SELF-ORGANIZED BIOLOGICAL DYNAMICS AND NONLINEAR CONTROL

The growing impact of nonlinear science on biology and medicine is fundamentally changing our view of living organisms and disease processes. This book introduces the application to biomedicine of a broad range of interdisciplinary concepts from non-linear dynamics, such as self-organization, complexity, coherence, stochastic resonance, fractals, and chaos.

The book comprises 18 chapters written by leading figures in the field. It covers experimental and theoretical research, as well as the emerging technological possibilities such as nonlinear control techniques for treating pathological biodynamics, including heart arrhythmias and epilepsy. The chapters review self-organized dynamics at all major levels of biological organization, ranging from studies on enzyme dynamics to psychophysical experiments with humans. Emphasis is on questions such as how living systems function as a whole, how they transduce and process dynamical information, and how they respond to external perturbations. The investigated stimuli cover a variety of different influences, including chemical perturbations, mechanical vibrations, thermal fluctuations, light exposures and electromagnetic signals. The interaction targets include enzymes and membrane ion channels, biochemical and genetic regulatory networks, cellular oscillators and signaling systems, and coherent or chaotic heart and brain dynamics. A major theme of the book is that any integrative model of the emergent complexity observed in dynamical biology is likely to be beyond standard reductionist approaches. It also outlines future research needs and opportunities ranging from theoretical biophysics to cell and molecular biology, and biomedical engineering.

JAN WALLECZEK is Head of the Bioelectromagnetics Laboratory and a Senior Research Scientist in the Department of Radiation Oncology at Stanford University School of Medicine. He studied biology at the University of Innsbruck, Austria, and then was a Doctoral Fellow and Research Associate at the Max-Planck Institute of Molecular Genetics in Berlin. Subsequently, he moved to California, where he was a Research Fellow in the Research Medicine and Radiation Biophysics Division at the Lawrence Berkeley National Laboratory, University of California, Berkeley, and at the Veterans Administration Medical Center in Loma Linda before founding the Bioelectromagnetics Laboratory at Stanford University in 1994. His recent publications include topics such as the nonlinear control of biochemical oscillators, coherent electron spin kinetics in magnetic field control of enzyme dynamics, nonlinear biochemical amplification, and stochastic resonance in biological chaos pattern detection. Jan Walleczek is a Founding Fellow of the Fetzer Institute, a Chair of the Gordon Research Conference on Bioelectrochemistry, and an Editorial Board member of the journal *Bioelectromagnetics*.

T0192235

SELF-ORGANIZED BIOLOGICAL DYNAMICS AND NONLINEAR CONTROL

*Toward Understanding Complexity, Chaos and Emergent Function
in Living Systems*

EDITED BY

JAN WALLECZEK

Department of Radiation Oncology, Stanford University

CAMBRIDGE
UNIVERSITY PRESS

CAMBRIDGE UNIVERSITY PRESS
Cambridge, New York, Melbourne, Madrid, Cape Town, Singapore, São Paulo

Cambridge University Press
The Edinburgh Building, Cambridge CB2 2RU, UK

Published in the United States of America by Cambridge University Press, New York

www.cambridge.org
Information on this title: www.cambridge.org/9780521624367

First published 2000
This digitally printed first paperback version 2006

A catalogue record for this publication is available from the British Library

Library of Congress Cataloguing in Publication data
Self-organized biological dynamics and nonlinear control: toward understanding
complexity, chaos and emergent function in living systems/edited
by Jan Walleczek.
p. cm.
Includes index.
ISBN 0 521 62436 3 (hb)
1. Cellular signal transduction. 2. Self-organizing systems. 3. Nonlinear systems.
I. Walleczek, Jan, 1964–

QP517.C45 S45 2000
571.6 – dc21   99-044949

ISBN-13 978-0-521-62436-7 hardback
ISBN-10 0-521-62436-3 hardback

ISBN-13 978-0-521-02607-9 paperback
ISBN-10 0-521-02607-5 paperback

Contents

List of contributors   *page* vii
Preface   xi

The frontiers and challenges of biodynamics research   *Jan Walleczek*   1

Part I   Nonlinear dynamics in biology and response to stimuli   13
1   External signals and internal oscillation dynamics: principal aspects
    and response of stimulated rhythmic processes   *Friedemann Kaiser*   15
2   Nonlinear dynamics in biochemical and biophysical systems: from
    enzyme kinetics to epilepsy   *Raima Larter, Robert Worth and
    Brent Speelman*   44
3   Fractal mechanisms in neuronal control: human heartbeat and gait
    dynamics in health and disease   *Chung-Kang Peng, Jeffrey
    M. Hausdorff and Ary L. Goldberger*   66
4   Self-organizing dynamics in human sensorimotor coordination and
    perception   *Mingzhou Ding, Yanqing Chen, J. A. Scott Kelso and
    Betty Tuller*   97
5   Signal processing by biochemical reaction networks
    Adam P. Arkin   112

**Part II   Nonlinear sensitivity of biological systems to electromagnetic
    stimuli**   145
6   Electrical signal detection and noise in systems with long-range
    coherence   *Paul C. Gailey*   147
7   Oscillatory signals in migrating neutrophils: effects of time-varying
    chemical and electric fields   *Howard R. Petty*   173
8   Enzyme kinetics and nonlinear biochemical amplification in
    response to static and oscillating magnetic fields   *Jan Walleczek
    and Clemens F. Eichwald*   193

9   Magnetic field sensitivity in the hippocampus   *Stefan Engström,*
    Suzanne Bawin and W. Ross Adey   216

**Part III   Stochastic noise-induced dynamics and transport in biological
    systems**   235
10 Stochastic resonance: looking forward   *Frank Moss*   236
11 Stochastic resonance and small-amplitude signal transduction in
    voltage-gated ion channels   *Sergey M. Bezrukov and*
    Igor Vodyanoy   257
12 Ratchets, rectifiers, and demons: the constructive role of noise in
    free energy and signal transduction   *R. Dean Astumian*   281
13 Cellular transduction of periodic and stochastic energy signals
    by electroconformational coupling   *Tian Y. Tsong*   301

Part IV   Nonlinear control of biological and other excitable systems 327
14 Controlling chaos in dynamical systems   *Kenneth Showalter*   328
15 Electromagnetic fields and biological tissues: from nonlinear
    response to chaos control   *William L. Ditto and Mark L. Spano*   341
16 Epilepsy: multistability in a dynamic disease   *John G. Milton*   374
17 Control and perturbation of wave propagation in excitable
    systems   *Oliver Steinbock and Stefan C. Müller*   387
18 Changing paradigms in biomedicine: implications for future
    research and clinical applications   *Jan Walleczek*   409

Index   421

Contributors

W. Ross Adey
Department of Biomedical Sciences, University of California at Riverside, Riverside, CA 92521, USA

Adam P. Arkin
Physical Biosciences Division, E. O. Lawrence Berkeley National Laboratory, Berkeley, CA 94720, USA

R. Dean Astumian
Departments of Surgery and of Biochemistry and Molecular Biology, University of Chicago, Chicago, IL 60637, USA

Suzanne Bawin
Research Service, Veterans Administration Medical Center, Loma Linda, CA 92357, USA

Sergey M. Bezrukov
Laboratory of Physical and Structural Biology, NICHD, National Institutes of Health, Bethesda, MD 20892-0924, USA

Yanqing Chen
Center for Complex Systems and Brain Sciences, Florida Atlantic University, Boca Raton, FL 33431-0991, USA

Mingzhou Ding
Center for Complex Systems and Brain Sciences, Florida Atlantic University, Boca Raton, FL 33431-0991, USA

William L. Ditto
Laboratory for Neural Engineering, Georgia Tech/Emory Biomedical Engineering Department, Georgia Institute of Technology, Atlanta, GA 30332-0535, USA

Clemens F. Eichwald
Bioelectromagnetics Laboratory, Department of Radiation Oncology, School of Medicine, Stanford University, Stanford, CA 94305-5304, USA

Stefan Engström
Department of Neurology, Vanderbilt University Medical Center, Nashville TN 37212-3375, USA

Paul C. Gailey
Department of Physics and Astronomy, Ohio University, Athens, OH 45701, USA

Ary L. Goldberger
Margret & H. A. Rey Laboratory for Nonlinear Dynamics in Medicine, Harvard Medical School, Beth Israel Deaconess Medical Center, Boston, MA 02215, USA

Jeffrey M. Hausdorff
Margret & H. A. Rey Laboratory for Nonlinear Dynamics in Medicine, Harvard Medical School, Beth Israel Deaconess Medical Center, Boston, MA 00215, USA

Friedemann Kaiser
Nonlinear Dynamics Group, Institute of Applied Physics, Technical University, Darmstadt, D-64289, Germany

J. A. Scott Kelso
Center for Complex Systems and Brain Sciences, Florida Atlantic University, Boca Raton, FL 33431, USA

Raima Larter
Department of Chemistry, Indiana University–Purdue University at Indianapolis, Indianapolis, IN 46202, USA

John G. Milton
Department of Neurology, University of Chicago Hospitals, Chicago, IL 60637, USA

Frank Moss
Laboratory for Neurodynamics, Department of Physics and Astronomy, University of Missouri at St Louis, St Louis, MO 63121, USA

Stefan C. Müller
Institut für Experimentelle Physik–Biophysik, Universitätsplatz 2, Otto-von-Guericke-Universität Magdeburg, Magdeburg, D-39106, Germany

Chung-Kang Peng
Margret & H. A. Rey Laboratory for Nonlinear Dynamics in Medicine, Harvard Medical School, Beth Israel Deaconess Medical Center, Boston, MA 02215, USA

Howard R. Petty
Department of Biological Sciences, Wayne State University, Detroit, MI 48202, USA

Kenneth Showalter
Department of Chemistry, West Virginia University, Morgantown, WV 26506-6045, USA

Mark L. Spano
Naval Surface Warfare Center, Silver Spring, MD 20817, USA

Brent Speelman
Department of Chemistry, Indiana University–Purdue University at Indianapolis, Indianapolis, IN 46202, USA

Oliver Steinbock
Institut für Experimentelle Physik–Biophysik, Universitätsplatz 2, Otto-von-Guericke-Universität Magdeburg, Magdeburg, D-39106, Germany

Tian Y. Tsong
Department of Biochemistry, Molecular Biology and Biophysics, University of Minnesota, St Paul, MN 55108, USA

Betty Tuller
Center for Complex Systems and Brain Sciences, Florida Atlantic University, Boca Raton, FL 33431-0991, USA

Igor Vodyanoy
Office of Naval Research Europe, 223 Old Marylebone Road, London, NW1 5TH, UK

Jan Walleczek
Bioelectromagnetics Laboratory, Department of Radiation Oncology, School of Medicine, Stanford University, Stanford, CA 94305-5304, USA

Robert Worth
Department of Neurosurgery, Indiana University–Purdue University at Indianapolis, Indianapolis, IN 46202, USA

Preface

The real voyage of discovery consists not in seeking new landscapes
but in having new eyes.   *Marcel Proust*

The tools and ideas from nonlinear dynamics such as the concept of self-
organization provide scientists with a powerful perspective for viewing living
processes in a new light. As in the physical sciences before, the nonlinear
dynamical systems approach promises to change scientific thinking in many
areas of the biomedical sciences. For example, two rapidly evolving branches
of nonlinear dynamics, popularly known as chaos and complexity studies,
which have opened up new vistas on the dynamics of the nonliving world, are
also beginning to impact deeply on our view of the living world. The key
concept at the core of this work states that complex nonlinear systems, under
conditions far from equilibrium, have a tendency to self-organize and to
generate complex patterns in space and time.

Living organisms are prime examples of nonlinear complex systems operating under far from equilibrium conditions and, hence, self-organization and
dynamical pattern formation is the hallmark of any living system. It thus
comes as no surprise that knowledge about the nonlinear dynamics of physical
systems can be successfully transferred to the study of biological systems. As a
result, previously difficult to explain biological phenomena can now be understood on a theoretical basis. Importantly, the nonlinear dynamical approach is
quickly leading to the discovery of novel biological behaviors and characteristics also. Many examples of often-unexpected biological insights, as a consequence of the nonlinear systems approach, and the emerging applications for
clinical diagnosis and therapy are among the topics discussed in this volume.

Motivated by the growing impact of nonlinear science on biomedicine I
proposed the organization of a workshop on 'Self-organized Biodynamics and
Control by Chemical and Electromagnetic Stimuli' from which the idea for this

volume originated. The workshop, which was jointly sponsored by the US Department of Energy and the Fetzer Institute, was held from 11 to 14 August 1996, at the Fetzer Institute in Kalamazoo, Michigan. Leading investigators, many of whom are the acknowledged authorities in their respective fields, met for three-and-a-half days to review current knowledge and to explore the most promising frontiers in this rapidly developing research field. The unifying theme was the nonlinear sensitivity of biological systems to weak external influences, and the development of novel methods that take advantage of this sensitivity in the study and nonlinear control of biological functions. Because of the demand generated by the first gathering, a second workshop was convened titled 'Towards Information-based Interventions in Biological Systems: From Molecules to Dynamical Diseases' from 23 to 26 August 1998. Between the two workshops a total of 38 stimulating presentations were given. Although this volume is not a workshop proceedings, the contributors, whose work is the subject of this volume, were drawn from the workshop speakers. Because of space constraints, several of the topics then discussed are not represented here, although I have made an effort in their selection to provide the broadest scope possible.

The interdisciplinary topics reflect the importance of the interplay between theoretical work and laboratory experiments in this new research area. While the book's primary goal is to provide an overview, the authors have tried to allow readers of diverse backgrounds to familiarize themselves with some of the details of the experimental and theoretical approaches presented. For example, chapters with a focus on experimental observations often provide important methodological information, so that the reader can better evaluate the challenges as well as opportunities of laboratory work in this area. In a similar fashion, the intent of the chapters that deal with the construction of theoretical models and the development of nonlinear analytical methods is to provide enough detail to enable the nonspecialist but technically oriented reader to follow the basic theoretical reasoning.

The use of concepts from nonlinear biological dynamics, or 'biodynamics' in short, to frame and solve critical research questions is rapidly expanding across many biological disciplines from cell and molecular biology to neuroscience. For example, the formation in 1998 of a program area on 'Quantitative Approaches to the Analysis of Complex Biological Systems' by the US National Institutes of Health is an indication that the nonlinear dynamical systems approach is near the threshold of entering the mainstream of biomedical research. I am convinced that it will be increasingly important for scientists in many biomedical disciplines to become familiar with the concepts outlined here. It is my hope that this book can serve as a useful guide to biodynamics for students and professionals and that it can provide them with a new framework for pursuing their own research interests.

Besides the 30 authors, who have generously given their time to write for this book, there are many other individuals whose support and contributions are directly responsible for making this book become a reality. In particular, this project would not have come to fruition without the enthusiasm and continuing support of the members of the Fetzer Institute's Board of Trustees. The task of planning and organizing the 1996 and 1998 workshops that provided the initial forum for evaluating the results and ideas presented here was carried out by the Fetzer Institute Task Force on 'Biodynamics', which was chaired by Bruce M. Carlson and whose other members included Paul C. Gailey, the late Kenneth A. Klivington, Harold E. Puthoff and myself. I thank my fellow task force members wholeheartedly for their excellent efforts. I also acknowledge the participation of Imre Gyuk, who provided the financial workshop support by the US Department of Energy, and I thank Frank Moss, who made the initial contact with Cambridge University Press. For valuable comments on the contributions written or co-written by me, I am grateful to Adam P. Arkin, Dean R. Astumian, Paul C. Gailey, Friedemann Kaiser, Susan J. Knox and Arnold J. Mandell. At Cambridge University Press, I wish to thank Simon Capelin and Sandi Irvine for patiently working with me to bring this book to completion.

Finally, I am indebted to George Hahn and Jeremy Waletzky for their important roles in the establishment of the Bioelectromagnetics Laboratory at Stanford, where I conducted most of my work in biodynamics. At the laboratory, I thank Jeffrey Carson, Clemens Eichwald, Pamela Killoran, Peter Maxim and Esther Shiu for their commitment to our work. I also want to express my gratitude to my parents and Lark, who were a source of inspiration and steady support throughout this project.

Jan Walleczek, Palo Alto

The frontiers and challenges of biodynamics research

JAN WALLECZEK

1 Background

As scientists unravel the secrets of the organization of life, an understanding of the temporal and spatiotemporal *dynamics* of biological processes is deemed crucial for creating a coherent, fully integrative picture of living organisms. In this endeavour, the basic challenge is to reveal how the coordinated, dynamical behavior of cells and tissues at the *macroscopic* level, emerges from the vast number of random molecular interactions at the *microscopic* level. This is the central task of modern biology and, traditionally, it has been tackled by focusing on the participating molecules and their microscopic properties, ultimately at the quantum level. Biologists often tacitly assume that once all the molecules have been identified, the complete functioning of the whole biological system can finally be derived from the sum of the individual molecular actions. This reductionistic approach has proven spectacularly successful in many areas of biological and medical research. As an example, the advances in molecular biology, which have led to the ability to manipulate DNA at the level of specific genes, will have a profound effect on the future course of medicine through the introduction of gene-based therapies. Despite this progress, however, the consensus is growing that the reductionist paradigm, by itself, may be too limiting for successfully dealing with fundamental questions such as (1) how living systems function *as a whole*, (2) how they transduce and process *dynamical information*, and (3) how they respond to *external perturbations*.[1]

2 Self-organization

The difficulties of addressing these questions by purely reductionistic approaches become immediately apparent when considering – from the

[1] For recent perspectives see Hess and Mikhailov (1994), Glanz (1997), Spitzer and Sejnowski (1997), Williams (1997), Coffey (1998) and Gallagher and Appenzeller (1999).

1

standpoint of physics – the following two defining features. (1) Living organisms are thermodynamically open systems; that is, they are in a state of permanent flux, continuously exchanging energy and matter with their environment. (2) They are characterized by a complex organization, which results from a vast network of molecular interactions involving a high degree of nonlinearity. Under appropriate conditions, the combination of these two features, *openness* and *nonlinearity*, enables complex systems to exhibit properties that are *emergent* or *self-organizing*. In physical and biological systems alike, such properties may express themselves through the spontaneous formation, from random molecular interactions, of long-range correlated, macroscopic dynamical patterns in space and time – the process of *self-organization*. The dynamical states that result from self-organizing processes may have features such as excitability, bistability, periodicity, chaos or spatiotemporal pattern formation, and all of these can be observed in biological systems.

Emergent or self-organizing properties can be defined as properties that are possessed by a dynamical system as a whole but not by its constituent parts. In this sense, the whole is *more* than the sum of its parts. Put in different terms, emergent phenomena are phenomena that are expressed at higher levels of organization in the system but not at the lower levels. One attempt to help to visualize the concept of self-organization is the sketch in Figure 1, which shows the dynamical interdependence between the molecular interactions at the microscopic level and the emerging global structure at the macroscopic level. The upward arrows indicate that, under nonequilibrium constraints, molecular interactions tend to spontaneously synchronize their behavior, which initiates the beginnings of a collective, macroscopically ordered state. At the same time, as indicated by the downward arrows, the newly forming macroscopic state acts upon the microscopic interactions to force further synchronizations. Through the continuing, *energy-driven* interplay between microscopic and macroscopic processes, the emergent, self-organizing structure is then stabilized and actively maintained.

Amongst the earliest examples for this behavior in a simple physical system is the spontaneous organization of long-range correlated macroscopic structures; that is, of convection cells (Bénard instability) in a horizontal water layer with a thermal gradient (e.g., Chandrasekhar, 1961). In this well-known case of hydrodynamic self-organization, the size of the emergent, global structures – that is, of spatiotemporal hexagonal patterns of the order of millimeters – is greater by many orders of magnitude than the size of the interacting water molecules. This implies that, when the thermal gradient has reached a critical value, the initially uncorrelated, random motions of billions of billions of

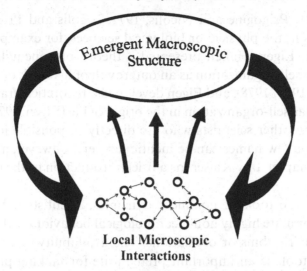

Figure 1. Sketch illustrating the dynamical interdependence between microscopic molecular interactions and the emerging global structure at the macroscopic level. The system under consideration is open to the flow of matter or energy. The upward arrows indicate that, under nonequilibrium constraints, molecular interactions tend to spontaneously synchronize their behavior, which initiates the beginnings of a macroscopic, ordered state. As indicated by the downward arrows, this newly forming state acts upon the microscopic interactions to force further synchronizations. Through the continuing, energy-driven interplay between microscopic and macroscopic processes, the emergent, self-organizing structure is stabilized and actively maintained.

molecules have synchronized spontaneously *without* any external instructions, hence, the term 'self-organization'.

3 Theoretical foundations and computer simulations

The above arguments reveal that the origins and dynamics of emergent, macroscopic patterns, including in biological systems, cannot be simply deduced from the sum of the individual actions of the system's microscopic elements. What is needed is an analysis of the system's collective, macroscopic dynamics, which results from the complex web of nonlinear interactions between the elements. During the early 1970s, general theoretical frameworks for this type of analysis, which are based on a branch of mathematics called *nonlinear dynamics*, became more widely available and recognized. In 1977, I. Prigogine was awarded a Nobel prize for the discovery that, in *apparent* contradiction to the second law of thermodynamics, physico-chemical systems far from thermodynamic equilibrium tend to self-organize by exporting entropy and form, what he termed, *dissipative structures* (Glansdorff and

Prigogine, 1971; Prigogine and Nicolis, 1971; Nicolis and Prigogine, 1977). Other pioneers in the physical or biological sciences, for example, include H. Haken and M. Eigen. Haken presented a theory of nonequilibrium phase transitions and self-organization as an outgrowth of his work on the theory of lasers (Haken, 1975, 1978), and Eigen developed a theoretical framework for a role of molecular self-organization in the origin of life (Eigen, 1971). There are, of course, many other scientists who are directly responsible for developing this field. Only a few names can be mentioned here, however, and the reader may consult Chapter 1 by Kaiser for a brief introduction to the history of this science.

Armed with the tools of nonlinear dynamics, scientists are now able to describe and simulate highly nonlinear biological behaviors such as biochemical and cellular rhythms or oscillations. The availability of the appropriate mathematical tools is an important prerequisite for making progress in the rapidly growing area of biological dynamics or 'biodynamics'. One specific reason stems from the fact that mechanistic explanations of self-organizing, biodynamical processes frequently defy intuition. This is due to the complexity of the dynamical interactions that underlie such processes, whose emergent properties cannot be readily grasped by the human observer (compare Figure 1). Thus, as is reflected in many of the contributions to this volume, scientists must rely heavily on computer simulations to explore complex biological dynamics and to make predictions about experimental outcomes. Common to all these approaches is the treatment of a biological system as an open system of nonlinearly interacting elements. Consequently, the field of *biodynamics* might be defined as *the study of the complex web of nonlinear dynamical interactions between and among molecules, cells and tissues, which give rise to the emergent functions of a biological system as a whole.*

4 Nonlinear dynamics moves into cell and molecular biology: cellular oscillators, biological signaling and biochemical reaction networks

Although the self-organization of macroscopic patterns, including temporal oscillations and spatiotemporal wave patterns, was first studied and theoretically understood in physical and chemical systems, numerous examples are now known at all levels of biological organization (for recent overviews, see Goldbeter, 1996; Hess, 1997). The most conspicuous examples of self-organizing biological activity are *biological rhythms* and *oscillations*. The formation of oscillatory dynamical states of different periodicities plays a fundamental role in living organisms. In humans, the observed oscillation periods cover a wide range from the subsecond time domain of neuronal oscillations to the 28-day period of the ovarian cycle. For instance, the perception of visual stimuli is associated with oscillatory synchronizations of neuronal assemblies at frequencies of 10s of hertz. At the cellular level, oscillatory signaling and metabolic processes such as oscillations in the intracellular concentration of calcium (Ca^{2+}), adenosine triphosphate (ATP) or nicotinamide adenine dinucleotide phosphate (NADPH) have periods of the order of seconds to minutes. For example, the activity of human neutrophils, a key component of cellular immune defense, involves oscillatory cell biochemical processes with periods on the order of 10 to 20 s. Finally, the cell cycle itself is a prime example of a biological oscillator: cell cycle progression is controlled by the mitotic oscillator whose oscillation periods may range from about 10 min to 24 h. Many more examples are known and several of them are covered in detail in this book.

The processes that underlie cellular oscillators are organized in complexly coupled *biochemical reaction networks*, wherein feedforward and feedback information flows provide the links between the different levels in the hierarchy of cell biochemical network organization. Such networks are also central components of the cellular machinery that controls biological signaling. Computer modeling has recently enabled scientists to investigate the properties of *biological signaling networks* such as their capacity to detect, transduce, process and store information. In these efforts, it was found that cellular signaling pathways may also exhibit properties of emergent complexity (for a recent example, see Bhalla and Iyengar, 1999). Such findings serve to demonstrate the difficulties that scientists face when they attempt to predict the dynamics of cellular signal transduction processes only on the basis of isolated signaling molecules and their individual microscopic actions. In order to develop an integrative, dynamical picture of biological signaling processes, therefore, it will be necessary to characterize the nonlinear relationships among the different molecular species making up the biochemical reaction networks, which control all aspects of cellular regulation as, for example, from RNA transcriptional control to cellular division. Theoretical models of biochemical reaction networks have been proposed that simulate, for example, cellular dynamics of Ca^{2+} oscillations (e.g., Goldbeter *et al.*, 1990), interactions between different cell signaling pathways (e.g., Weng *et al.*, 1999), genetic regulatory circuits (e.g., McAdams and Arkin, 1998), cellular control networks for DNA replication (Novak and Tyson, 1997) and cellular division (Borisuk and Tyson, 1998). Such theoretical work is not limited, however, to the analysis of normal cell function. Nonlinear modeling has been applied, for example, to pathological cell signaling involved in cancer formation (Schwab and Pienta, 1997).

From this and related work the perspective is developing that biological cells can be viewed as highly sophisticated information-processing devices that can discern complex patterns of extracellular stimuli (Bray, 1995). In line with this view is the finding that, in analogy to electrical circuits, biochemical reaction networks can perform computational functions such as switching, amplification, hysteresis, or band-pass filtering of frequency information (e.g., Arkin and Ross, 1994). The development of the theoretical and computational tools for deducing the function of complex biochemical reaction and nonlinear signaling networks will become even more important for biologists now that many genome projects are nearing completion. The ambitious goal of these projects is to provide researchers with a complete list of all cellular proteins and genetic regulatory systems. The daunting task that biologists face is to functionally integrate the massive amount of data from these projects. Clearly, this will require an approach that can account for the emergent, collective properties of the vast network of nonlinear biochemical reactions that underlie the biocomplexity of cells, tissues and of the whole organism.

5 Biological interactions with external stimuli and nonlinear control

The nonlinear dynamical nature of living processes turns out to be crucial for understanding how biological systems interact with the external environment. Specifically, the intrinsic nonlinearity of living systems is of great significance to scientists who study the response of cells, tissues and whole organisms to natural or artificial stimuli. The reason is that the response behavior of a nonlinear system may differ drastically from that of a linear system. In a linearly behaving system, the response magnitude to an applied stimulus is proportional to the strength of the stimulus. In contrast, disproportionately large changes may result in a nonlinear system. The inherent *amplification properties* of nonlinear systems thus represent one critical aspect that defines the system's sensitivity and the magnitude of its response to external perturbations.

Another aspect concerns the capacity of complex, nonlinear systems to detect and process information contained in incoming signals. For instance, the response of nonlinear systems can depend, in a highly nonlinear fashion, on the frequency information contained in an oscillating external perturbation. For these and other reasons discussed further below, the response of nonlinear processes such as may occur in biological systems may lead to unexpected sensitivities and complex response patterns. Knowledge about this behavior is not only of significance for revealing the mechanistic basis of stimulus–response effects but, importantly, can be exploited for the *nonlinear control* of dynamical biological processes for practical purposes.

Within the context of this volume, nonlinear control refers to mechanisms or methods that control chemical, biochemical or biological processes by exploiting the nonlinear dynamical features that underlie these processes. For example, the goals of nonlinear control may be to cause excitation or suppression of oscillations, entrainment and synchronization, or transitions from chaotic to periodic oscillations and vice versa. Specifically, the term 'control' refers to the modification of the behavior of a nonlinear system by variation of one or more of the *control parameters* that govern the system's macroscopic dynamics. This may be achieved by variations that are caused either by processes within the system or by appropriately designed external perturbations. In this approach, global macroscopic dynamics, rather than microscopic kinetics, thus provides the critical information for system control.

6 Purpose and contents

This volume provides an introduction to the application of a broad range of concepts from nonlinear dynamics such as self-organization, emergent phenomena, stochastic resonance, coherence, criticality, fractals and chaos to biology and medicine. The selected contributions cover nonlinear self-organized dynamics at all major levels of biological organization, ranging from studies on enzyme dynamics to psychophysical experiments with humans. The emphasis is on work from (1) experimentalists who study the response of nonlinear dynamical states in biological and other excitable systems to external stimuli and (2) theorists who create predictive models of nonlinear stimulus–response interactions. The investigated stimuli cover a variety of different influences, including chemical perturbations, electromagnetic signals, mechanical vibrations, light stimuli or combinations thereof. The interaction targets include cyclical, excitable and oscillatory behavior in biological and related systems. They include membrane ion channels and pumps, biochemical reaction networks, oscillatory chemical or enzyme activity, oscillations in cellular metabolites, Ca^{2+} oscillations, genetic regulatory networks, excitable states in neurons and sensory cells, and chaotic or periodic heart and brain tissue dynamics. This volume's two main purposes are: (1) to introduce the reader to the present state of theoretical and experimental knowledge in this rapidly expanding field of interdisciplinary research, and (2) to outline the future research needs and opportunities from the perspective of the different disciplines, from theoretical physics to biomedical engineering.

The individual contributions summarize a wide range of experimental and theoretical investigations by biologists, neuroscientists, chemists, physicists, bioengineers and medical researchers. This selection emphasizes (1) the need for cross-disciplinary dialogue and (2) the importance of the interplay between theoretical modeling and laboratory experiments. It also reflects the responsibility of the recent focus on collaborations between theorists and experimentalists for the increasing progress in understanding complex stimulus–response interactions in biosystems. This volume covers both the basic research aspects as well as the emerging technological dimensions. Basic research includes the interplay between theory development, laboratory experimentation and computer simulations. The promise of future technologies comes from the development of techniques that exploit the self-organized dynamics intrinsic to living systems for diagnostic, prognostic and therapeutic purposes. Special attention is given to three interconnected components of the stimulus–response paradigm:

(1) The often surprising sensitivity of biological and related excitable systems to weak external influences, whether they are chemical, mechanical or electromagnetic in nature, is illustrated by many examples. The stimuli are either time-varying or constant. Time-dependent stimuli include periodic oscillations and random fluctuations. These are applied to systems that generate deterministic temporal or spatiotemporal behavior by methods that in some cases involve feedback control. A major focus is the application of electromagnetic stimuli as a specific, minimally invasive tool for influencing biodynamical systems, which is the subject of the research area known as *bioelectromagnetics*. This volume includes important information regarding (a) the theoretical limits of the interaction of electromagnetic field signals with chemical, biochemical and biological systems and (b) the laboratory evidence for electric or magnetic field interactions with isolated enzymes, cells and tissues. The targets of electromagnetic fields may be any physicochemical processes that are sensitive to these fields and that play a role in the generation or maintenance of self-organizing dynamics. The fundamental physical constraints that govern these interactions are explained for both the initial energy transduction step in the presence of thermodynamic noise and for the responsiveness of the dynamical state to a weak perturbation.

(2) The recognition of deterministic macroscopic dynamics in biological systems also opens up unanticipated opportunities for probing biological systems. For example, information regarding the intrinsic dynamics of a biological system can be obtained by analyzing its response to an applied stimulus. Computer simulations have long shown that the imposition of weak stimuli on systems with complex dynamics, including living systems, may induce responses that depend not only on the intensity of the stimulus but, importantly, on its temporal pattern as well as the initial state of the system. State dependence and sensitivity to the temporal characteristics of the applied stimulus is a fundamental feature of self-organized biological activity. There now exists experimental evidence that is in excellent agreement with the predictions from theoretical modeling: an increasing number of laboratories report that excitable systems, including chemical, biochemical and biological systems, display complex responses with nonlinear dependence on imposed tem-

poral patterns. For example, resonance-like responses to coherently oscillating stimuli that depend strictly on the frequency of the imposed stimulus have been observed in many experiments in recent years. They include excitable chemical reactions, isolated enzymes and membrane ion transporters, biochemical reaction networks, Ca^{2+}-dependent gene expression, and neuronal or heart muscle cell activity in single cells and tissue preparations.

(3) The identification of self-organized dynamical states in living systems and the knowledge about their sensitivity has also paved the way for developing new strategies for influencing or controlling biological dynamics. Importantly, it was found that the sensitivity of biodynamical systems to appropriately designed stimuli could be exploited for practical purposes, like the ability to shift the dynamics of biological activity from an unwanted state to a desired one. The discovery of deterministic biological chaos, for example, offers novel strategies for therapeutic interventions. Here, chaos does not refer to disordered, random processes, but rather characterizes hidden dynamical order within apparent disorder. As this book illustrates, methods initially developed to control chaos in physical systems have also been found to be effective in controlling chaotic dynamics in chemical and biological systems. Dramatic demonstrations of this possibility are experiments in which chaos control techniques were applied to heart and brain tissue preparations and, most recently, to human heart patients. This book discusses implications of these possibilities for treating so-called 'dynamical diseases' such as heart arrhythmias and epileptic seizures.

A new research area that is critical to each of the three components of the stimulus–response paradigm is the exploration of the constructive role that intrinsic or external random fluctuations may play in physiological functions. Consequently, one part of the volume is devoted to theory and experimentation on the previously unsuspected, beneficial role of stochastic noise in controlling or influencing nonlinear dynamic and transport phenomena in living systems. At first glance this notion seems counterintuitive, but established physical concepts, including the ones known as *stochastic resonance* and *fluctuation-driven transport*, make such phenomena theoretically plausible (see, e.g., Astumian and Moss, 1998). This volume covers both the applied and basic research dimensions of noise-assisted biochemical and biological processes. It summarizes theoretical and experimental evidence demonstrating a beneficial or even necessary role for noise in biological signaling, including neuronal information processing. The developing technological applications, which are based on the principle of stochastic resonance, are also addressed. This work includes the modulation of biological signal transmission through the controlled addition of noise to diagnose or to improve human sensory perception.

7 Frontiers and outlook

What is the physical basis of biological self-organization? How do the basic elements of biological activity interact to give rise to the function of a living organism as a whole? How do biodynamical systems respond to weak biochemical or electromagnetic perturbations? How can the nonlinear features of biodynamical processes be put to practical use, for example, in clinical diagnosis and therapy? These are among the questions that are explored in the 18 chapters that follow. The presented ideas and experimental observations demonstrate that many important features of the dynamics of living processes can be understood on a theoretical basis. Importantly, the validity of this knowledge is confirmed by the success of the emerging biomedical applications that have already resulted from this work, for example by employing fractal time series analysis and nonlinear control methods.

In summary, the perspective of a living system as a self-organizing, complex, far-from-equilibrium biochemical state allows physics to enter the study of dynamical biological functions in a quantitative, predictive manner. While the nonlinear dynamical systems approach does not yet, however, represent a physical theory for the organization of life, its broad scope and power suggests that it will be a crucial building block in the construction of any such theory in the future. At a minimum, biodynamics research is revealing how complex, sophisticated and remarkably sensitive living processes really are. Finally, this work suggests that any integrated understanding of the functional complexity observed in dynamical biology is probably beyond the scope of standard reductionistic approaches. It is our hope that the reader can share the excitement of discovery conveyed in the following chapters and thus will be motivated to view and explore biological processes from a new perspective.

References

Arkin, A. P. and Ross, J. (1994) Computational functions in biochemical reaction networks. *Biophys. J.* **67**: 560–578.

Astumian, R. D. and Moss, F. (1998) Overview: the constructive role of noise in fluctuation driven transport and stochastic resonance. *Chaos* **8**: 533–538.

Bhalla, U. S. and Iyengar, R. (1999) Emergent properties of networks of biological signaling pathways. *Science* **283**: 381–387.

Borisuk, M. T. and Tyson, J. J. (1998) Bifurcation analysis of a model of mitotic control in frog eggs. *J. Theor. Biol.* **195**: 69–85.

Bray, D. (1995) Protein molecules as computational elements in living cells. *Nature* **376**: 307–312.

Chandrasekhar, S. (1961) *Hydrodynamic and Hydromagnetic Stability.* Oxford: Oxford University Press.

Coffey, D. S. (1998) Self-organization, complexity and chaos: the new biology for medicine. *Nature Med.* **4**: 882–885.

Eigen, M. (1971) Molecular self-organization and the early stages of evolution. *Quart. Rev. Biophys.* **4**: 149–212.

Gallagher, R. and Appenzeller, T. (1999) Beyond reductionism. *Science* **284**: 79.

Glansdorff, P. and Prigogine, I. (1971) *Thermodynamic Theory of Structure, Stability and Fluctuations.* New York: Wiley.

Glanz, J. (1997) Mastering the nonlinear brain. *Science* **277**: 1758–1760.

Goldbeter, A. (1996) *Biochemical Oscillations and Cellular Rhythms.* Cambridge: Cambridge University Press.

Goldbeter, A., Dupont, G. and Berridge, M. J. (1990) Minimal model for signal-induced Ca^{2+}-oscillations and for their frequency encoding through protein phosphorylation. *Proc. Natl. Acad. Sci. USA* **87**: 1461–1465.

Haken, H. (1975) Cooperative effects in systems far from thermal equilibrium and in nonphysical systems. *Rev. Mod. Phys.* **47**: 67–121.

Haken, H. (1978) *Synergetics: An Introduction.* Berlin: Springer-Verlag.

Hess, B. (1997) Periodic patterns in biochemical reactions. *Quart. Rev. Biophys.* **30**: 121–176.

Hess, B. and Mikhailov, A. (1994) Self-organization in living cells. *Science* **264**: 223–224.

McAdams, H. H. and Arkin, A. P. (1998) Simulation of prokaryotic genetic circuits. *Annu. Rev. Biophys. Biomol. Struct.* **27**: 199–224.

Nicolis, G. and Prigogine, I. (1977) *Self-organization in Nonequilibrium Systems.* New York: Wiley.

Novak, B. and Tyson, J. J. (1997) Modeling the control of DNA replication in fission yeast. *Proc. Natl. Acad. Sci. USA* **94**: 9147–9152.

Prigogine, I. and Nicolis, G. (1971) Biological order, structure and instabilities. *Quart. Rev. Biophys.* **4**: 107–148.

Schwab, E. D. and Pienta, K. J. (1997) Explaining aberrations of cell structure and cell signaling in cancer using complex adaptive systems. *Adv. Mol. Cell Biol.* **24**: 207–247.

Spitzer, N. C. and Sejnowski, T. J. (1997) Biological information processing: bits of progress. *Science* **277**: 1060–1061.

Weng, G., Bhalla, U. S. and Iyengar, R. (1999) Complexity in biological signaling systems. *Science* **284**: 92–96.

Williams, N. (1997) Biologists cut reductionist approach down to size. *Science* **277**: 476–477.

Part I
Nonlinear dynamics in biology and response to stimuli

Part I introduces the terminology and definitions of key concepts in nonlinear dynamics and provides examples of their application at different levels of physiological organization. The examples show how common principles from nonlinear dynamics can be applied in the study of systems that differ greatly in terms of their material composition, scale of organization, and biological function. Chapter 1 by Friedemann Kaiser first reviews the reasons why nonlinear dynamics is critical to understanding biological function and order, and also provides a historical background. The chapter then introduces basic concepts and mathematical definitions that are essential to theoretical analyses of nonlinear biological phenomena, with a focus on model construction and responses to stimuli. Chapter 2 by Raima Larter and co-workers begins with a description of a nonlinear enzyme oscillator, the peroxidase–oxidase system, which is the best-characterized biochemical *in vitro* reaction showing diverse dynamics such as periodicity and bifurcation into chaos. Insights into the dynamical principles that govern the enzyme oscillator are then related to development of a model of neuroelectrical oscillations during epileptic brain activity. Work from the laboratory of Ary Goldberger is reviewed in Chapter 3, which demonstrates that neuronal control processes underlying heart and gait dynamics are characterized by long-range power law correlations. This chapter introduces the use of the concept of fractal dimensionality in biology and shows how fractal time series analysis can be put to use in clinical diagnosis and prognosis. Chapter 4, written by Mingzhou Ding and collaborators, continues with the theme of fractal analysis and discusses results obtained from psychophysical studies with humans. These experiments provide evidence for self-organized dynamics in human sensorimotor coordination and speech perception. The final chapter in this part, Chapter 5, returns to the cellular and subcellular levels of biological organization. Adam Arkin explains how engineering principles from electric circuit analysis can be

13

employed in the modeling of computational functions of biochemical reaction networks that are involved in nonlinear cell signaling networks, cellular oscillators and genetic regulatory circuits.

1

External signals and internal oscillation dynamics: principal aspects and response of stimulated rhythmic processes

FRIEDEMANN KAISER

1.1 Introduction

The description of order and function in biological systems has been a challenge to scientists for many decades. The overwhelming majority of biological order is functional order, often representing *self-organized dynamical states* in living matter. These states include spatial, temporal and spatiotemporal structures, and all of them are ubiquitous in living as well in nonliving matter. Prominent examples are patterns (representing static functions), oscillatory states (rhythmic processes), travelling and spiraling waves (nonlinear phenomena evolving in space and time).

From a fundamental point of view, biological function must be treated in terms of dynamic properties. Biological systems exhibit a relative stability for some modes of behavior. In the living state, these modes remain very far from thermal equilibrium, and their stabilization is achieved by nonlinear interactions between the relevant biological subunits. The functional complexity of biological materials requires the application of macroscopic concepts and theories, the consideration of the motion of individual particles (e.g., atoms, ions, molecules) is either meaningless or not applicable in most cases.

The existence and stabilization of far-from-equilibrium states by nonlinear interactions within at least some subunits of a physical, chemical or biological system are intimately linked with cooperative processes. Besides the well-known strong equilibrium cooperativity, thermodynamically metastable states and nonequilibrium transitions in cooperatively stabilized systems can occur, provided a certain energy input is present. In equilibrium, an entire subunit or a domain of a macromolecular system reacts as a unit, which means that it transforms as a whole. Responsible for equilibrium phase transitions are physical changes and chemical transformations of macrovariables. Well-known examples are transitions from liquid to solid and from para- to

15

ferroelectric states, changes of crystalline structures, superfluidic and super-conducting systems. In nonequilibrium situations, nonlinearities create dissi-pative elements that lead to new states, including trigger action, threshold effects and hysteresis. Additional interactions of these nonequilibrium states with external stimuli increase in a dramatic way the number and types of specific modes of behavior.

In recent years it has become clear that nonlinear phenomena and their interactions with *external signals* (fields and forces of electromagnetic, mech-anical or chemical nature) are abundant in Nature. Examples range from mechanics (anharmonic oscillators), hydrodynamics (pattern formation and turbulence), electronics (Josephson junctions), nonlinear optics (laser, optical bistability, information processing and storage), acoustics (sonoluminescence), chemistry (oscillating reactions and spiral waves), biochemistry (glycolytic and Ca^{2+} oscillations) to biology (large spectrum of rhythms). Only a few fields of research and some examples are mentioned.

Already more than six decades ago, membrane phenomena in living matter were considered as steady states instead of equilibrium states (Hill, 1930). The importance was stressed that cells are open systems, the steady states of which are created and stabilized by the flux of energy and matter through the system. These considerations led to the concept '*Fliessgleichgewicht*' ('dynamical equi-librium') for nonequilibrium states (Von Bertalanffy, 1932). First mathematical modeling approaches (Rashevsky, 1938) and quite general theoretical con-siderations on a strong physical basis (Schrödinger, 1945) were first attempts to describe biological order and function with existing concepts and laws of physics, and to look for the essential properties separating living from nonliv-ing matter.

A simple, nonlinear two-variable model, including diffusion, revealed that inherent temporal and spatial instabilities can create different spatiotemporal structures, thus offering a simple chemical basis for morphogenesis (Turing, 1952). Later on it was stressed that 'the cell cannot have a steady state unless it is accompanied by oscillations' (Bernhard, 1964). This statement implies that in order to achieve a stable oscillatory condition the system must lose as much energy as it gains on average over one oscillation cycle. From a modern point of view, these oscillations are sustained oscillations of the *limit-cycle* type, representing temporal dissipative structures in nonlinear systems.

Nonlinear phenomena require the investigation of nonlinear dynamical models with only a few relevant degrees of freedom. Nonlinear dynamics is a very old problem, originating in studies of planetary motion. Henri Poincaré was first to investigate the complex behavior of simple mathematical systems by applying geometrical methods and studying topological structures in *phase space* (Poincaré, 1892). He discovered that the strongly deterministic equations for the motion of planets and other mechanical systems could display an *irregular* or *chaotic motion*. Some years later, a mathematical basis for this behavior was given (Birkhoff, 1932). Only in 1963 in a 'computer experiment' of a model of boundary layer convection was it discovered that a system of three first-order nonlinear differential equations can exhibit a chaotic behavior (Lorenz, 1963). Contrary to Poincaré's example (deterministic chaos in conservative or Hamiltonian deterministic systems), Lorenz discovered *deterministic chaos* in *dissipative systems*.

The Lorenz model may be viewed as the prototype example for many nonlinear dynamical systems, e.g., for the biologically motivated studies on limit cycles in populations (May, 1972) and in brain function (Kaiser, 1977). The development of the digital computer offered an additional tool to study aspects of nonlinear behavior that were previously considered to be too complex.

The essential results of Lorenz are: (1) oscillations with a pseudo-random time behavior (now called *chaotic*); (2) trajectories that oscillate chaotically for a long time before they run into a static or periodic stable stationary state (*preturbulence*); (3) some trajectories alternating between chaotic and stable periodic oscillations (*intermittency*); (4) for certain parameter values trajectories appearing chaotic, although they stay in the neighborhood of an unstable periodic oscillation (*noisy periodicity*).

It took another 10 years before the importance of these results was recognized. Since then the number of both theoretical and experimental studies on the complex behavior of simple systems has rapidly increased in all scientific disciplines. Synonyms for complex behavior of nonlinear systems via spatiotemporal instabilities are *cooperative effects* and *long-range coherence* (Fröhlich, 1969), *dissipative structures* (Nicolis and Prigogine, 1977), *self-organization* and *synergetics* (Haken, 1978), *coherent* and *emergent phenomena* (Hameroff, 1987), and *local activity* (Chua, 1998).

As a general result one may conclude the following. New concepts are developing in an emerging field of interdisciplinary research. The common basis is nonlinearity and the temporal evolution and spatiotemporal instabilities are similar in all nonlinear systems, thus permitting a unified description. These concepts comprise fascinating phenomena: irregular or chaotic motion originating from simple steady states, or regular oscillations as well as regular and turbulent spatiotemporal structures originating from spatially homogeneous states. Some problems remain unsolved. For example, the identification of the chaotic attractor in the mathematical theory (Birkhoff, 1932) and in the simulation 'experiment' (Lorenz, 1963) has not yet succeeded, i.e., the chaoticity of the numerically computed results has not been proven in the strong mathematical sense. Furthermore, it should be emphasized that chaos is only a part of the fascinating behavior that nonlinear systems can exhibit. Since already, for regular motion, a huge number of states and bifurcations exist, the pursuit of modern trends and the concern only with chaotic motion should be avoided. In particular, with respect to externally excited systems, regular motion and bifurcations need to be considered as well.

In this chapter, the terminology of nonlinear dynamics and basic concepts needed to describe nonlinear states, complex phenomena and possible transitions are presented, as well as their *response to external stimuli*. To keep the presentation self-contained, only the temporal evolution of excited nonlinear systems is discussed, with models representing simple mechanical oscillators and complex biological rhythms. Mathematical details are omitted as well as spatially irregular and spatiotemporally irregular structures, i.e., fractals and turbulent states, respectively.

1.2 Nonlinear dynamics

1.2.1 Basic concepts

The theoretical analysis of nonlinear phenomena and their stimulation is performed with the help of nonlinear evolution equations. These model equations describe the dynamic behavior; that is, the evolution in time of the system under consideration. Two kinds of modeling approach are appropriate, a continuous description ($\cdot = \mathrm{d}/\mathrm{d}t$, time derivative)

$$\dot{x} = F(x,\mu,t) \tag{1}$$

and a discontinuous description ($n + 1 = n + \Delta t$, Δt scaled to 1)

$$x_{n+1} = f(x_{n,\mu}). \tag{2}$$

The state of the system is $x = (x_1(t),\ldots,x_m(t))$ or $x_n = (x_{1,n},\ldots,x_{m,n})$ with the state variables $x(t), x_n \in R^m$, i.e., m-dimensional systems ($m/2$ degrees of freedom) are considered. F and f are nonlinear vector functions of the variables. These functions depend on a whole series of parameters μ. Equation (1) consists of m nonlinear and coupled ordinary first-order differential equations, whereas Equation (2) represents m nonlinear one-dimensional maps. For our general considerations discrete systems and systems with additional delay terms are neglected (for effects of delay terms see Milton, Chapter 16, this volume). Continuous systems are closer to physical and biological reality.

The variables x span the state space, where the trajectories (starting at initial conditions) advance in time toward limit sets called *attractors*. In principle, four types of attractors exist and they may all coexist with their respective basins of attraction (Figure 1). Attractors represent asymptotically stable solutions. In order to obtain information about the system's behavior and the types of solution, standard methods of nonlinear dynamics have to be applied (see e.g., Kaiser, 1988; Schuster, 1988). Besides *oscillation* and *phase plane diagrams* (Figure 1a,b), *power spectra* (by fast Fourier transformations) yield information regarding the system's stable steady states. For complex periodic, quasiperiodic and chaotic states, highly sophisticated methods have to be employed to yield a clear distinction between the different attractors. Whereas *stable fixed points* (static attractors) are determined by a simple – in most cases – linear stability analysis, the three other types of attractor require a detailed analytical and numerical investigation of the complete nonlinear models.

A series of problems and questions arises. (1) Is an apparently aperiodic state exhibited in a time series from numerical calculations or from experiments really chaotic, or is it quasiperiodic or complex periodic? (2) How can one separate noise and uniform randomness from deterministic irregularities (chaos)? (3) What is the origin of deterministic chaoticity, and how can one measure the strength of chaos? Meanwhile, some methods have been developed that allow for some answers. These methods are a direct continuation of the standard methods (i.e., fixed points, stability analysis, oscillation and phase plane diagrams, power spectra). To keep the discussion within a reasonable range, only the most characteristic measures are given.

(1) *Lyapunov exponent:* This measures the convergence or divergence of nearby trajectories. Equation (1) has m exponents; a stable fixed point has m negative exponents; a stable periodic (quasiperiodic) motion has at least one (two) exponent(s) equal to zero, the others being negative; a chaotic attractor is represented by at least one positive exponent; while, in addition, at least two nonpositive exponents have to exist. Besides this dynamic measure for which the long-term behavior of the system is needed, static measures are adequate to separate chaotic from nonchaotic motion. Two examples are discussed.

(2) *Fractal dimension d_F:* Different dimensions can be defined for strange attractors, including the Hausdorff, information, capacity and correlation dimensions. All these have noninteger values for chaotic states and can be calculated by embedding and box-counting methods.

(3) *Kolmogorov entropy K:* A fundamental measure for chaotic motion, representing the average rate at which information about the state of a dynamical system is lost in time. For a regular motion, K becomes zero; for random systems it is infinite. Deterministic chaos exhibits a small, positive K-value. Meanwhile an enormous amount of literature exists where both the mathematical background and the details regarding applications of the methods are described (e.g., Ruelle, 1989).

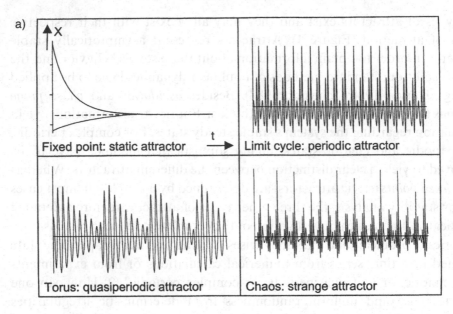

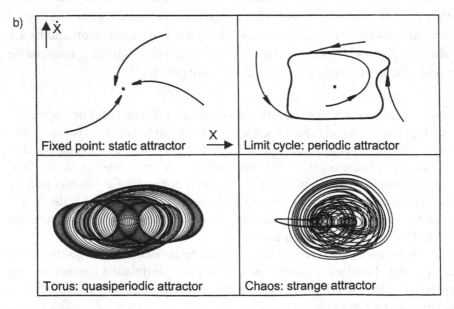

Figure 1. The four types of asymptotically stable solutions (attractors) existing in dissipative nonlinear systems. (a) *Oscillation diagrams* (amplitude x versus time t); (b) *phase plane diagrams* (variable $y = \dot{x}$ versus amplitude x). The system's behavior is governed by one of these steady state solutions, the trajectories finally join these stable states. The diagrams represent typical examples; a different model is chosen for each of the eight diagrams.

Measured time signals are one dimensional and discrete in most cases. However, in nonlinear systems this single coordinate contains information about the other variables. Certain procedures have been developed to construct the attractor from the data set by embedding and delay techniques and to extract the inherent information (Schuster, 1988). The fractal dimension, d_F, and the embedding dimension provide a strong indication of the number of relevant degrees of freedom in the dynamics of the system.

Chaotic states are characterized by their *initial state sensitivity*, leading to a loss of final state predictability. This behavior results from the divergence of nearby trajectories in at least one direction of phase space. In the latter case one Lyapunov exponent is positive. Chaotic motion is irregular and complex, yet it is stable, spatially coherent and completely deterministic. Refined methods have been developed to distinguish and to extract random from chaotic motion. These problems as well as the methods to control chaotic motion are discussed in several chapters of this volume, for example see Ditto and Spano, Chapter 15, this volume.

1.2.2 Principal aspects of driven nonlinear systems

Externally driven nonlinear systems exhibit an enormous variety of behaviors. If at least one internal or external parameter is changed, the system undergoes continuous or discontinuous changes and transitions from one attractor to another at some critical value. The relevant and determining parameter is called *control* or *bifurcation parameter*. The transitions via instabilities are *bifurcations* of steady-state solutions of the dynamical system.

There are three types of local bifurcation. (1) *Hopf bifurcation*: typical examples are transitions form a static attractor (fixed point) to a periodic attractor (limit cycle) and from the latter to the motion on a torus. (2) *Saddle-node* or *tangent bifurcations*: transitions from a limit cycle to a new one, discontinuous transitions in hysteresis, and transitions from quasiperiodic to phase-locked periodic states are the dominating bifurcations. (3) *Period-doubling bifurcations*: a limit cycle of period, T, bifurcates into an oscillation with period $2T$, which, in many cases, is followed by a whole cascade of further period-doubling bifurcations to states with periods $4T, 8T, \ldots, 2^n T$. Period-tripling and multiplying bifurcations can also occur.

Hopf and saddle-node bifurcations can create new frequencies, either in an incommensurate ratio to the original one, or as subharmonics of the external driver frequency, ω_{ext}, for an external signal, $F(t) = F_1 \cos(\omega_{ext} t)$, the latter being harmonic for simplicity. All three types of local bifurcation can terminate in chaotic motion, representing the three generic *routes to chaos* (Schuster,

F. Kaiser

1988). Having in mind the external stimulation of biological rhythmicity, we restrict the discussion to externally driven limit cycles and their respective responses. A limit cycle (periodic attractor) represents a *self-sustained oscillation*, the period and amplitude of which is completely determined by the internal parameters and no external forcing is required (*active* oscillator). Figure 2 shows essential aspects of the nonlinear response of externally perturbed systems with main emphasis on *sub-* and *superharmonic resonances*.

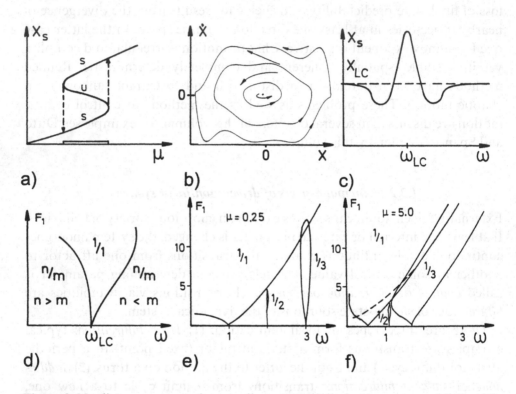

Figure 2. Nonlinear response of externally perturbed dissipative systems. (a) *Bistability*: steady state amplitude x versus bifurcation parameter μ; the arrows show the transitions for increasing and decreasing μ, leading to a hysteretic behavior (u is the unstable branch). (b) *Multi-limit-cycle system* (Equation 3): phase plane diagram ($\dot{x}$ versus x) containing one unstable fixed point, two stable limit cycles and an unstable one in between. (c) *Steady-state response of an externally driven limit cycle*: amplitude x_s versus frequency ω for a fixed driver strength, x_{LC} (ω_{LC}) is the amplitude (frequency) of the unperturbed limit cycle. (d) *Resonance diagram*: principal response of a driven limit cycle to the external driver strength, F_1, with frequency, ω ($F(t) = F_1 \cos(\omega t)$). Besides the main resonance (1/1 Arnold tongue), a large number of subharmonic ($n < m$, $\omega > \omega_{LC}$) and superharmonic ($n > m$, $\omega < \omega_{LC}$) resonances exist. (e–f) *Resonance diagrams for a driven Van der Pol oscillator* (Equation 3 with $\alpha, \beta = 0$, only three tongues are shown), indicating that an increasing internal dissipation (parameter μ) leads to strong changes in the resonance structure.

A prototypical limit-cycle oscillator is the Van der Pol oscillator. Its generaliz-ed version is given by the equation (Kaiser, 1980, 1981)

$$\ddot{x} + \mu(x^2 - 1 + \alpha x^4 + \beta x^6)\dot{x} + x = F(t) \tag{3}$$

(where `¨` denotes the second differential with respect to time) or equivalently as a first-order system (Equation (1))

$$\dot{x} = y$$
$$\dot{y} = -\mu(x^2 - 1 + \alpha x^4 + \beta x^6)y - x + F(t), \tag{4}$$

where μ is a measure for the internal dissipation. The case $\alpha = 0$ and $\beta = 0$ represents the Van der Pol oscillator, $F(t) = F_1 \cos(\omega t)$. Figure 3 depicts a detailed resonance diagram of the large-amplitude limit cycle (Equation (3) and Figure 2b; Kaiser and Eichwald, 1991). The sequence of resonances is Farey ordered, between the resonances, called *resonance horns* or *Arnold tongues*; many additional resonances (periodic states) with decreasing width

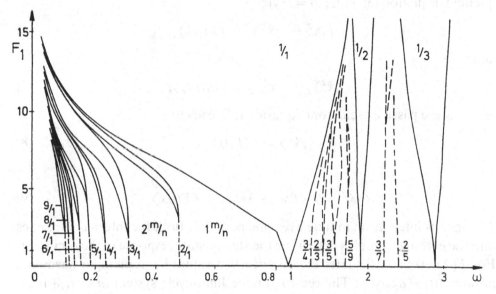

Figure 3. Resonance diagram. Response of the large-amplitude limit cycle of Figure 2b to an external periodic drive with frequency, ω, and strength, F_1. For $0 < \omega \leq 0.8$ the frequency scale is expanded by a factor of 4 to enlarge the superharmonic resonances. All resonances are ordered by the Farey construction rule: between tongues a/b and c/d one finds $(a + c)/(b + d)$, e.g., between 1/2 and 2/5 one gets the 3/7 resonance etc. The denominator determines the periodicity of the complex periodic oscillations. The *Farey sequences* are found in most nonlinear continuous systems, whereas another ordering principle, the U-sequence, is dominant in the small regions where resonance horns (Arnold tongues) overlap (see Figure 2f). The latter sequence is general in discrete systems (nonlinear maps; Equation (2)).

exist. The states in between are *quasiperiodic*. Only within very restricted areas in the F_1–ω plane – mainly where resonance horns overlap – chaotic states do exist. Their measure tends to zero. This is a general feature of driven limit cycles, which is completely different from driven passive nonlinear oscillators. The latter represent driven fixed points. In the latter case only a few resonance-like structures exist with chaotic states in between. Quasiperiodic states can exist only when at least two external incommensurate frequencies are applied. The Duffing oscillator is a prototypical passive oscillator, its equation reads

$$\alpha\ddot{x} + \gamma\dot{x} + \beta x + \delta x^3 = F(t). \tag{5}$$

It represents, for example, the linearly damped motion in a doublewell poten-tial ($\beta < 0$, $\delta > 0$) under the influence of an external stimulus. The principal difference between an *active* and a *passive* oscillator can easily be demon-strated. From the energy balance per period (action balance per period), i.e., multiplying the oscillator equation by $\dot{x}$ (by x) and integrating over one unknown period T, one gets the steady-state amplitude (steady-state fre-quency). Equation (3) with $\alpha, \beta = 0$ yields

$$\langle\mu(x^2 - 1)\dot{x}^2\rangle_T = \langle F(t)\dot{x}\rangle_T \tag{6}$$

and

$$\langle\dot{x}^2\rangle_T = \langle x^2\rangle_T + \langle F(t)x\rangle_T, \tag{7}$$

whereas the passive oscillator (Equation (5)) leads to

$$\langle\gamma\dot{x}^2\rangle_T = \langle F(t)\dot{x}\rangle_T \tag{8}$$

and

$$\langle\alpha\dot{x}^2\rangle_T = \langle\beta x^2 + \delta x^4\rangle_T + \langle F(t)x\rangle_T. \tag{9}$$

$\langle\;\rangle_T$ means integration in time over one period T. Both the internal dynamics and the external stimulus determine the steady-state response of the oscillator. For $F(t) = 0$ one calculates the unperturbed values by applying a harmonic ansatz: $x(t) = a\cos\omega t$. The results for the limit-cycle system ($a = x_{LC}$) read: $x_{LC} = 2$, $\omega_{LC} = 1$, whereas for the double-well system ($a = x_s$) $x_s = 0$, $\omega^2 = \beta + \frac{3}{4}\delta x_s^2$. The amplitude dependence of ω only occurs for $x_s \neq 0$; the oscillator must be driven ($F(t) \neq 0$).

Limit cycles require at least a two-variable system, i.e., two nonlinear first-order differential equations are necessary. Sufficient conditions, e.g., in chemical reaction systems involving Hopf bifurcations (i.e., LC behavior) are: (1) at least a three-molecular step, e.g., representing a quadratic autocatalytic process, or an autocatalytic step plus a nonlinear production rate for a two-dimensional system; (2) at least one two-molecular step for $d = 3$. Three coupled first-order equations are necessary for an autonomous system (no external drive) to become chaotic. However, this is not sufficient in many cases.

Figure 4 displays the *Farey construction principle* governing the response of driven limit cycles. Knowing the parents, the topology of all daughters can be deduced. Having the information of one parent (e.g., 1/1) and one daughter (e.g., 3/4), the structure of the other parent (2/3) is determined. This property exhibits a specific method of information encoding and its subsequent decoding. Figure 5 depicts the response of the Van der Pol oscillator, showing the essential aspects contained in *oscillation* and *phase plane diagrams* and in *power spectra*. Only the external frequency is varied in the diagrams.

1.2.3 Consequences for the system's behavior

Besides sub- and superharmonic resonances, externally driven limit cycles exhibit coexistence, leading to multistability and hysteretic behavior. In addition, global bifurcations create crisis-induced intermittent states and merging

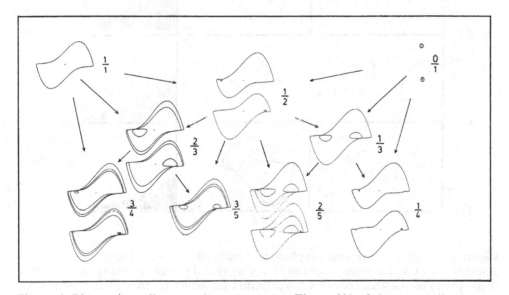

Figure 4. Phase plane diagrams ($\dot{x}$ versus x, see Figure 2b) of the externally driven, strongly dissipative Van der Pol oscillator (Equation 3) exhibiting the Farey construction principle, here in the subharmonic resonance regime. The Farey parents (0/1 and 1/1) and the three first Farey generations (the Farey daughters 1/2, 2/3 + 1/3, 3/4 + 3/5 + 2/5 + 1/4, respectively) are shown. Only n/m = odd/odd leads to inversion symmetric oscillations, 1/1, 1/3 and 3/5 in the example and both parents of an inversion-symmetric daughter have to be noninversion symmetric. The 4/5 and 1/5 oscillations in Figure 5 belong to the fourth generation: 4/5, 5/7, 5/8, 4/7, 3/7, 3/8, 2/7, 1/5.

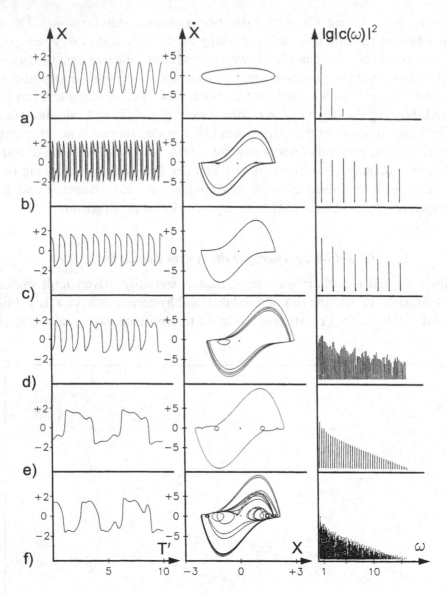

Figure 5. Oscillation diagrams (amplitude x versus time T', scaled to the period of the external drive), phase plane diagrams ($\dot{x}$ versus x) and power spectra (power spectral density $c(\omega)$ versus frequency ω, same arbitrary log-scale), for the driven Van der Pol oscillator. (a) Small dissipation, nearly harmonic 1/1 oscillation, $\omega_{LC} = 1$ dominates. (b–f) Strong dissipation, relaxation-type oscillations, showing the 5/1 superharmonic resonance, the 1/1, 4/5, 1/5 subharmonic resonances, and chaotic response, respectively. Many lines in the power spectra are strong, the dominating line is in all cases the resonance line, n/m. For inversion-symmetric oscillations only odd super- and subharmonic lines occur. In the chaotic spectrum some of the discrete frequencies are exhibited above the 'noisy' background, which, however, is completely deterministic. The external frequency decreases from (b) to (f).

attractors. Nevertheless, resonant, i.e., frequency-dependent responses will still dominate. Sharp resonances and synchronization lead to frequency selectivity and to a multifrequency response. Frequency and intensity windows determined by the resonance horns may lead to threshold and saturation effects, provided a certain n/m resonance may be related to a specific functional state. Transitions from limit cycles to quasiperiodic and chaotic states provide additional behavior, where different bifurcations can presumably lead to different states of information storage and transfer. Knowledge regarding the physical criteria that enable bifurcations to other states is also highly desirable. At least for some cases, there is a stabilization of those states, because the dissipated energy per period is minimal and decreases with increasing driver strength (Kaiser, 1987).

Very slow external signals ($\omega_{ext} \ll \omega_{LC}$) can synchronize the system's fast motion to the slow drive, whereas very fast signals ($\omega_{ext} \gg \omega_{LC}$) create frequency-locked states in the far subharmonic region. Figure 6 represents a quite general scheme. The system either can be in one of three fixed-point states or in a limit-cycle oscillation. The graph shows the system's principal response to weak (no crossing of bifurcation lines) and strong (crossing of bifurcation lines)

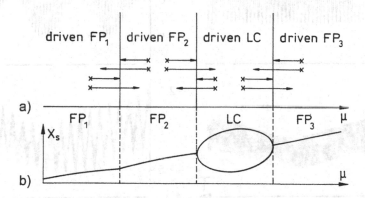

Figure 6. General scheme for an externally driven nonlinear system. The internal or external time-independent bifurcation parameter μ determines the system's steady state, one of the three fixed points (FP) or a limit-cycle oscillation (LC). (a) The strength of a periodic signal (given by the arrows) applied to a steady state (marked by the crosses) determines whether the system is partially driven to a neighboring steady state or remains within the same state. The frequency of the external drive determines whether the system is for many internal oscillations in the neighboring steady state ($\omega_{ext} \ll \omega_{LC}$) or spends there only a small or vanishing part during one cycle ($\omega_{ext} \gg \omega_{LC}$). (b) A typical realization of the unperturbed scheme. With increasing μ, the steady state value x_s increases, FP_1 becomes unstable and FP_2 is stabilized, which, in turn, bifurcates via a Hopf bifurcation into a stable limit cycle. An inverse Hopf bifurcation leads back to a new, stable nonoscillating situation, FP_3. Many biochemical systems exhibit this kind of behavior.

stimuli. The frequency relation ω_{ext}/ω_{LC} determines the system's preferred state, a driven limit cycle or a driven fixed point. In Figure 7 the principal responses of a *passive* and an *active* oscillator to a very *slow* (and weak) and to a very *fast* (and much stronger) stimulus are compared. The figure reveals the pronounced differences in the response of the two systems. These differences may also be deduced from experimental time series.

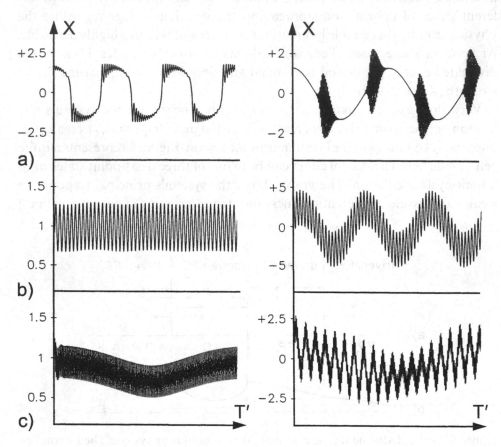

Figure 7. Comparison of the response of a nonlinear passive oscillator (double-well oscillator, Equation 5, with internal frequency ω_{int}, left column) and of an active oscillator (Van der Pol oscillator, Equation 3, frequency ω_{LC}, right column) to an external periodic stimulus with frequency ω_{ext}. (a) $\omega_{ext} \ll \omega_{int}, \omega_{LC}$: the system is entrained even by a weak external signal; that is, it partially decouples within one external cycle and behaves like a damped oscillator (fixed point, left) or a free limit cycle (right). (b) $\omega_{ext} \gg \omega_{int}, \omega_{LC}$: the passive system (driven fixed point, left) oscillates with ω_{ext} within one of the two minima even for rather strong drives; the limit cycle (right) performs free oscillations, its amplitude is high-frequency modulated. (c) Response to the combined influence of the weak, slow and the strong, fast signals. The passive system follows both drives, whereas the limit cycle exhibits all three frequencies. Three external periods are shown in (a), 60 in (b) and 1 in (c).

1.2.4 *Combined influence of very fast and very slow signals*

The response of a driven system to an external stimulus depends on the signal's strength and on its frequency. Detailed investigations have shown that the critical driver strength, $F_{1,c}$, required to synchronize passive and active oscillators in the 1/1 resonance is at least one order of magnitude smaller in the superharmonic case ($\omega_{ext} \ll \omega_{LC}, \omega_{int}$) compared to the subharmonic resonances. Figure 8 displays such a situation, where $F_{1,c}$ ($F_2 = 0$) at least is two orders of magnitude smaller than $F_{2,c}$ for $F_1 = 0$. The combined influence of both stimuli reveals a phase transition-like behavior, small F_1 values ($F_1 < F_2$) lead to bifurcations into the large period-one states (P1$_L$) with values $F_2 < F_{2,c}$. Thresholds, being relevant for monochromatic fast (slow) excitations can dramatically be lowered by an additional slow (fast) signal. Figure 9 shows examples of externally driven oscillators for both, very slow and very fast stimuli, in comparison with the internal dynamics or mechanics and for the combined influence of both. The passive system behaves like a driven fixed point, whereas for the active system the internal limit-cycle frequency is always present, together with ω_1 and/or ω_2.

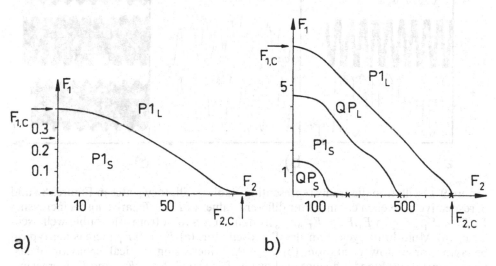

Figure 8. Response of nonlinear oscillators to the combined influence of a fast and a slow external signal, $F(t) = F_1 \cos(\omega_1 t) + F_2 \cos(\omega_2 t)$. (a) Transition of a double-well oscillator (Equation 5) from an *intra*-well oscillation (small, periodic oscillation, P1$_S$) to an *inter*-well oscillation (large, periodic oscillation, P1$_L$) as a function of both F_1 and F_2. (b) Transitions of a multi-limit cycle oscillator (Equation 3; see Figure 2b) from quasiperiodic small oscillations (QP$_S$) via periodic small (P1$_S$) and quasiperiodic large (QP$_L$) to periodic large oscillations (P1$_L$). Note that the F_1 scales are expanded by a factor of 100 compared with F_2, indicating that in the superharmonic region ($\omega_{ext} \ll \omega_{LC}, \omega_{int}$) the critical driver strength required for the transition is much smaller than in the subharmonic regions, $\omega_2 = 1000\omega_1$ in the example.

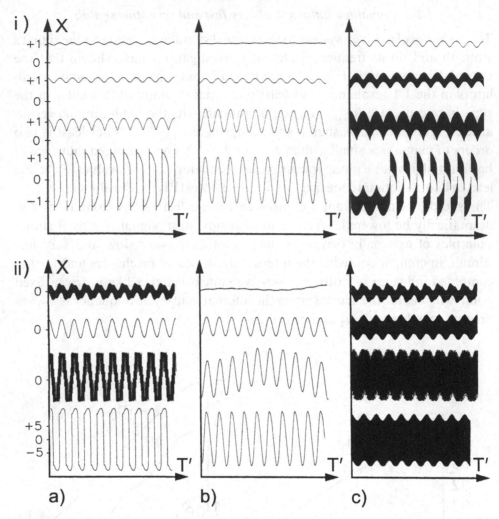

Figure 9. Oscillation diagrams representing stable oscillatory states in Figures 8a and b, respectively. In each column four different values of F_1 or F_2 are chosen, increasing $F_1,F_2 < F_{1,c},F_{2,c}$ to $F_1,F_2 > F_{1,c},F_{2,c}$ from from top to bottom. (i) Double-well oscillator; (ii) Multi-limit-cycle Van der Pol oscillator. (a) $F_2 = 0, F_1$ increasing: typical behavior for a slow excitation; (b) $F_1 = 0, F_2$ increasing: typical behavior of fast excitation; (c) combined influence of slow and fast signal, $F_1 < F_{1,c}$ and F_2 increasing, but $F_2 < F_{2,c}$. The small arrows in Figure 8a indicate the lines along which the series were taken. All diagrams within (i) and (ii) have the same scaling of the x-axis. Ten slow external periods are shown in (a) and (c); 10 fast external periods in (b); $\omega_2 = 1000\omega_1$ for both oscillators.

1.2.5 Combined influence of static, periodic and noisy signals

In physical, chemical and biological systems *noise* is always present. Besides the *internal* noise the *external* signal may exhibit an additional noisy component (from the environment or from neighboring subsystems). Specific states within a system are determined by the internal dynamics, its parameters and by internal noise, the latter being irrelevant for asymptotically stable steady states far enough away from bifurcation lines.

Biological systems are neither completely deterministic nor completely random. The influence of noise is well documented for both experimental results and theoretical investigations. Whereas usually – at least in linear systems – noise leads to a randomization of processes, to a broadening of frequency lines, to a destruction of spatial structures, to thermalization etc., noise may exhibit a constructive role in nonlinear systems. These noise-induced transitions include order-to-order and chaos-to-order transitions, in a way similar to chaos-induced processes such as the well-known three types of crisis, showing enlargement, merging or destruction of attractors (Gassmann, 1997). Noise can induce coherence and coherence resonance (Pikovsky and Kurths, 1997). In coupled systems, an increase in noise can create synchronous firing of stochastically responding model neurons, whereas for stronger noise levels coherence is lost again (Kurrer and Schulten, 1995). Furthermore, a very weak external signal (periodic or aperiodic) can be amplified by constructive interference with noise. This mechanism is well known as *stochastic resonance* (Moss et al., 1994; see also Moss, Chapter 10, this volume). Small-signal amplification and extreme sensitivity to specific frequencies near bifurcation points by a periodic signal have been stressed as further relevant processes (e.g., Wiesenfeld and McNamara, 1986; Kaiser, 1988).

A rather general statement for the influence of noise on nonlinear systems can be given: a driven fixed point (passive oscillator) is rather sensitive to noise, whereas a driven limit cycle is nearly insensitive to noise (e.g., Kurrer and Schulten, 1991; Eichwald and Kaiser, 1995). The situation can change dramatically, if the limit cycle (fixed point) partially becomes a driven fixed point (limit cycle), when a very slow signal is applied to the limit cycle (fixed point) near a bifurcation line (see Figure 6). Such a situation is present in many biological oscillators.

The combined influence of static, periodic and noisy signals, given by

$$F(t) = F_0 + F_1 \cos{(\omega t)} + \xi(t) \tag{10}$$

can have dramatic influences on the system's behavior. To keep the discussion within a reasonable limit, the following set of equations is considered (a special case of Equation (1))

$$\dot{x} = A + f(x) + \alpha xy + F(t)$$
$$\dot{y} = B + g(x) - \alpha xy$$

(11)

with a prototypical nonlinearity xy and general nonlinear functions $f(x)$ and $g(x)$. A second-order differential equation can be derived by eliminating the variable y. It reads ($' = d/dx$)

$$\ddot{x} + \frac{1}{x}(A + \alpha x^2 + f(x) - \dot{x} - f'(x)x + F(t))\dot{x} - \alpha(B + A + g(x) + f(x)$$
$$+ F(t))x = \dot{F}(t)$$

(12)

or in an abbreviated version

$$\ddot{x} + \tilde{g}(x,\dot{x},F(t))\dot{x} + \tilde{f}(x,F(t))x = \tilde{F}(t),$$

(13)

where $\tilde{g}, \tilde{f}$ and $\tilde{F}$ can be extracted directly from Equation (12). This equation represents the prototype of a nonlinear driven oscillator (see Equations (3) and (5)). The external force influences parametrically, both, the nonlinear dissipation (function $\tilde{g}(x,\dot{x},F(t))$) and the amplitude-dependent 'frequency' (function $\tilde{f}(x,F(t))$). Even a static stimulus ($F(t) = F_0$) can perform both influences. The stochastic component operates in an additive (via $\dot{F}(t)$) and in a multiplicative (via $F(t)$) manner. If, for example, Equations (11) describe chemical reactions, both the kinetics and the dynamics are influenced in a deterministic and in a stochastic fashion.

In principle, external stimuli can be even more complicated. Besides a contribution not varying in time, the time-dependent, deterministic component can be periodic, pulsatile or aperiodic, it can be fast or slow compared to the internal dynamics and it can include amplitude or frequency modulation. In this case the response then not only depends on one frequency and the related amplitude as well as the internal state of the system, it also depends on the specific *temporal pattern* of the signal. Nevertheless, general trends can be deduced, because in most cases a few frequencies will dominate. New temporal structures and temporal organizations emerge beyond a critical point of instability of a nonlinear steady state. The same holds for spatial structures, occurring beyond a critical point of instability of a homogeneous state. The combination of both types of instability and subsequent steady states leads to spatiotemporal structures. Therefore much may be learned from simple non-linear models, because the behavior for many systems is generic and the bifurcations and their characterizations obey universal laws.

1.3 Biophysical rhythmicity

Rhythmic phenomena play an important role in many aspects of biological order and function. The involved systems range from the molecular to the macroscopic level, the corresponding periods cover the submicrosecond time domain to hours (molecular and cellular oscillators), days (circadian rhythms) and even months (population cycles). Oscillation dynamics define biological clock functions and are of essential importance in intra- and intercellular signal transmission and in cellular differentiation. The importance of non-linear dynamical concepts in biology is stressed by the following facts: (1) sustained autonomous oscillations do exist (Kaiser, 1980, 1988; Goldbeter, 1996); (2) sudden changes in the system's dynamics indicate bifurcations of nonlinear systems; (3) chaotic dynamics, meaning fractal behavior in the temporal domain, is exhibited by many biochemical and biological systems (Degn *et al.*, 1988; Goldbeter, 1996). The functional role of chaos, however, is still a matter of discussion (e.g., see Ditto and Spano, Chapter 15, this volume). In the present contribution, no biological models will be discussed. Only essential features of nonlinear dynamics are presented by considering certain properties of physical or biochemical oscillators.

1.3.1 Requirements and concepts for a modeling approach

'Things should be made as simple as possible – but not simpler.' A. Einstein

Endogenous biological rhythms often exhibit stable periodic oscillations. These oscillations are modeled by nonlinear differential equations exhibiting self-sustained oscillations (limit cycles). Both the relevant variables (e.g., concentrations of the reacting molecules) and the nonlinear processes (e.g., chemical and enzymatic reactions) have to be known. For a concrete biological situation, for example, a complete reaction chain within a cell, the resulting large set of equations would include many variables and processes. However, if too much information is included, the set of equations cannot be analyzed sufficiently. Restricting to an explanation of a phenomenon and, consequently, to a description of a certain mechanism instead of the whole 'biological reality', one obtains a reduced set of variables that contains the governing nonlinear dynamics. The result is a *minimal model*, containing all the essential elements of a specific process. This is the common procedure for developing models in theoretical physics.

A simple demonstration can explain the method. Assume that a process is described by the following set of equations

$$\dot{q} = F(q) = A_{\underline{q}} + B\tilde{F}(q), \tag{14}$$

$F(q)$ is separated into a linear $(A_{\underline{q}})$ and a nonlinear $(B\tilde{F}(q))$ part, where the vector $q = (\ldots s,t,u \ldots x,y,z \ldots)$ represents n variables. Equation (14) thus consists of n differential equations, that are coupled and nonlinear. The governing dynamics is assumed to be determined by only a few variables. Then all the other variables can be replaced by their steady-state values. By an adiabatic elimination procedure for all the fast (irrelevant) variables, only a few equations result, an example being Equations (11), when only the variables x and y are relevant. This procedure is the basis for the concept of self-organization and synergetics (Haken, 1978). It can equally well be applied to spatiotemporal problems and to delay systems (e.g., see Milton, Chapter 16, this volume).

Subsequently, the methods and concepts of nonlinear dynamics have to be applied to the resulting equations. Furthermore, in order to investigate the response of a biological system, or of parts of it, to an external stimulus, the modeling approach has to be extended to include the additional dynamics. Quite generally, three steps are indispensable: (1) the external signal couples to a molecular target (primary physical mechanisms), (2) a complicated series of internal transduction and transmission processes is activated, (3) the translation via a causal link of biochemical steps creates the system's response (secondary biological mechanisms; see Walleczek, 1995; Kaiser, 1996). Chemical signal transduction mechanisms across cellular membranes within single cells or cell-to-cell communication processes are relevant candidates for biological pathways (see also Walleczek and Eichwald, Chapter 8, this volume).

1.3.2 A paradigmatic model

Many important aspects of stimulated rhythms are already contained in models of externally driven limit cycles. At this point in the modeling approach, the type of the external stimulus (mechanical, chemical, hormonal, electromagnetic, etc.) is irrelevant. It is the information contained in the signal that is significant, however, especially its frequencies and amplitudes and its temporal pattern. These characteristics of the input step must be within a useful range of the subsequent cycle to which the input couples. Then, the original information is encoded, e.g., by rate- or temporal-encoding procedures as part of the secondary interaction mechanism. At least for the second messenger calcium (Ca^{2+}), frequency encoding instead of amplitude encoding seems to be relevant for information processing. For example, experimental evidence exists showing that proteins can decode intracellular Ca^{2+} oscillations (De Koninck and Schulman, 1998). Finally, the information is transmitted through the pathway, and the relevant features are extracted by the final cycle for further processing. This encoding procedure has to be fast and efficient to be able to induce an unequivocal response.

To keep the presentation within a reasonable limit, a model is discussed that contains essential elements of a coupled and stimulated system (Figure 10). Each model component can be exchanged with any other passive or active oscillator. The model consists of seven elements, five passive oscillators, coupled in a symmetric array, and two active oscillators. The passive system's output $x_5(t)$ is a complex signal, containing information of the external signal, the noise and the internal dynamics. Both active oscillators have been chosen because of their specific properties that are essential for many biological oscillators: a fixed point for weak and strong stimuli; a limit cycle in between, allowing for threshold and excitability properties (compare Figure 6). The first oscillator is a Ca^{2+} oscillator based upon the Goldbeter–Dupont–Berridge minimal model (GDB; Goldbeter *et al.*, 1990). The second oscillator is represented by the FitzHugh–Nagumo model (FHN), originally introduced to describe the onset of nerve pulses (FitzHugh, 1961). Coupled arrays of FHN oscillators represent important model systems to describe signal transduction in sensory neuronal systems, e.g., indicating that noise may play a functional role via stochastic resonance (Collins *et al.*, 1995). Besides the ever-present internal noise, the system, or a part of it, is exposed to an external noise source, acting synchronously. This means that σ_c can be viewed as spatially coherent noise, whereas σ_i ($i = 1 \dots 7$) is spatially incoherent.

In general, noise is temporally incoherent. Instead of white noise, the numerical simulations use exponentially correlated colored noise $\zeta_i(t,\sigma_i)$, generated through an integral algorithm using a higher-dimensional Orenstein–Uhlenbeck process. σ_i^2 is the variance of the Gaussian-distributed noise amplitudes. Different time scales govern the system's motion, the fluctuations being the fastest, that is, the internal frequencies ($\omega_{int}, \omega_{LC}$) are in the region of 1 hertz = s^{-1}, and the external signal oscillates more slowly, i.e., by two orders of magnitude. The model equations in Figure 10 and the results in Figure 11 represent typical examples for representing the system's dynamics. For simplicity, delay is neglected in all Langevin systems and we take equal coupling strengths ε_i. The passive oscillators (x_1 and x_5) perform transitions from one well to the other one in a nonperiodic fashion. Transitions can occur only when both stimuli, the coherent signal and the noise, are present (subthreshold situation). The Ca^{2+} oscillator exhibits its fast limit-cycle oscillations when system x_5 is in the right-sided potential well, and it acts as a driven fixed point (below limit cycle threshold) when system x_5 is in the left-sided well. The FHN oscillator is a driven fixed point (above limit cycle threshold) when the Ca^{2+}

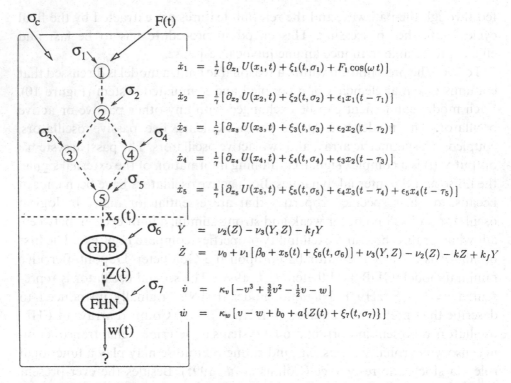

$$\dot{x}_1 = \tfrac{1}{\gamma}[\partial_{x_1} U(x_1, t) + \xi_1(t, \sigma_1) + F_1 \cos(\omega t)]$$

$$\dot{x}_2 = \tfrac{1}{\gamma}[\partial_{x_2} U(x_2, t) + \xi_2(t, \sigma_2) + \epsilon_1 x_1(t - \tau_1)]$$

$$\dot{x}_3 = \tfrac{1}{\gamma}[\partial_{x_3} U(x_3, t) + \xi_3(t, \sigma_3) + \epsilon_2 x_2(t - \tau_2)]$$

$$\dot{x}_4 = \tfrac{1}{\gamma}[\partial_{x_4} U(x_4, t) + \xi_4(t, \sigma_4) + \epsilon_3 x_2(t - \tau_3)]$$

$$\dot{x}_5 = \tfrac{1}{\gamma}[\partial_{x_5} U(x_5, t) + \xi_5(t, \sigma_5) + \epsilon_4 x_3(t - \tau_4) + \epsilon_5 x_4(t - \tau_5)]$$

$$\dot{Y} = \nu_2(Z) - \nu_3(Y, Z) - k_f Y$$

$$\dot{Z} = \nu_0 + \nu_1[\beta_0 + x_5(t) + \xi_6(t, \sigma_6)] + \nu_3(Y, Z) - \nu_2(Z) - kZ + k_f Y$$

$$\dot{v} = \kappa_v[-v^3 + \tfrac{3}{2}v^2 - \tfrac{1}{2}v - w]$$

$$\dot{w} = \kappa_w[v - w + b_0 + a\{Z(t) + \xi_7(t, \sigma_7)\}]$$

Figure 10. A paradigmatic model. Five double-well systems (1–5) are coupled in a symmetric configuration. Each of them represents a bistable passive state. The single oscillator equation is given by Equation (5); $\alpha = 0$ is reasonable for $\gamma > 0.5$ (strong damping allows for the adiabatic limit, leading to Langevin systems). $\partial_{xi} U_i = -\beta x_i - \delta x_i^3$. Only the first oscillator is stimulated by a harmonic force $F(t)$. The output of one oscillator acts as the input of the next one, $\varepsilon_i x_i(t - \tau_i)$, with coupling strength ε_i and time delay τ_i. The complex output of the double-well system 5 is coupled to an oscillator, which is in a steady state for both weak and strong stimuli and performs limit-cycle oscillations in between (compare Figure 6). As an example the Goldbeter–Dupont–Berridge (GDB) Ca^{2+} oscillator minimal model is employed. Its output, in turn, is injected into a second active oscillator with the same properties (either a fixed point or a limit cycle). As an example the FitzHugh–Nagumo (FHN) model has been chosen. All seven systems are selected for their steady-state properties and without any intention of implementing a realistic biological system. Each system is subjected to its own noise term $\xi_i(t, \sigma_i)$ with noise strength σ_i; σ_c is an additional noise source acting synchronously, i.e., spatially coherent, on systems 1–5 or 1–7, respectively. The parameters are chosen that systems 1–5 are double-well oscillators, the GDB–(FHN) system is a limit cycle (fixed point above limit cycle) for vanishing inputs. Details of the different models are irrelevant at this stage of discussion. The ν_i are linear ($i = 0,1$) and nonlinear ($i = 2,3$) fluxes; the variables Y and Z denote the calcium concentration of an intracellular calcium store and the intracellular calcium concentration, respectively; v represents a fast-switching variable (e.g., a voltage in the case of a membrane); w is a slow recovery variable.

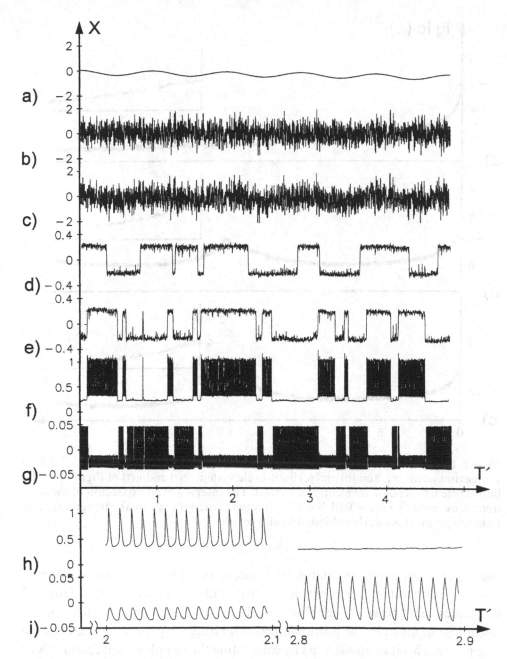

Figure 11. Oscillation diagrams for (a) $(F_1(t) = F_1 \cos(\omega t)$, (b) noise $\xi(t)$, (c) $F_1(t) + \xi(t)$, (d) + (e) double-well systems $x_1(t)$ and $x_5(t)$, (f) $z(t)$, (g) $w(t)$. Time is scaled to the external slow harmonic stimulus. (h) and (i) show the limit-cycle oscillations and the driven fixed points of (f) and (g) on an extremely expanded time scale. $\sigma_i\,(i = 1 \text{ to } 7) \equiv \sigma_{inc} = 0.6;\ \sigma_c = 0,\ \omega = 0.01,\ F_1 = 0.3.$

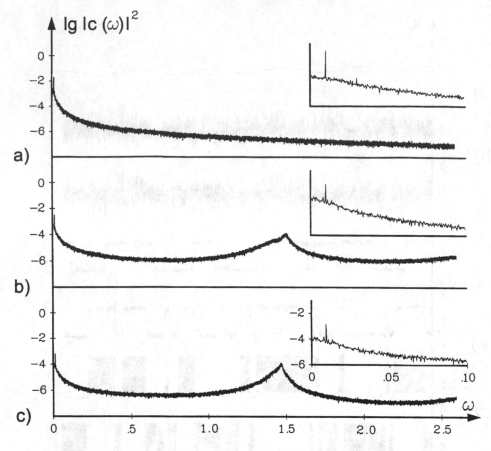

Figure 12. Fourier spectra (power spectral density $c(\omega)$ on a log-scale) for (a) fifth Langevin system, x_5, and (b) and (c) limit-cycle systems $z(t)$ and $w(t)$ of Figure 11. The limit-cycle frequencies are clearly expressed. The inserts are an expansion of the low-frequency parts. The $\omega = 0.01$ line (low-frequency signal with weak strength applied only to system 1) is clearly exhibited in all spectra.

oscillator is an active oscillator, and vice versa. The weak external signal, $\cos(\omega t)$, cooperates in a constructive way with the noisy inputs. Figure 12 illustrates that the weak signal is transduced to all systems. Furthermore, it becomes amplified by the noise sources, exhibiting the properties of stochastic resonance. This is demonstrated by calculating the signal-to-noise ratio (SNR) for systems 5–7. The results in Figure 13 reveal that either σ_{inc} or σ_c can amplify the weak coherent signal, whereas the combined influence of both noise sources has dramatic consequences for signal amplification, similar to the phase transition-like behavior presented in Figure 8. For example, if the system operates below its amplification maximum, a weak uniform noise

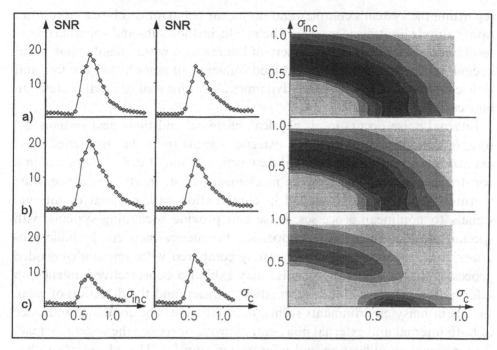

Figure 13. Signal-to-noise ratio (SNR) versus noise strength σ. (a) Langevin system $x_5(t)$, (b) GDB oscillator $z(t)$, (c) FHN oscillator $w(t)$. Left column: SNR as a function of σ_{inc}, spatially incoherent noise is applied to systems 1–7 and $\sigma_c = 0$. Middle column: SNR as a function of σ_c, spatially coherent noise applied to systems 1–5 or 1–7 (which makes no difference) and $\sigma_{inc} = 0$. Right column: gray-scale plot of the SNR (from white to black as the SNR increases) as a function of both σ_c and σ_{inc}. Other parameters are as in Figure 11.

contribution (spatially coherent noise σ_c) can enhance signal amplification, whereas a system that is already operating in its optimal mode, may be influenced only in a destructive way by a small σ_c contribution.

Much information may be obtained from coupled nonlinear models such as the one presented here. The inclusion of delay terms and different coupling strengths (with and without backcoupling) offers additional extraordinary motion and shows the way to proceed from simple models to the complex biological behavior relevant for rhythmicity, information transfer and, hence, final biological function.

1.4 Conclusions

This overview demonstrates that the concepts of nonlinear dynamics in general, and of externally driven active oscillators in particular, are indispensable to the description and analysis of biological rhythms. Much information regarding the system's complex and nonlinear behavior can be deduced from rather simple limit-cycle models. The results include sub- and superharmonic resonances as well as a large variety of bifurcations within regular states and irregular, chaotic ones. The combined influence of stimuli that are fast and slow compared with the internal dynamics allows for a substantial decrease in bifurcation thresholds.

Internal noise occurs in all physical, chemical and biological systems. Its role, in combination with weak external signals (periodic or aperiodic), is extraordinary. Instead of its usual destructive action, it can also operate in a constructive manner, e.g., by the mechanisms of stochastic resonance. Furthermore, when noise is applied in combination with external or internal signals to nonlinear processes, noise can provide oscillating systems with special information-encoding properties. Frequency-encoded rhythmic processes seem to operate more accurately compared with amplitude-encoded processes. In those situations noise may exhibit a constructive influence on information processing and information transfer, and the detection of weak signals in noisy environments seems, thus, possible. The combined influence of both internal and external noise can improve or reduce the system's capacity for signal amplification and information transfer. This phenomenon has been demonstrated by the results with the paradigmatic model in Section 1.3.2. The constructive role of internal noise in information transfer via an increase in *external* noise has been shown quite recently as well (Gailey *et al.*, 1997).

Temporal structures may display a huge amount of diverse patterns. Irregular temporal dynamics exhibits scaling and universal behavior. Chaotic states contain a very large number of unstable period orbits, to which the system may be stabilized by chaos control techniques (for details, see Showalter, Chapter 14, this volume). For many problems in physics, chemistry and biology, the spatial variations have to be analyzed as well. Many nonlinear spatiotemporal systems are governed by a few universal equations, the most prominent of which are real or complex Ginzburg–Landau equations. Equations of this type result when a small-amplitude expansion near a bifurcation point (Hopf bifurcation in many cases) is performed. The typical structure, when restricting to one complex variable A, is given by the amplitude equation

$$\partial_t A = \mu A + \alpha |A|^2 A + \beta \Delta A, \tag{15}$$

where μ, α, β are complex parameters, ∂_t denotes derivation with respect to time, and Δ denotes the second spatial derivations perpendicular to the direction of propagation, describing diffusion and diffraction. Equation (15) exhibits many regular spatiotemporal solutions (rolls, hexagons, squares, traveling waves etc.). Disordered structures occur via spatiotemporal instabilities. These spatiotemporal chaotic or 'turbulent' states may be controlled or suppressed by refined chaos control techniques, since they contain a very large number of unstable patterns to which the system may be stabilized in a way similar to temporal chaos control. Furthermore, with very weak control signals, preselected regular patterns can be generated.

Finally, two fundamental questions remain. (1) How do nonlinear oscillations arise in a concrete biological system or in a functional subunit making up part of the system? (2) What is the function of these oscillations, i.e., how does functional order translate into biochemical signaling and biological function? Answers to the first question are accumulating for many different systems. With respect to the second question, most answers are more speculative in nature. New experiments have already demonstrated, however, that specifically the *frequency* of cellular Ca^{2+} oscillations could indeed determine the specificity and magnitude of gene expression in cells (Dolmetsch *et al.*, 1998; Li *et al.*, 1998). The further exploration of these questions is the subject of many of the remaining chapters in this volume.

References

Bernhard, R. (1964) Survey of some biological aspects of irreversible thermodynamics. *J. Theor. Biol.* 7: 532–557.

Birkhoff, G. D. (1932) Sur quelques courbes fermées remarquables. *Bull. Soc. Math. Fr.* **60**: 1–26.

Chua, A. (1998) *CNN: A Paradigm for Complexity*. Singapore: World Scientific.

Collins, J. J., Chow, C. C. and Imhoff, T. T. (1995) Stochastic resonance without tuning. *Nature* **376**: 236–237.

De Koninck, P. and Schulman, H. (1998) Sensitivity of CaM kinase II to the frequency of Ca^{2+} oscillations. *Science* **279**: 227–230.

Degn, H., Holden, A. V. and Olsen, L. F. (eds.) (1988) *Chaos in Biological Systems*. New York: Plenum Press.

Dolmetsch, R. E., Xu, K. and Lewis, R. S. (1998) Calcium oscillations increase the efficiency and specificity of gene expression. *Nature* **392**: 933–936.

Eichwald, C. and Kaiser, F. (1995) Model for external influences on cellular signal transduction pathways including cytosolic calcium oscillations. *Bioelectromagnetics* **16**: 75–85.

FitzHugh, R. (1961) Impulses and physiological states in theoretical models of nerve membrane. *Biophys. J.* **1**: 445–466.

Fröhlich, H. (1969) Quantum mechanical concepts in biology. In *Theoretical Physics and Biology* (ed. M. Marois), pp. 13–22. Amsterdam: North-Holland.

Gailey, P. C., Neiman, A., Collins, J. J. and Moss, F. (1997) Stochastic resonance in ensembles of nondynamical elements: the role of internal noise. *Phys. Rev. Lett.* **79**: 4701–4704.

Gassmann, F. (1997) Noise-induced chaos–order transitions. *Phys. Rev. E* **55**: 2215–2221.

Goldbeter, A. (1996) *Biochemical Oscillations and Cellular Rhythms.* Cambridge: Cambridge University Press.

Goldbeter, A., Dupont, G. and Berridge, M. J. (1990) Minimal model for signal-induced Ca^{2+}-oscillations and for their frequency encoding through protein phosphorylation. *Proc. Natl. Acad. Sci. USA* **87**: 1461–1465.

Haken, H. (1978) *Synergetics: An Introduction.* Berlin: Springer-Verlag.

Hameroff, S. R. (1987) *Ultimate Computing: Biomolecular Consciousness and Nanotechnology.* Amsterdam: North-Holland.

Hill, A. V. (1930) Membrane-phenomena in living matter. *Trans. Faraday Soc.* **26**: 667–678.

Kaiser, F. (1977) Limit cycle model for brain waves. *Biol. Cybern.* **27**: 155–163.

Kaiser, F. (1980) Nonlinear oscillations (limit cycles) in physical and biological systems. In *Nonlinear Electromagnetics* (ed. P. L. E. Uslenghi), pp. 343–389. New York: Academic Press.

Kaiser, F. (1981) Coherent modes in biological systems – perturbations by external fields. In *Biological Effects of Nonionizing Radiation* (ed. K. H. Illinger), pp. 219–241. Washington, DC: American Chemical Society.

Kaiser, F. (1987) The role of chaos in biological systems. In *Energy Transfer Dynamics* (eds. W. Barett and H. Pohl), pp. 224–236. New York: Springer-Verlag.

Kaiser, F. (1988) Theory of nonlinear excitations. In *Biological Coherence and External Stimuli* (ed. H. Fröhlich), pp. 25–48. New York: Springer-Verlag.

Kaiser, F. (1996) External signals and internal oscillation dynamics: biophysical aspects and modeling approaches for interactions of weak electromagnetic fields at the cellular level. *Bioelectrochem. Bioenerg.* **41**: 3–18.

Kaiser, F. and Eichwald, C. (1991) Bifurcation structure of a driven, multi-limit-cycle Van der Pol oscillator. *Int. J. Bifurc. Chaos* **1**: 485–491.

Kurrer, C. and Schulten, K. (1991) Effect of noise and perturbations on limit cycle systems. *Physica D* **50**: 311–320.

Kurrer, C. and Schulten, K. (1995) Noise-induced synchronous neural oscillations. *Phys. Rev. E* **51**: 6213–6218.

Li, W. H., Llopis, J., Whitney, M., Zlokarnik, G. and Tsien, R. Y. (1998) Cell-permeant caged InsP$_3$ ester shows that Ca^{2+} spike frequency can optimize gene expression. *Nature* **392**: 936–941.

Lorenz, E. N. (1963) Deterministic nonperiodic flow. *J. Atmosph. Sci.* **20**: 130–141.

May, R. M. (1972) Limit cycles in predator–prey communities. *Science* **177**: 900–902.

Moss, F., Pierson, D. and O'Gorman, D. (1994) Stochastic resonance. *Int. J. Bifurc. Chaos* **4**: 1283–1297.

Nicolis, G. and Prigogine, I. (1977) *Self-organization in Non-equilibrium Systems.* New York: Wiley.

Pikovsky, A. and Kurths, J. (1997) Coherence resonance in a noise-driven excitable system. *Phys. Rev. Lett.* **78**: 775–778.

Poincaré, H. (1892) *Les Méthods Nouvelles de la Mécanique Céleste.* Paris: Gauthier-Villars.

Rashevsky, N. (1938) *Mathematical Biophysics.* Chicago: Chicago University Press.

Ruelle, D. (1989) *Chaotic Evolution and Strange Attractors.* Cambridge: Cambridge University Press.

Schrödinger, E. (1945) *What is Life?* Cambridge: Cambridge University Press.

Schuster, H. G. (1988) *Deterministic Chaos.* Weinheim: Physik Verlag.

Turing, A. M. (1952) The chemical basis of morphogenesis. *Phil. Trans. Roy. Soc. Lond. B* **237**: 37–72.

Von Bertalanffy, L. (1932) *Theoretische Biologie I*. Berlin: Borntraeger.

Walleczek, J. (1995) Magnetokinetic effects on radical pairs: a paradigm for magnetic field interactions with biological systems at lower than thermal energy. In *Electromagnetic Fields: Biological Interactions and Mechanisms* (ed. M. Blank), *Advances in Chemistry*, No. 250, pp. 395–420. Washington, DC: American Chemical Society.

Wiesenfeld, K. and McNamara, B. (1986) Small-signal amplification in bifurcating dynamical systems. *Phys. Rev. A* **33**: 629–642.

2

Nonlinear dynamics in biochemical and biophysical systems: from enzyme kinetics to epilepsy

RAIMA LARTER, ROBERT WORTH AND
BRENT SPEELMAN

2.1 Introduction

Biological systems provide many examples of well-studied, self-organized nonlinear dynamical behavior including biochemical oscillations, cellular or tissue-level oscillations or even dynamical diseases (Goldbeter, 1996). The latter include such phenomena as cardiac arrhythmias, Parkinson's disease and epilepsy. Part of the reason for progress in understanding these phenomena has been the willingness of investigators to communicate and share insights across disciplinary boundaries, even when this communication is hampered by differing jargon or concepts unfamiliar to the nonexpert. The common language of nonlinear systems theory has helped to facilitate this cross-disciplinary conversation as well as to provide a new definition of what it means to say that two things are dynamically 'similar' or even 'the same'.

In this chapter, we compare the dynamics of a well-studied biochemical oscillator, the peroxidase–oxidase reaction, with that of epilepsy, a dynamical disease (Milton and Black, 1995). We are so accustomed to the normal way of reasoning in science that it seems wrong, somehow, to point out the similarities between, on the one hand, the oscillations in substrate concentration during an enzyme-catalyzed reaction and, on the other, the regular oscillations in the electroencephalography (EEG) signal observed during certain types of epileptic seizure. While these two systems could not be more different in terms of their material nature, they are actually quite similar *dynamically*. By noticing the dynamical similarities between these two systems, we are able to apply the insights from a thorough and long-term study of the enzyme system to the search for possible mechanisms of the origin and spread of partial seizures. These insights into the complex disease of epilepsy would have been much more elusive without the cross-disciplinary comparison of these two phenomena, an approach typical of nonlinear dynamics.

The type of investigation described in this chapter was recently highlighted in a Research News article in *Science* that reviewed theoretical and clinical studies of epilepsy (Glanz, 1997). As pointed out in this article, 'Unlike traditional neuroscience, which often focuses on the details of the brain – neurotransmitters, receptors, and neurons, alone or in small groups – nonlinear dynamics aims to identify the large-scale patterns that emerge when neurons interact *en masse*'. As we will describe below, these large-scale patterns are dynamically similar to those commonly observed in oscillatory chemical reactions. Thus, it makes sense to look to the kinetics of these reactions for clues about the mechanistic source of epileptic dynamics.

Another example of the fruitful interplay between disciplines that occurs in the field of nonlinear science involves the recent flurry of activity in studying calcium (Ca^{2+}) oscillations (Cuthbertson and Cobbold, 1991; Berridge, 1993). The rapidity with which these oscillations, observed in different cell types (muscle, liver, oocyte, glial, etc.), were identified as being dynamically similar can be attributed directly to the comparisons made between these latter observations and well-known chemical oscillators that had already been studied for decades. Once dynamical similarity had been noted, a suggestion was made to look for Ca^{2+} waves, since the similarity to chemical systems suggested that spiral waves ought, also, to exist in cellular systems. Spiral waves of Ca^{2+} activity were, indeed, observed in some of the cell types studied (Lechleiter *et al.*, 1991). The observation of these waves and their similarity to both chemical waves (Epstein, 1991) and waves of electrical excitation in excitable cardiac tissue (another biological system whose understanding has been greatly facilitated by comparison with chemical systems) lent further support to the notion that the Ca^{2+} oscillations observed in different cell types were *dynamically similar*. This type of investigation was very helpful in extracting order from a broad set of observations that otherwise would not have been deemed 'similar' by most investigators.

In this chapter, we review the research, both theoretical and experimental, that has been carried out on a well-studied enzyme oscillator, the peroxidase–oxidase reaction, and briefly describe how the insights from our research with this system have allowed us to propose a mechanism for the origin and spread of epileptic seizures. The procedure uses the notion of dynamical similarity, again, by noting the feedback characteristics of certain subnetworks in the hippocampal region of the brain and modeling the subnetwork dynamics using an approach drawn from enzyme kinetics.

2.2 The peroxidase–oxidase reaction

Since the observation of oscillations in the horseradish peroxidase-catalyzed oxidation of NADH over 30 years ago (Yamazaki *et al.*, 1965; Nakamura *et al.*, 1969), extensive studies of this system have been carried out. This reaction has recently been reviewed both in terms of its dynamical features (Larter *et al.*, 1993) and its constituent chemistries (Scheeline *et al.*, 1997). When this reaction takes place in a flow system with reduced nicotinamide adenine dinucleotide (NADH) as the reductant, the concentrations of reactants (oxygen and NADH) as well as some enzyme intermediates, can be seen to oscillate with periods ranging from several minutes to about an hour, depending on the experimental conditions. The peroxidase–oxidase (PO) reaction exhibits many complex dynamical behaviors including *bistability*, *birhythmicity, quasiperiodicity, complex oscillatory behavior* and *chaos*. As the only single enzyme system to exhibit *in vitro* oscillations in homogeneous, stirred solutions, the PO reaction is the simplest nonlinear biochemical system known. In more complex biochemical systems, such as the glycolysis reaction, metabolic control features such as allosteric enzyme kinetics are also operative. Because this represents an additional means of regulation not available to the PO system or to purely inorganic chemical oscillators, the PO system has been said to be intermediate in type between chemical oscillators and the much more complex, but highly regulated, biochemical oscillators (Goldbeter, 1996).

The PO reaction occurs as the first step in a sequence of reactions in plants that eventually culminates in the production of lignin (Mäder and Amberg-Fisher, 1982; Mäder and Füssl, 1982); it also is involved in the important processes of the photosynthetic dark reactions (Pantoja and Willmer, 1988). At this point, it is not known whether the oscillations observed in the flow system have any bearing on behavior *in vivo*; however, recent studies with horseradish cell extracts revealed the existence of damped oscillations, indicating that oscillations are possible *in vivo* as well (Møller *et al.*, 1998). Oscillatory behavior is known to occur in other biochemical settings and seems to be a ubiquitous phenomenon at many levels in biological systems. The existence of chaotic behavior in the PO reaction brings to mind many intriguing questions, for example if chaos can occur in a single enzyme reaction (such as the PO reaction) does the existence of multiple enzyme networks of reactions make chaos inevitable *in vivo*? Is chaos a sign of health or disease?

The overall reaction that we refer to as the PO reaction is:

$$2NADH + O_2 + 2H^+ \rightarrow 2NAD^+ + 2H_2O \tag{1}$$

Figure 1 shows typical experimental results that can be obtained from the *in vitro* study of this reaction as the concentration of the coenzyme, 2,4-dichlorophenol (DCP) is varied. In the second set of panels in the figure, a comparison is made with simulations using one of the more recent proposed mechanisms for the PO reaction. As can be seen from this figure, the theoretical studies are in quite good agreement with the experimental findings, indicating that recently proposed mechanisms are beginning to converge on reality.

A great deal of experimental and theoretical work has been devoted to determining the mechanism by which oscillations and chaos arises in the PO reaction. Although some controversy still exists regarding the mechanistic basis of *complex* oscillatory behavior, including chaos in this system, the origin of *simple* oscillatory behavior is solidly established. The part of the reaction mechanism widely agreed to give rise to simple oscillatory behavior in the PO reaction is illustrated in the reaction network shown in Figure 2. The free radical species, NAD$^{\bullet}$, plays a central role by being involved in a positive feedback loop, i.e., an autocatalytic reaction. The autocatalysis leads to an explosion in the NAD$^{\bullet}$ concentration, which would proceed unchecked, were it not for the free radical termination reactions that also occur. The latter may involve bimolecular radical–radical dimerization or, perhaps, unimolecular deactivation via collisions with the container wall. Either way, the oscillatory behavior may be understood as an alternation between an explosive production of free radicals and a rapid termination, followed by another cycle of production, then termination, etc.

In addition to oscillatory and chaotic behavior, the PO reaction exhibits bistability. Although this phenomenon has not attracted as much interest as have the more exotic dynamical effects of oscillations and chaos, it is mentioned here because of its suspected kinetic source: the inhibition of the enzyme by molecular oxygen. Thus, in the reaction mechanism for this system we have both autocatalysis and inhibition, i.e., positive and negative feedback. In the following section we will see how these same dynamical features arise in models of brain dynamics. In the latter types of model, positive and negative feedback exist in the form of excitatory and inhibitory neuronal connections, respectively. There is very little, if any, difference between the dynamics of autocatalysis in enzyme kinetics and excitatory feedback in neuronal dynamics. Similarly, inhibition in neuronal systems is modeled in a fashion identical with that of inhibition in enzyme kinetics.

A simple model for the PO reaction, proposed in 1979, provides a good example of the use of nonlinear dynamical techniques in elucidating the origin of certain dynamical features of this system (Degn *et al.*, 1979). It is a

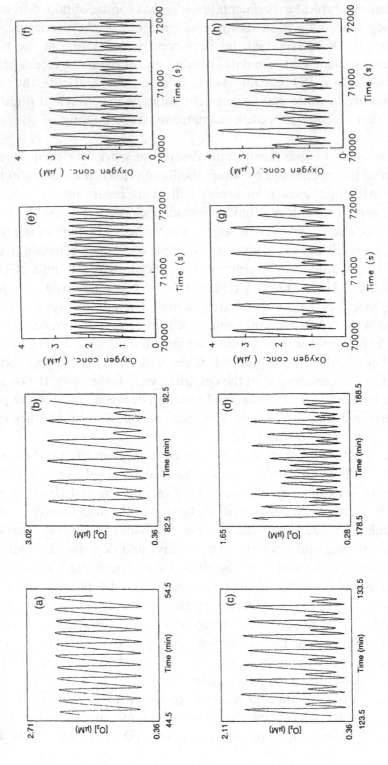

Figure 1. Comparison of experimental and theoretical results for the peroxidase–oxidase reaction. Panels (a)–(d) correspond to a variation of 2,4-dichlorophenol, while panels (e)–(h) correspond to variation in a rate constant in one model of this reaction (reprinted, with permission, from Scheeline *et al.*, 1997, Copyright 1997 American Chemical Society).

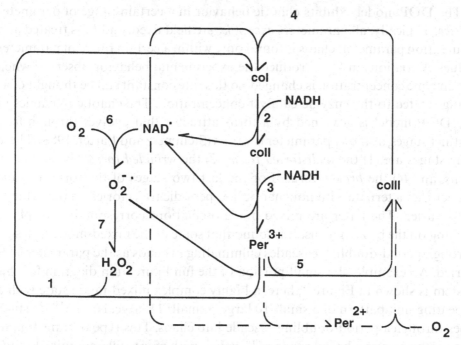

Figure 2. Steps in the peroxidase–oxidase reaction that lead to oscillatory behavior co, compound. A key species is NAD˙, which is involved in an autocatalytic feedback loop (reprinted, with permission, from Scheeline *et al.*, 1997, Copyright 1997 American Chemical Society).

four-variable model (now known as the DOP model after Degn, Olsen and Perram) and is given by the following system of rate equations:

$$\dot{A} = -k_1 ABX - k_3 ABY + k_7 - k_{-7}A$$

$$\dot{B} = -k_1 ABX - k_3 ABY + k_8$$

$$\dot{X} = k_1 ABX - 2k_2 X^2 + 2k_3 ABY - k_4 X + k_6$$

$$\dot{Y} = -k_3 ABY + 2k_2 X^2 - k_5 Y,$$

$$(2)$$

where A is the concentration of dissolved O_2, B is the concentration of NADH, and X and Y are concentrations of two critical intermediates, X and Y. The dot over each variable denotes the derivative of this concentration variable with respect to time. From many comparisons of simulations and experiment, it has been determined that X mimics the probable dynamics of a free radical species, NAD˙, while Y corresponds to an enzyme–substrate complex known as compound III, which consists of a molecule of oxygen bound to a reduced form of the enzyme known as Per^{2+}; the native enzyme is Per^{3+}.

The DOP model exhibits chaotic behavior in a certain range of parameter values. Typically, all parameters except k_1 are held fixed, and k_1 is treated as a bifurcation parameter; chaos is found only within a certain range of parameter values. Variations in k_1 reproduce the experimental behavior observed when the enzyme concentration is changed, so this rate constant can be thought of as being related to the enzyme catalyst concentration. The chaotic dynamics in the DOP model is governed by a torus attractor that evolves through four distinct stages, as the k_1 parameter is varied (Steinmetz and Larter, 1988). These four stages are (1) the *undistorted torus*, (2) the *wrinkled torus*, (3) the *fractal torus*, and (4) the *broken torus*. For the last two stages of the torus, chaotic trajectories alternate with nonchaotic, i.e., periodic trajectories, as the value of k_1 is varied. The latter are mixed-mode oscillations corresponding to phase locking on the broken torus. It is found that some of the mixed-mode states go through period-doubling cascades culminating in chaos as the parameter k_1 is varied. An example of a small portion of the full bifurcation diagram for this system is shown in Figure 3; here, a highly complex mixed-mode state with a repeating unit pattern of 5 small/10 large/5 small/ 11 large, i.e., a $5^{10}5^{11}$ state, goes through a period-doubling cascade into chaos. This type of transition to chaos is seen over a broad range of k_1 values with many different mixed-mode states.

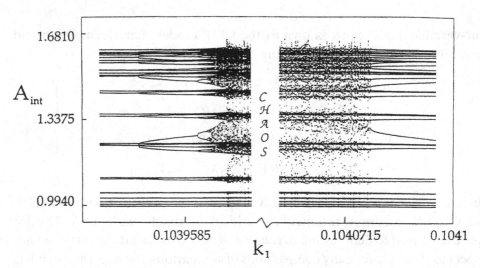

Figure 3. Bifurcation diagram showing the value of the variable A_{int} in the Poincaré section as a function of the parameter k_1. Chaotic dynamics are observed to arise from period-doubling cascades from complex mixed-mode periodic states. Here, the complex state at low k_1 values which undergoes the cascade is the $5^{10}5^{11}$ state (reprinted, with permission, from Steinmetz, 1991).

This example, then, shows that mixed-mode oscillations, while arising from a torus attractor that bifurcates to a fractal torus, give rise to chaos via the familiar period-doubling cascade in which the period becomes infinite and the resulting chaotic orbit consists of an infinite number of unstable periodic orbits. Recent experiments on the PO reaction have revealed the existence of mixed-mode oscillations that occur in a period-adding sequence (Hauser and Olsen, 1996). Although the simple model discussed here clearly does not contain as much information as the more detailed model in Figure 2, it does include the critical part of the mechanism that gives rise to this and other dynamical features of this reaction. The study of this and another simple, abstract model (Olsen, 1983) has helped to guide the elucidation of the more detailed mechanism by focusing the investigators' attention on the essential features that lead to oscillatory and complex dynamics.

We now turn to a discussion of epilepsy, which has been described as a dynamical disease (Milton and Black, 1995) and will frame our discussion by comparing dynamic processes in networks of neurons to dynamics in chemical reaction networks. For discussion of dynamical diseases, in particular of epilepsy, see also Milton (Chapter 16, this volume).

2.3 Epilepsy

One of the primary motivations for studying central nervous system (CNS) disorders from the perspective of nonlinear dynamical theory is that it allows one to consider phenomena such as epilepsy that involve the brain in a global manner (Kelso and Fuchs, 1995). Such an approach runs counter to the reductionist tendencies of much contemporary neuroscience. Focusing on small details of the CNS has produced some spectacular successes, some of which are discussed below, but does not provide much information on processes such as memory or seizures, many of whose characteristics are emergent and are not understandable below the level of ensembles of neurons (see also Ding *et al.*, Chapter 4, this volume). Consequently, it is critical that the correct level of organization be chosen to model a given neuronal phenomenon. Figure 4 illustrates some of the different levels of organization at which neuronal modeling can be attempted.

In a general way, epilepsy can be thought of as a situation characterized by an abnormal coherence of neuronal oscillations in both the temporal and spatial domains. Thus, when investigating the dynamics of epilepsy, the most appropriate level of description would seem to be a population or group of neurons. The incidence of epilepsy in developed nations is about 7 in 1000 persons (García-Flores *et al.*, 1998). The most common type of adult epilepsy is

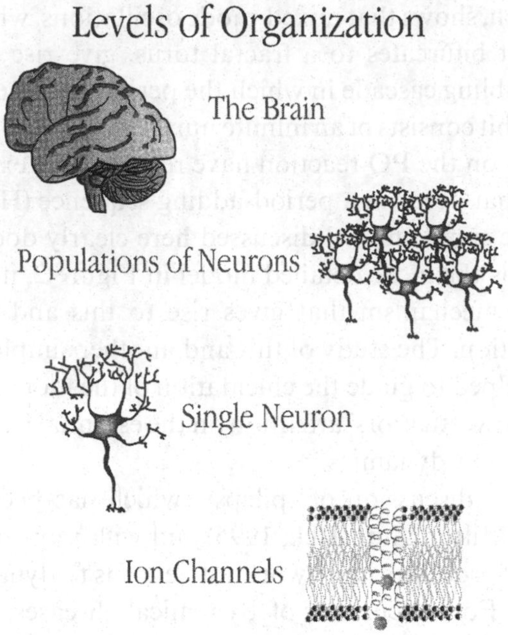

Levels of Organization

The Brain

Populations of Neurons

Single Neuron

Ion Channels

Figure 4. Levels of organization in the brain.

that classed as *partial seizures* (Hauser, 1993; Hauser *et al.*, 1993). These result from a focal area of structurally and physiologically abnormal neurons, most commonly located in the temporal lobe. The focal area generates abnormal activity of regular periodicity, which can then spread to recruit adjacent regions that are presumably anatomically and functionally normal; the end result is a behaviorally manifest seizure (Jefferys, 1990; Luciano, 1993). As many as 5–10% of patients with epilepsy will eventually become medically uncontrollable (Wyllie, 1993); however, many in this latter group can be helped by operative intervention to remove the focal generator site. This surgery can render such a group seizure free in up to 90% of cases in properly selected patients (Salanova *et al.*, 1998). It is still not clearly understood, though, why the removal of the focus should work so well to stop seizures, nor why it fails in the 10% of intractable cases. Thus it is of great clinical importance to understand more thoroughly the dynamics of seizure generation and spread.

Seizures are usually monitored by EEG, which measures the spatially averaged electrical potential produced by populations of neurons by using arrays of scalp electrodes (Lopes da Silva, 1991; Niedermayer, 1993). Seizures are generally characterized by a high degree of synchronization across the electrode array and an abnormal degree of periodic regularity (see Figure 5 for an example). It is this latter feature of the dynamics of epilepsy that we seek to

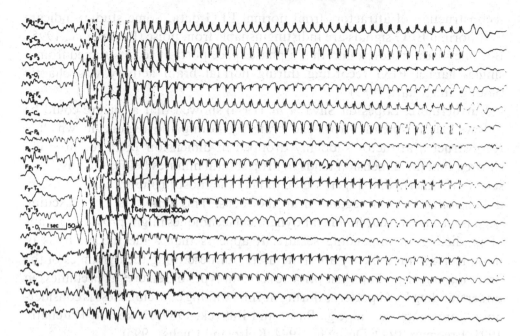

Figure 5. EEG tracings from a patient undergoing *petit mal* seizure. Notice that the irregular tracings in normal brain state prior to the seizure are replaced by more orderly, nearly periodic and highly synchronized firing during the seizure (reprinted, with permission, from Niedermeyer, 1993, in *Electroencephalography: Basic Principles, Clinical Applications and Related Fields*, third edition, Copyright Williams & Wilkins, 1993).

explain with the tools of nonlinear dynamics. Other groups have sought to apply nonlinear dynamical techniques to much more sweeping questions involving not only epileptic EEG signals but normal EEG as well. For example, Babloyantz and Destexhe (1986; Destexhe and Babloyantz, 1991) used techniques from nonlinear dynamics to analyze the difference in voltage between two electrodes placed at different positions on the head; they claimed that the dynamics of the EEG time series was governed by a *strange attractor* for the normal, awake state and a seizure state. This attractor could be visualized by reconstructing it in a phase space of dimension D equal to approximately 4 for the awake state and 2 for a generalized, *petit mal* seizure state; the latter value would be expected from the general appearance of Figure 5. Thus, the *petit mal* seizure state can be thought of as a periodic limit-cycle oscillation; this conclusion seems quite solid and is probably noncontroversial. The general applicability of their result of $D = 4$ for a normal awake state is somewhat more questionable, however, as other investigators have found much higher dimensions when attempting to create phase space reconstruction of attractors from normal EEG data. Some studies indicate that normal EEG signals are indistinguishable from noise (Belair *et al.*, 1995). But one investigation found that a phase space portrait reconstructed from a single-channel EEG recording during normal brain activity was chaotic (Soong and Stuart, 1989).

Furthermore, Lopes da Silva *et al.* (1994) showed that a seizure could be followed as it spread across the brain by noting the regions in which a reconstructed attractor suddenly dropped in dimension D. A recent study showed that the dimension tends to drop several minutes before the onset of a seizure (Lehnertz and Elger, 1998); in fact, the most pronounced decrease in dimension occurs in regions near the focus. This report is quite exciting because it suggsts that nonlinear dynamical techniques might provide a means of predicting an impending seizure. It appears that while the methods of nonlinear dynamic theory are quite useful for describing the dynamics of the brain *during* (or perhaps just prior to) certain types of seizure, its usefulness in understanding EEG patterns during normal waking consciousness is considerably more controversial (e.g., Basar and Bullock, 1989; Duke and Pritchard, 1991; Freeman, 1992; Destexhe, 1994; Kelso and Fuchs, 1995).

If nonlinear dynamics can be used to understand dynamical diseases such as epilepsy, it has great potential for clinical applications. One example of such an approach is the work of Schiff *et al.* (1994), who, with the use of small, correctly timed electrical perturbations, could coax the dynamics of a synchronously firing, i.e., 'seizing', hippocampal slice into a more normal chaotic regime. The hippocampus is the structure in the medial temporal lobe in which focal seizures most commonly originate. To a first approximation, its circuitry is organized in a 'lamellar' pattern orthogonal to the long axis of the temporal lobe. Thus thin slices in planes parallel to these lamellae preserve most of the important intercellular connections and can be used in the laboratory as an important experimental system. By creating a first-return map of the interspike interval measured in the hippocampal slice, these investigators demonstrated that the dynamics could be controlled with small electrical perturbations, which maintained the trajectory near an unstable period-1 limit cycle. They were also able to steer the brain slice into a more normal, chaotic regime. This technique, called *anti-control*, may have possible clinical application in the future for controlling seizures in human patients, since the onset of periodic dynamics is associated with a seizure (for details, see Ditto and Spano, Chapter 15, this volume).

Our intent in the studies summarized below was to investigate epilepsy by considering the dynamics of a small population of neurons known as a *subnetwork*. A subsequent investigation involved coupling together several units of a subnetwork to investigate the spread of regular periodic firing of the subnetwork as might occur during a seizure (Speelman, 1997; Larter *et al.*, 1999). The basic equations we start with in building the model are very commonly used to describe the biophysics of nerve conduction. A necessary but not sufficient condition for neuronal excitability is that the interior of the nerve cell is electrically negative relative to the outside. This is due to the fact that the neuronal membrane is *selectively permeable* to different ions, chiefly Na^+ and K^+. Sufficiency is conferred by the property that these membrane permeabilities, i.e., *conductances*, vary in response to the state of the neuron and its rate of change. This mechanism for the generation of the neuronal action potential is elegantly described by a four-dimensional system of non-linear differential equations initially put forth in a series of remarkably prescient papers by Hodgkin and Huxley in 1952 (Hodgkin and Huxley, 1952a,b). A number of excellent reviews of these equations and their mathematical properties are available; one particularly good one is Cronin (1987).

Once excited, the neuron conducts an electrical impulse, the action potential, in a nondecremental fashion along its axon to a terminal, where it communicates with another neuron at a synapse. Synaptic transmission involves electrical–chemical–electrical transduction and the influence on the secondary neuron can be either *excitatory* (making it more likely to fire) or *inhibitory* (less likely). Two important chemical transmitters in the CNS are glutamate, which is excitatory, and γ-aminobutyric acid (GABA), which is inhibitory. A time-honored axiom called *Dale's Law* holds that a given neuron is either excitatory or inhibitory, but not both, although recent evidence suggests hat this is not strictly true in every case.

As noted above, partial seizures are thought to be *initiated* by a focal area of structural and functional abnormality in the hippocampus. The exact pathological anatomy is not yet clear but it is likely that the biophysical malfunction involves either excess excitation, decreased inhibition (including disinhibition) or both (Schwartzkroin, 1993; Sloviter, 1994; Holmes, 1995; Dichter, 1998). This type of dysfunction is considered in the simulations using the model described below.

2.4 Modeling of neuronal dynamics

Computational modeling of neurons can reasonably be considered to have begun with the work of Hodgkin and Huxley (1952a,b) discussed above. By making reasonable assumptions about the kinetics of the conductance variables, FitzHugh determined that the essential dynamics could be represented by a reduced set of two differential equations (FitzHugh, 1960). Although

FitzHugh sacrificed the ability to calculate exact quantitative values, he did create a simple model that recreated the essential qualitative dynamics of the Hodgkin–Huxley equations. Morris and Lecar later created a hybrid two-variable model for the membrane potential of neurons in a mollusk (Morris and Lecar, 1981). It is this particular variation of the Hodgkin–Huxley equations that we start with in constructing a model for the subnetwork of interest in the CA3 region of the hippocampus, i.e., the subnetwork thought to be responsible for seizure initiations.

The goal of the subnetwork model study is to simulate the recruitment and synchronization of neurons and to determine through this simulation those parameters that affect the initiation and propagation of a seizure. However, rather than connecting together a large number of explicitly modeled neurons, typical of some approaches (Traub *et al.*, 1982; Traub and Miles, 1991), we collectively model *groups* or *populations* of neurons. Thus, by defining a group, or population, of neurons as a dynamical system, i.e., one described by a few differential equations, the emergent behavior of this population or group can be studied. This dynamical system, consequently, describes the behavior of an important subnetwork in a more complex network constituting the real system of the hippocampus. Similar models using interconnected excitatory and inhibitory elements have been previously studied (Wilson and Cowan, 1972; Kaczmarek, 1976; Plant, 1981; Mackey and an der Heiden, 1984; Castelfranco and Stech, 1987; Milton *et al.*, 1990). The interpretation of the variables in our model is that they describe average properties of populations of neurons, i.e., a prototypical or stereotypical neuron, rather than actual single neurons as the basic elements of the network. In this work, then, we are taking an approach similar to that in chemical kinetics in which a mass-action rate law, derived by considering the behavior of prototypical single molecules undergoing collision, is reinterpreted on the macroscopic scale in terms of concentrations, i.e., variables that describe the average properties of very large numbers of molecules. As has been pointed out by others (Golomb and Rinzel, 1993, 1994), what one always loses in approaches like this is information about the distribution of states that might exist on the microscopic scale; what we gain, of course, is the ability to simulate macroscopic or collective processes which may not have meaning at the microscopic level.

The subnetwork model we consider consists of three differential equations that describe the qualitative behavior of the relevant subnetwork, a population of neurons in region CA3 in the hippocampus. Two types of neuron are included in this simple model: pyramidal cells with membrane potential V and inhibitory interneurons with membrane potential V_I, interconnected by synapses and fed by current from the excitatory, i.e., perforant, pathway (see Figure 6).

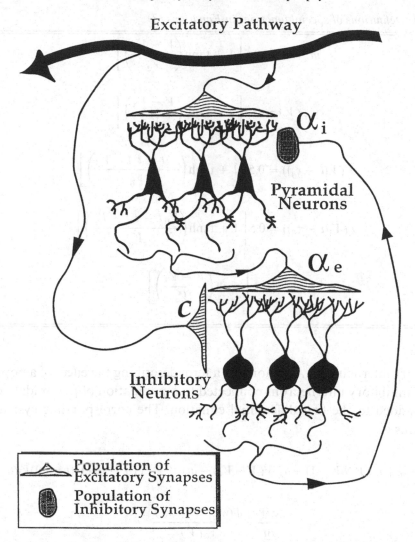

Figure 6. Schematic drawing of the subnetwork model associated with Equations (3). For symbols, see the text (reprinted, with permission, from Speelman, 1997).

We have described the dynamical behavior of this subnetwork by a system of three differential equations based on the two-variable Morris–Lecar model. Rinzel and Ermentrout (1989) studied the dynamical features of this two-variable model and found that it was generally applicable to many neuronal systems. Here, the two-variable Morris–Lecar model is taken to describe the dynamics of a population of pyramidal cells. To model the behavior of the subnetwork, we have added a third equation to the Morris–Lecar model to simulate the effect of a population of inhibitory interneurons connected to the pyramidal cells and the effect of an excitatory pathway connected to the

Table 1. *Definitions of special functions in Equations (3)*

$$m_\infty(V) = 0.5\left[1 + \tanh\left(\frac{V - V_1}{V_2}\right)\right]$$

$$w_\infty(V) = 0.5\left[1 + \tanh\left(\frac{V - V_3}{V_4}\right)\right]$$

$$\alpha_e(V(t - \tau_d)) = 0.5\left[1 + \tanh\left(\frac{V(t - \tau_d) - V_5}{V_6}\right)\right]$$

$$\alpha_i(V_I(t - \tau_d)) = 0.5\left[1 + \tanh\left(\frac{V_I(t - \tau_d) - V_7}{V_6}\right)\right]$$

$$\tau_w(V) = \left[\cosh\left(\frac{V - V_3}{2V_4}\right)\right]^{-1}$$

inhibitory interneurons. Additionally, a term describing the effect of a population of inhibitory interneurons connected to a population of pyramidal neurons is added to the first differential equation. The corresponding system of equations is:

$$\frac{dV}{dt} = -g_{Ca}m_\infty(V)(V - 1) - g_K W(V - V_K) - g_L(V - V_L) + i - \alpha_i(V_I(t - \tau_d))V_I(t - \tau_d)$$

$$\frac{dW}{dt} = \frac{\phi(w_\infty(V) - W)}{\tau_w(V)} \tag{3}$$

$$\frac{dV_I}{dt} = b(ci + \alpha_e(V(t - \tau_d))V(t - \tau_d)),$$

where V is the average membrane potential of a typical pyramidal type neuron in the CA3 region of the hippocampus and W is a relaxation factor which is essentially the fraction of open K^+ channels in these pyramidal cells. V_I is the potential of the inhibitory interneuron and the g_i represents the total conductances for the $i = Ca^{2+}$, K^+ and leakage channels. The other parameters in these equations are defined in Table 1, and typical values used in our simulations are given in Table 2.

The functions α_e and α_i are hyperbolic functions that describe the collective

Table 2. *Definitions and typical values of the parameters in Equations (3) and the special functions in Table 1*

Parameter	Description	Value
V_1	Threshold value for m_∞	-0.01
V_2	Steepness parameter for m_∞	0.15
V_3	Threshold value for w_∞	0.0
V_4	Steepness parameter for w_∞	0.30
V_5	Threshold value for α_e	0.0
V_6	Steepness parameter for α_e and α_i	Variable
V_7	Threshold value for α_i	0.0
g_{Ca}	Conductance of population of Ca^{2+} channels	1.1
g_K	Conductance of population of K^+ channels	2.0
g_L	Conductance of population of leakage channels	0.5
V_K	Equilibrium potential of K^+	-0.7
V_L	Equilibrium potential of leakage channels	-0.5
τ_d	Delay	Usually zero
i	Applied current	0.30
b	Time constant scaling factor	Variable
c	Strength of feedforward inhibition	Varied (Figure 7)
ϕ	Temperature scaling factor	0.7

activity of a population of synapses. Each individual synaptical connection, whether inhibitory or excitatory, is assumed to act like an 'on–off' switch between the pyramidal cell and the inhibitory interneuron. For example, when the pyramidal cell potential, V, becomes larger than V_5, a fixed threshold potential, the connection between the pyramidal cell and the interneuron (α_e) is opened. The interneuron has an inhibitory effect on the pyramidal cell through the negative sign preceding α_i, the connection from the inhibitory interneurons to the pyramidal neurons. A simple distribution of synapses is assumed, which results in a smooth sigmoidal shape for the functions α_e and α_i. This assumption is similar to that originally taken by Wilson and Cowan (1972) to describe the response of individual populations of excitatory and inhibitory cells in response to an average level of excitation. Their model describes recurrently connected small populations of excitatory and inhibitory cells, whereas the current model describes small populations of recurrently connected excitatory and inhibitory cells. In spirit, then, the current work is more similar to that of Kaczmarek (1976) and Plant (1981).

The system of Equations (3) was numerically solved using the function NDSolve included in Mathematica (Wolfram Research) on an Indigo II Silicon Graphics workstation. Figure 7 shows a summary of these solutions as a bifurcation diagram created by plotting the maxima in the V time series while varying the parameter c. This parameter c is a measure of the current strength

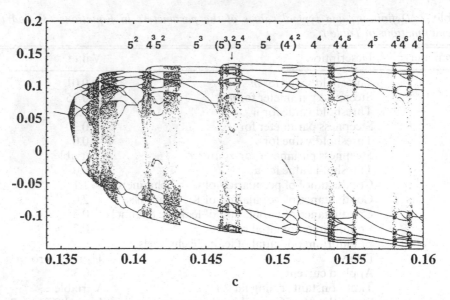

Figure 7. Bifurcation diagram showing the typical behavior of the subnetwork model as parameter c is varied. Parameter c corresponds to the degree of inhibition in the subnetwork of Figure 6 (reprinted, with permission, from Speelman, 1997).

flowing from the excitatory pathway directly into the inhibitory neurons and, thus, its value can be interpreted as the degree of inhibition in the subnetwork. A large variety of mixed-mode states are seen to be interspersed with regions of apparent chaotic behavior. The mixed-mode states are found to be phase-locked periodic states on a fractal torus attractor (Speelman, 1997); many of these states appear in Farey-sequence order and the intervening chaotic states arise via period-doubling cascades from the mixed-mode states. The similarity between the bifurcation diagram for this neuronal subnetwork model (Figure 7) and the DOP model for the PO reaction (Figure 3) is striking but not unexpected. Further details regarding the dynamical behavior of this model, a coupled lattice derived from it to model propagation of a seizure and the implications of these simulations in elucidating the mechanism of complex partial seizures can be found in Speelman (1997) and Larter *et al.* (1999).

2.5 Conclusions

The dynamical features of a particular system (such as a malfunctioning hippocampus in an epileptic patient) often provide clues to the type of mechanism that might be operative in that system. These clues come from direct observation and are enhanced by experience with dynamical mechanisms in other well-studied systems (e.g., population biology or chemical oscillators). The role of *feedback processes* is central in all these types of system and, as we have seen in the epilepsy example described here, any imbalances between positive and negative forms of feedback can lead to serious disorders, if these feedback processes occur in critical physiological systems.

The reduction of a population of neurons to two or three differential equations may seem drastic, but this approach is not too different from the original approach of Hodgkin and Huxley (1952a,b), who fit macroscopic experimental measurements to a four-variable system of differential equations. The general form of Hodgkin and Huxley's equations are still used today in the program GENESIS (Koch and Segev, 1989), which breaks up neurons into small, discrete compartments in order to deal with the simulation of neuronal behavior and applies Hodgkin–Huxley-type equations to each compartment. The compartments can represent different areas of the neuron such as the dendrite or axon. With this compartment modeling, different ion channels can be incorporated into each of the different compartments. The compartments then become analogous to atoms used in molecular dynamics simulations. Recently, reduced compartment models have been used which is an approach similar to using models that involve atom types in a molecular dynamics simulation, such as a one pseudo-atom representation of the methyl group in molecular mechanics modeling (Bush and Sejnowski, 1993; Pinsky and Rinzel, 1994).

One possible avenue for future exploration involves the success of the new technique of vagal nerve stimulation for seizure control. It is possible that the success of this new technique might be at least partially explained by experiments of the type Schiff and co-workers carried out in which a hippocampal slice undergoing seizures was coaxed back to a chaotic regime with pulses of current (Schiff *et al.*, 1994). Another important question is the mechanism by which the seizure propagates to other areas of the brain, recruiting presumably normal neurons. Clinical observation reveals that patients who have epilepsy with anatomically and physiologically stable brain abnormalities do not have seizures on a constant basis. It is likely that an area larger than that of the stable abnormalities must malfunction in order to produce a behaviorally manifest seizure. This, then, leads naturally to the question of what prevents the spatial spread of abnormally periodic oscillations to physiologically normal tissue. This second issue was looked at using a coupled-lattice model (Speelman, 1997) and has been described elsewhere (Larter *et al.*, 1999).

There are additional similarities between neuronal modeling and molecular dynamics. Modeling the entire neuron with one set of differential equations as was done early on, for example by Kaczmarek (1976), is similar to considering an amino acid as one unit in simulations of large proteins. Finally, modeling a population of neurons, where only the average behavior of the population is considered would be similar to using stochastic boundary conditions in molecular dynamics where a 'bath' of molecules is represented as a mean field. This flexibility available to investigators in computational modeling illustrates the extent of artistry involved in deciding both the appropriate level at which to model a problem and the simplification and assumptions that can be justified. The similarity to the problems inherent in modeling the dynamics of the brain is a good example of the progress that can be achieved in understanding a complex system when cross-disciplinary fertilization of ideas occurs.

Acknowledgments

The authors acknowledge support of this work by the National Science Foundation under grant CHE-9307549 and by a Research Venture Award from the Research Investment Fund of IUPUI.

References

Babloyantz, A. and Destexhe, A. (1986) Low dimensional chaos in an instance of epilepsy. *Proc. Natl. Acad. Sci. USA* **83**: 3513–3517.

Basar, E. and Bullock, T. H. (eds.) (1989) *Brain Dynamics: Progress and Perspectives.* New York: Springer-Verlag.

Belair, J., Glass, L., an der Heiden, U. and Milton, J. (1995) Dynamical disease: identification, temporal aspects and treatment strategies of human illness. *Chaos* **5**: 1–7.

Berridge, M. J. (1993) Inositol triphosphate and calcium signalling. *Nature* **361**: 315–325.

Bush, P. C. and Sejnowski, T. J. (1993) Reduced compartmental models of neocortical pyramidal cells. *J. Neurosci. Meth.* **46**: 159–166.

Castelfranco, A. M. and Stech, H. W. (1987) Periodic solutions in a model of recurrent neural feedback. *SIAM J. Appl. Math.* **47**: 573–588.

Cronin, J. (1987) *Mathematical Aspects of Hodgkin–Huxley Neural Theory.* New York: Cambridge University Press.

Cuthbertson, K. S. R. and Cobbold, P. H. (eds.) (1991) *Oscillations in Cell Calcium: Collected Papers and Review. Cell Calcium* **12**: 61–268.

Degn, H., Olsen, L. F. and Perram, J. W. (1979) Bistability, oscillations, and chaos in an enzyme reaction. *Ann. NY Acad. Sci.* **316**: 623–637.

Destexhe, A. (1994) Oscillations, complex spatiotemporal behavior, and information transport in networks of excitatory and inhibitory neurons. *Phys. Rev. E* **50**: 1594–1606.

Destexhe, A. and Babloyantz, A. (1991) Pacemaker-induced coherence in cortical networks. *Neural Comput.* **3**: 145–154.

Dichter, M. (1998) Overview: the neurobiology of epilepsy. In *Epilepsy: A Comprehensive Textbook* (ed. J. Engel and T. A. Pedley), pp. 233–235. Philadelphia: Lippincott-Raven.

Duke, D. W. and Pritchard, W. S. (eds.) (1991) *Measuring Chaos in the Human Brain*. Singapore: World Scientific.

Epstein, I. R. (1991) Perspective: spiral waves in chemistry and biology. *Science* **252**: 67.

FitzHugh, R. (1960) Thresholds and plateaus in the Hodgkin–Huxley nerve equations. *J. Gen. Physiol.* **43**: 867–896.

Freeman, W. J. (1992) Tutorial on neurobiology: from single neurons to brain chaos. *Intl. J. Bifurc. Chaos* **2**: 451–458.

García-Flores, E., Farías, R. and García-Almaguer, E. (1998) Epidemiology of epilepsy in North America. In *Textbook of Sterotactic and Functional Neurosurgery* (ed. P. L. Gildenberg and R. Tasker), pp. 1775–1779. New York: McGraw-Hill.

Glanz, J. (1997) Mastering the nonlinear brain. *Science* **277**: 1758–1760.

Goldbeter, A. (1996) *Biochemical Oscillations and Cellular Rhythms: The Molecular Bases of Periodic and Chaotic Behavior*. Cambridge: Cambridge University Press.

Golomb, D. and Rinzel, J. (1993) Dynamics of globally coupled inhibitory neurons with heterogeneity. *Phys. Rev. E* **48**: 4810–4814.

Golomb, D. and Rinzel, J. (1994) Clustering in globally coupled inhibitory neurons. *Physica D* **72**: 259–282.

Hauser, W. A. (1993) The natural history of seizures. In *The Treatment of Epilepsy: Principles and Practice* (ed. E. Wyllie), pp. 165–170. Philadelphia: Lea and Febiger.

Hauser, W. A., Annegers, J. F. and Kurland, L. T. (1993) The incidence of epilepsy and unprovoked seizures in Rochester, Minnesota, 1935–1984. *Epilepsia* **34**: 453–468.

Hauser, M. and Olsen, L. F. (1996) Mixed-mode oscillations and homoclinic chaos in an enzyme reaction. *J. Chem. Soc. Faraday Trans.* **92**: 2857–2863.

Hodgkin, A. L. and Huxley, A. F. (1952a) The components of membrane conductance in the giant axon of *Loligo*. *J. Physiol.* **116**: 473–496.

Hodgkin, A. L. and Huxley, A. F. (1952b) A quantitative description of membrane current and its application to conduction and excitation in nerve. *J. Physiol.* **117**: 500–544.

Holmes, G. L. (1995) Pathogenesis of epilepsy: the role of excitatory amino acids. *Clev. Clinic J. Med.* **62**: 240–247.

Jefferys, J. G. R. (1990) Basic mechanisms of focal epilepsies. *Exp. Physiol.* **75**: 127–162.

Kaczmarek, L. K. (1976) A model of cell firing patterns during epileptic seizures. *Biol. Cybern.* **22**: 229–234.

Kelso, J. A. S. and Fuchs, A. (1995) Self-organizing dynamics of the human brain: critical instabilities and Sil'nikov chaos. *Chaos* **5**: 64–69.

Koch, C. and Segev, I. (1989) *Methods in Neuronal Modeling*. Cambridge, MA: MIT Press.

Larter, R., Olsen, L. F., Steinmetz, C. G. and Geest, T. (1993) Chaos in biochemical systems: the peroxidase reaction as a case study. In *Chaos in Chemical and Biochemical Systems* (ed. R. J. Field and L. Györgyi), pp. 175–224. Singapore: World Scientific Press.

Larter, R., Speelman, B. and Worth, R. M. (1999) A coupled ordinary differential equation lattice model for the simulation of epileptic seizures. *Chaos* **9**: 795–804.

Lechleiter, J., Girard, S., Peralta, E. and Clapham, D. (1991) Spiral calcium wave propagation and annihilation in *Xenopus laevis* oocytes. *Science* **252**: 123–126.

Lehnertz, K. and Elger, C. E. (1998) Can epileptic seizures be predicted? Evidence from nonlinear time series analysis of brain electrical activity. *Phys. Rev. Lett.* **80**: 5019–5022.

Lopes da Silva, F. (1991) Neural mechanisms underlying brain waves: from neural membranes to networks. *Electroencephalogr. Clin. Neurophysiol.* **79**: 81–93.

Lopes da Silva, F. H., Pijn, J. and Wadman, W. J. (1994) Dynamics of local neuronal networks: control parameters and state bifurcations in epileptogenesis. *Progr. Brain Res.* **102**: 359–370.

Luciano, D. (1993) Partial seizures of frontal and temporal origin. *Neurologic Clin.* **4**: 805–822.

Mackey, M. C. and an der Heiden, U. (1984) The dynamics of recurrent inhibition. *J. Math. Biol.* **19**: 211–225.

Mäder, M. and Amberg-Fisher, V. (1982) Role of peroxidase in lignification of tobacco cells. I. *Plant Physiol.* **70**: 1128–1131.

Mäder, M. and Füssl, R. (1982) Role of peroxidase in lignification of tobacco cells. II. *Plant Physiol.* **70**: 1132–1134.

Milton, J. G., an der Heiden, U., Longtin, A. and Mackey, M. C. (1990) Complex dynamics and noise in simple neural networks with delayed mixed feedback. *Biomed. Biochim. Acta* **49**: 697–707.

Milton, J. and Black, D. (1995) Dynamic diseases in neurology and psychiatry. *Chaos* **5**: 8–13.

Møller, A. C., Hauser, M. J. B. and Olsen, L. F. (1998) Oscillations in peroxidase-catalyzed reactions and their potential function in vivo. *Biophys. Chem.* **72**: 63–72.

Morris, C. and Lecar, H. (1981) Voltage oscillations in the barnacle giant muscle fiber. *Biophys. J.* **35**: 193–213.

Nakamura, S., Yokota, K. and Yamazaki, I. (1969) Sustained oscillations in a lactoperoxidase, NADPH and O_2 system. *Nature* **222**: 794.

Niedermayer, E. (1993) Epileptic seizure disorders. In *Electroencephalography: Basic Principles, Clinical Applications and Related Fields*, third edition (ed. E. Niedermayer and F. Lopes da Silva), pp. 461–564. Baltimore, MD: Williams & Wilkins.

Olsen, L. F. (1983) An enzyme reaction with a strange attractor. *Phys. Lett.* **94A**: 454–457.

Pantoja, O. and Willmer, C. M. (1988) Redox activity and peroxidase activity associated with the plasma membrane of guard-cell protoplasts. *Planta* **174**: 44–50.

Pinsky, P. F. and Rinzel, J. (1994) Intrinsic and network rhythmogenesis in a reduced Traub model for CA3 neurons. *J. Comput. Neurosci.* **1**: 39–60.

Plant, R. E. (1981) A FitzHugh differential-difference equation modeling recurrent neural feedback. *SIAM J. Appl. Math.* **40**: 150–151.

Rinzel, J. and Ermentrout, G. B. (1989) Analysis of neural excitability and oscillations. In *Methods in Neuronal Modeling* (ed. C. Koch and I. Segev), pp. 135–169. Cambridge, MA: MIT Press.

Salanova, V., Markand, O., Worth, R. M., Smith, R., Wellman, H., Hutchins, G., Park, H., Ghetti, B. and Azzarelli, B. (1998) FDG-PET and MRI in temporal lobe epilepsy: relationship to febrile seizures, hippocampal sclerosis and outcome. *Acta Neurol. Scand.* **97**: 146–153.

Scheeline, A., Olson, D., Williksen, E., Horras, G., Klein, M. L. and Larter, R. (1997) The peroxidase–oxidase reaction and its constituent chemistries. *Chem. Rev.* **97**:

739–756.

Schiff, S. J., Jerger, K., Duong, D. H., Chang, T., Spano, M. L. and Ditto, W. L. (1994) Controlling chaos in the brain. *Nature* **370**: 615–620.

Schwartzkroin, P. A. (1993) Basic mechanisms of epileptogenesis. In *The Treatment of Epilepsy* (ed. E. Wyllie), pp. 83–98. Philadelphia: Lea and Febiger.

Sloviter, R. S. (1994) The functional organization of the hippocampal dentate gyrus and its relevance to the pathogenesis of temporal lobe epilepsy. *Neurolog. Progr.* **35**: 640–654.

Soong, A. C. K. and Stuart, C. I. J. M. (1989) Evidence of chaotic dynamics underlying the human alpha-rhythm electroencephalogram. *Biol. Cybern.* **62**: 55–62.

Speelman, B. (1997) A dynamical systems approach to the modeling of epileptic seizures. Ph.D. thesis. Indiana University–Purdue University.

Steinmetz, C. G. (1991) Chaos in the peroxidase–oxidase reaction. Ph.D. thesis. Indiana University–Purdue University.

Steinmetz, C. G. and Larter, R. (1988) The quasiperiodic route to chaos in a model of the peroxidase–oxidase reaction. *J. Chem. Phys.* **94**: 1388–1396.

Traub, R. D. and Miles, R. (1991) *Neuronal Networks of the Hippocampus.* Cambridge: Cambridge University Press.

Traub, R. D., Miles, R. and Wong, R. K. S. (1982) Cellular mechanisms of neuronal synchronization in epilepsy. *Science* **216**: 745–747.

Wilson, H. R. and Cowan, J. D. (1972) Excitatory and inhibitory interactions in localized populations of model neurons. *Biophys. J.* **12**: 1–24.

Wyllie, E. (1993) *The Treatment of Epilepsy: Principles and Practice.* Philadelphia: Lea and Febiger.

Yamazaki, I., Yokota, K. and Nakajima, R. (1965) Oscillatory oxidations of reduced pyridine nucleotide by peroxidase. *Biochem. Biophys. Res. Comm.* **21**: 582–586.

3

Fractal mechanisms in neuronal control: human heartbeat and gait dynamics in health and disease

CHUNG-KANG PENG, JEFFREY M. HAUSDORFF
AND ARY L. GOLDBERGER

3.1 Introduction

Clinical diagnosis and basic investigations are critically dependent on the ability to record and analyze physiological signals. Examples include heart rate recordings of patients at high risk of sudden death (Figure 1), electroencephalographic (EEG) recordings in epilepsy and other disorders, and fluctuations of hormone and other molecular signal messengers in neuroendocrine dynamics. However, the traditional bedside and laboratory analyses of these signals have not kept pace with major advances in technology that allow for recording and storage of massive data sets of continuously fluctuating signals. Surprisingly, although these typically complex signals have recently been shown to represent processes that are nonlinear, nonstationary, and nonequilibrium in nature, the tools to analyze such data often still assume linearity, stationarity and equilibrium-like conditions. Such conventional techniques include analysis of means, standard deviations and other features of histograms, along with classical power-spectrum analysis. An exciting recent finding is that such complex data sets may contain *hidden information*, defined here as information not extractable with conventional methods of analysis. Such information promises to be of clinical value (forecasting sudden cardiac death in ambulatory patients, or cardiopulmonary catastrophes during surgical procedures), as well as to relate to basic mechanisms of healthy and pathological function. Fractal analysis is one of the most promising new approaches for extracting such hidden information from physiological time series. This is partly due to the fact that the absence of characteristic temporal (or spatial) scales – the hallmark of fractal behavior – may confer important biological advantages, related to the adaptability of response (Goldberger *et al.*, 1990; Bassingthwaighte *et al.*, 1994; Bunde and Havlin, 1994; Goldberger, 1996; Iannaconne and Khokha, 1996; Goldberger, 1997).

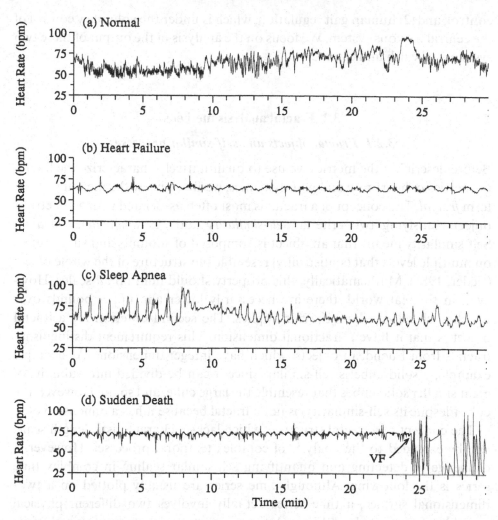

Figure 1. Representative complex physiological fluctuations. Heart rate (normal sinus rhythm) time series of 30 min from (a) a healthy subject at sea level, (b) a subject with congestive heart failure, (c) a subject with obstructive sleep apnea, and (d) a sudden cardiac death subject who sustained a cardiac arrest with ventricular fibrillation (VF). Note the highly nonstationary and 'noisy' appearance of the healthy variability, which is related in part to fractal (scale-free) dynamics. In contrast, pathological states may be associated with the emergence of periodic oscillations, indicating the emergence of a characteristic timescale. bpm, beats per minute.

In this chapter, we present some recent progress in applying fractal analysis to human physiology. We begin with a definition of *fractal dynamics*, followed by an introduction to some special problems posed by physiological time series. We then discuss the analysis of the output from two model systems: (1) human heartbeat regulation, which is under involuntary (neuroautonomic) control; and (2) human gait regulation, which is under the voluntary control of the central nervous system. We focus on the analysis of the output of these two systems in health and disease.

3.2 Fractal analysis methods

3.2.1 *Fractal objects and self-similar processes*

Before describing the metrics we use to quantitatively characterize the fractal properties of heart rate and gait dynamics, we first review the meaning of the term *fractal*. The concept of a fractal is most often associated with geometrical objects satisfying two criteria: *self-similarity* and *fractional dimensionality*. Self-similarity means that an object is composed of subunits and subsubunits on multiple levels that (statistically) resemble the structure of the whole object (Feder, 1988). Mathematically, this property should hold on all scales. However, in the real world, there are necessarily lower and upper bounds over which such self-similar behavior applies. The second criterion for a fractal object is that it have a fractional dimension. This requirement distinguishes fractals from Euclidean objects, which have integer dimensions. As a simple example, a solid cube is self-similar, since it can be divided into subunits of eight smaller solid cubes that resemble the large cube, and so on. However, the cube (despite its self-similarity) is not a fractal because it has a dimension of 3.

The concept of a fractal structure, which lacks a characteristic length scale, can be extended to the analysis of complex temporal processes. However, a challenge in detecting and quantifying self-similar scaling in complex time series is the following. Although time series are usually plotted on a two-dimensional surface, a time series actually involves two different physical variables. For example, in Figure 1, the horizontal axis represents 'time', while the vertical axis represents the value of the variable that changes over time (in this case, heart rate). These two axes have independent physical units, minutes and beats/minute, respectively. (Even in cases where the two axes of a time series have the same units, their intrinsic physical meaning is still different.) This situation is different from that of geometrical curves (such as coastlines and mountain ranges) embedded in a two-dimensional plane, where both axes represent the same physical variable. To determine whether a two-dimensional curve is self-similar, we can do the following test: (1) take a subset of the object and rescale it to the same size as the original object, using the same magnification factor for both its width and height; and then (2) compare the statistical properties of the rescaled object with the original object. In contrast, to properly compare a subset of a time series with the original data set, we need two magnification factors (along the horizontal and vertical axes), since these two axes represent different physical variables.

To put the above discussion into mathematical terms: a time-dependent process (or time series) is self-similar if

$$y(t) \overset{\mathrm{d}}{=} a^{\alpha} y\left(\frac{t}{a}\right), \tag{1}$$

where $\overset{\mathrm{d}}{=}$ means that the statistical properties of both sides of the equation are identical. In other words, a self-similar process, $y(t)$, with a parameter α has the identical probability distribution as a properly rescaled process, $a^{\alpha} y(t/a)$, i.e., a time series which has been rescaled on the x-axis by a factor a ($t \to t/a$) and on the y-axis by a factor of a^{α} ($y \to a^{\alpha} y$). The exponent α is called the *self-similarity parameter*.

In practice, however, it is impossible to determine whether two processes are statistically identical, because this strict criterion requires them to have identical distribution functions (including not just the mean and variance, but all higher moments as well)[1]. Therefore, one usually approximates this equality with a weaker criterion by examining only the means and variances (first and second moments) of the distribution functions for both sides of Equation (1).

Figure 2a shows an example of a self-similar time series. We note that with the appropriate choice of scaling factors on the x- and y-axes, the rescaled time series (Figure 2b) resembles the original time series (Figure 2a). The self-similarity parameter α as defined in Equation (1) can be calculated by a simple relation

$$\alpha = \frac{\ln M_y}{\ln M_x} \tag{2}$$

where M_x and M_y are the appropriate magnification factors along the horizontal and vertical direction, respectively.[2]

In practice, we usually do not know the value of the α exponent in advance. Instead, we face the challenge of extracting this scaling exponent (if one does exist) from a given time series. To this end, it is necessary to study the time series on observation windows with different sizes and adopt the weak criterion of self-similarity defined above to calculate the exponent α. The basic idea is illustrated in Figure 2. Two observation windows in Figure 2a, *window 1* with horizontal size n_1 and *window 2* with horizontal size n_2, were arbitrarily

---

[1] Equation (1) also requires that the joint probability functions (covariance and all higher-order correlations) are the same.

[2] Note that the variable, t/a, on the right hand side of Equation (1) actually represents a magnification factor of $M_x = a$ in a graphical representation.

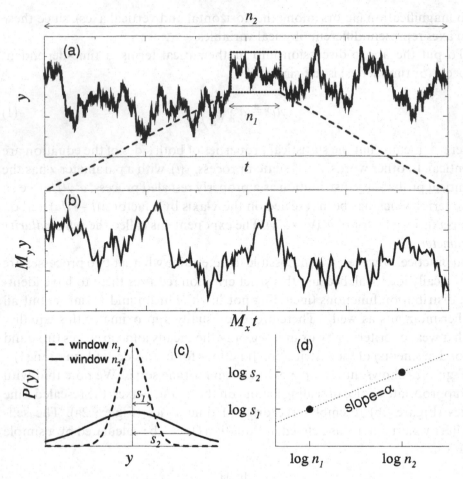

Figure 2. Illustration of the concept of self-similarity for a simulated random walk. (a) Two observation windows, with time scales n_1 and n_2, are shown for a self-similar time series $y(t)$. (b) Magnification of the smaller window with time scale n_1. Note that the fluctuations in (a) and (b) look similar, provided that two different magnification factors, M_x and M_y, are applied on the horizontal and vertical scales, respectively. (c) The probability distribution, $P(y)$, of the variable y for the two windows in (a), where s_1 and s_2 indicate the standard deviations for these two distribution functions. (d) Log–log plot of the characteristic scales of fluctuations, s, versus the window sizes, n.

selected to demonstrate the procedure. The goal is to find the correct magnification factors such that we can rescale window 1 to resemble window 2. It is straightforward to determine the magnification factor along the horizontal direction, $M_x = n_2/n_1$. But for the magnification factor along the vertical direction, M_y, we need to determine the vertical characteristic scales of windows 1 and 2. One way to do this is by examining the probability distributions (histograms) of the variable y for these two observation windows (Figure 2c). A reasonable estimate of the characteristic scales for the vertical heights, i.e., the typical fluctuations of y, can be defined by using the standard deviations of these two histograms, denoted as s_1 and s_2, respectively. Thus, we have $M_y = s_2/s_1$. Substituting M_x and M_y into Equation (2), we obtain

$$\alpha = \frac{\ln M_y}{\ln M_x} = \frac{\ln s_2 - \ln s_1}{\ln n_2 - \ln n_1}. \tag{3}$$

This relation is simply the slope of the line that joins these two points, (n_1, s_1) and (n_2, s_2), on a log–log plot (Figure 2d).

In analyzing 'real-world' time series, we perform the above calculations using the following procedures. (1) For any given size of observation window, the time series is divided into subsets of independent windows of the same size. To obtain a more reliable estimation of the characteristic fluctuation at this window size, we average over all individual values of s obtained from these subsets. (2) We then repeat these calculations, not just for two window sizes (as illustrated above), but for many different window sizes. The exponent α is estimated by fitting a line on the log–log plot of s versus n across the relevant range of scales.

3.2.2 Mapping 'real-world' time series to self-similar processes

For a self-similar process with $\alpha > 0$, the fluctuations grow with the window size in a power-law way. Therefore the fluctuations on large observation windows are significantly larger than those of smaller windows. As a result, the time series is *unbounded*. However, most physiological time series of interest, such as heart rate and gait, are *bounded* – they cannot have arbitrarily large amplitudes no matter how long the data set is. This practical restriction causes further complications for our analyses. Consider the case of the heart rate time series shown in Figure 3a. If we zoom in on a subset of the time series, we notice an apparently self-similar pattern. To visualize this self-similarity, we do not need to rescale the y-axis ($M_y = 0$) – only rescaling of the x-axis is needed. Therefore, according to Equation (3), the self-similarity parameter is zero – not an informative result. Consider another example where we randomize the sequential order of the original heart rate time series, generating a completely uncorrelated 'control' time series (Figure 3b) – white noise. The white noise data set also has a self-similarity parameter of zero. However, it is obvious that the patterns in Figure 3a and b are quite different. An immediate problem, therefore, is how to distinguish the trivial parameter zero in the latter case of uncorrelated noise, from the non-trivial parameter zero computed for the original data.

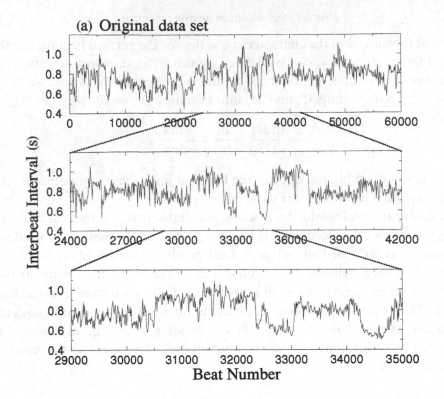

(a) Original data set

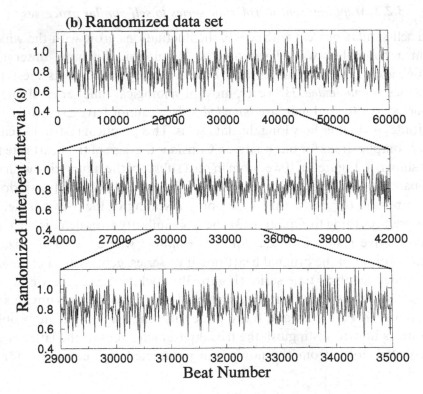

(b) Randomized data set

Figure 3 (*opposite*). A cardiac interheartbeat interval (inverse of heart rate) time series is shown in (a) and a randomized control is shown in (b). Successive magnifications of the subsets show that both time series are self-similar with a trivial exponent $\alpha = 0$ (i.e., $M_y = 1$), albeit the patterns are very different in (a) and (b).

Physicists and mathematicians have developed an innovative solution for this central problem in time series analysis (Hurst, 1951; Kolmogorov, 1961). The 'trick' is to study the fractal properties of the accumulated (integrated) time series, rather than those of the original signals (Feder, 1988; Beran, 1994). One well-known physical example with relevance to biological time series is the dynamics of Brownian motion. In this case, the random force (noise) acting on particles is bounded, as with physiological time series. However, the trajectory (an integration of all previous forces) of the Brownian particle is not bounded and exhibits fractal properties that can be quantified by a self-similarity parameter. When we apply fractal-scaling analysis to the integrated time series of Figure 3a and b, the self-similarity parameters are indeed different in these two cases, providing meaningful distinctions between the original and the randomized control data sets. The details of this analysis are discussed in the next section.

In summary, mapping the original bounded time series to an integrated signal is a crucial step in fractal time series analysis. In the rest of this chapter, therefore, we apply fractal analysis techniques after integration of the original time series.

3.2.3 Detrended fluctuation analysis

As discussed above, a bounded time series can be mapped to a self-similar process by integration. However, another challenge facing investigators applying this type of fractal analysis to physiological data is that these time series are often highly nonstationary[3] (Figure 1a). The integration procedure will further exaggerate the nonstationarity of the original data.

To overcome this complication, we have introduced a modified root mean square analysis of a random walk – termed *detrended fluctuation analysis* (DFA)[4] – to the analysis of biological data (Peng *et al.*, 1994a, 1995). Advantages of DFA over conventional methods (e.g., spectral analysis and Hurst

---

[3] A simplified and general definition characterizes a time series as stationary if the mean, standard deviation and higher moments, as well as the correlation functions are invariant under time translation. Signals that do not obey these conditions are nonstationary.

[4] The DFA computer program is available at *http://reylab.bidmc.harvard.edu* without charge.

analysis) are that it permits the detection of intrinsic self-similarity embedded in a seemingly nonstationary time series, and also avoids the spurious detection of apparent self-similarity, which may be an artifact of extrinsic trends[5]. This method has been successfully applied to a wide range of simulated and physiological time series in recent years (Buldyrev *et al.*, 1993; Hausdorff *et al.*, 1995b, 1996; Ossadnik *et al.*, 1994; Peng *et al.*, 1994a, 1995).

To illustrate the DFA algorithm, we use the interbeat time series shown in Figure 3a as an example. First, the interbeat interval time series (of total length N) is integrated,

$$y(k) = \sum_{i=1}^{k} [(B(i) - B_{ave}],$$

where $B(i)$ is the ith interbeat interval and B_{ave} is the average interbeat interval. As discussed above, this integration step maps the original time series to a self-similar process. Next, we measure the vertical characteristic scale of the integrated time series. To do so, the integrated time series is divided into boxes of equal length, n. In each box of length n, a least-squares line is fit to the data (representing the *trend* in that box; see Figure 4). The y-coordinate of the straight-line segments is denoted by $y_n(k)$. Next we detrend the integrated time series, $y(k)$, by subtracting the local trend, $y_n(k)$, in each box. For a given box size n, the characteristic size of fluctuation for this integrated and detrended time series is calculated by

$$F(n) = \sqrt{\frac{1}{N} \sum_{k=1}^{N} [y(k) - y_n(k)]^2}. \qquad (4)$$

(This quantity F is similar to but not identical with the quantity s measured in the previous section.)

This computation is repeated over all time scales (box sizes) to provide a relationship between $F(n)$ and the box size n. Typically, $F(n)$ will increase with box size n. A linear relationship on a double log graph indicates the presence of scaling (self-similarity) – the fluctuations in small boxes are related to the fluctuations in larger boxes in a power-law fashion. The slope of the line relating log $F(n)$ to log n determines the scaling exponent (self-similarity parameter), α, as discussed before.

---

[5] The DFA algorithm works better for certain types of nonstationary time series (especially slowly varying trends). However, it is not designed to handle all possible nonstationarities in real-world data (Peng *et al.*, 1995).

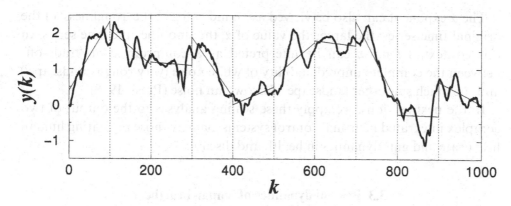

Figure 4. The integrated time series: $y(k) = \sum_{i=1}^{k} [B(i) - B_{ave}]$, where $B(i)$ is the i-th interbeat interval shown in Figure 3a. The vertical dotted lines indicate boxes of size $n = 100$, and the solid straight line segments represent the 'trend' estimated in each box by a linear least-squares fit. (From Peng *et al.*, 1995.)

3.2.4 *Relationship between self-similarity and autocorrelation functions*

The self-similarity parameter of an integrated time series is related to the more familiar autocorrelation function, $C(\tau)$, of the original (nonintegrated) signal. Briefly:

(1) For white noise, where the value at one instant is completely uncorrelated with any previous values, the integrated value, $y(k)$, corresponds to a random walk and therefore $\alpha = 0.5$ (Montroll and Shlesinger, 1984; Feder, 1988). The autocorrelation function, $C(\tau)$, is 0 for any τ (time lag) not equal to zero.

(2) Many natural phenomena are characterized by short-term correlations with a characteristic time scale, τ_0, and an autocorrelation function, $C(\tau)$, that decays exponentially, i.e., $C(\tau) \sim \exp(-\tau/\tau_0)$. The initial slope of F_n versus $\log n$ may be different from 0.5, but α will approach 0.5 for large window sizes.

(3) An α greater than 0.5 and less than or equal to 1.0 indicates *persistent* long-range power-law correlations, i.e., $C(\tau) \sim \tau^{-\gamma}$. The relation between α and γ is $\gamma = 2 - 2\alpha$. Note also that the power spectrum, $S(f)$, of the original (nonintegrated) signal is also of a power-law form, i.e., $S(f) \sim 1/f^{\beta}$, because the power spectrum density is simply the Fourier transform of the autocorrelation function, $\beta = 1 - \gamma = 2\alpha - 1$. The case of $\alpha = 1$ is a special one, which has interested physicists and biologists for many years – it corresponds to $1/f$ noise ($\beta = 1$).

(4) When $0 < \alpha < 0.5$, power-law *anti-correlations* are present such that large values are more likely to be followed by small values and vice versa (Beran, 1994).

(5) When $\alpha > 1$, correlations exist but cease to be of a power-law form; $\alpha = 1.5$ indicates brown noise, the integration of white noise.

The α exponent can also be viewed as an indicator of the 'roughness' of the original time series: the larger the value of α, the smoother the time series. In this context, $1/f$ noise can be interpreted as a compromise or 'trade-off' between the complete unpredictability of white noise (very rough 'landscape') and the much smoother landscape of Brownian noise (Press, 1978).

In the next sections, we apply these scaling analyses to the output of two complex integrated neuronal control systems, namely those regulating human heart rate and gait dynamics in health and disease.

3.3 Fractal dynamics of human heartbeat

Clinicians have traditionally described the normal activity of the heart as 'regular sinus rhythm'. However, contrary to subjective impression and clinical assumption, cardiac interbeat intervals normally fluctuate in a complex, apparently erratic manner, even in individuals at rest (Figure 1a; Kitney and Rompelman, 1980; Goldberger *et al.*, 1990). This highly irregular behavior defies conventional analyses that require 'well-behaved' (stationary) data sets. Fractal analysis techniques developed above are good candidates for studying this type of time series where fluctuations on multiple time scales appear to occur.

3.3.1 *Is the healthy human heartbeat fractal?*

To test whether heartbeat time series exhibit fractal behavior, we can apply the DFA algorithm to the full, 24-hour data sets excerpted in Figure 3. Figure 5 compares the DFA analysis of the interbeat interval time series for the healthy subject with the randomized control time series. For the healthy subject, DFA analysis shows scaling behavior with exponent $\alpha = 1$ over three decades, consistent with $1/f$-type of dynamics as previously reported (Kobayashi and Musha, 1982; Peng *et al.*, 1993b).[6] As expected, the randomized control data set shows a trivial exponent $\alpha = 0.5$, indicating uncorrelated randomness. Power spectrum analysis confirms the DFA results. The β exponent derived from the power spectrum, however, is less accurate because the stationarity requirement for Fourier analysis is not satisfied in this case.

---

[6] One alternative method to reduce the effects of nonstationarity in heart rate time series is to study the first difference of the original time series. In that case, the interbeat interval *increments* exhibit long-range anti-correlations (Peng *et al.*, 1993b).

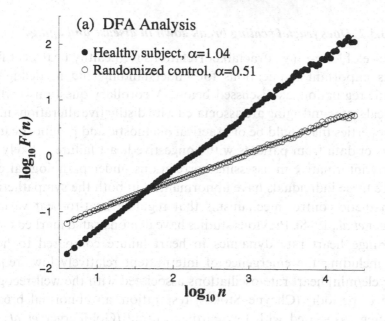

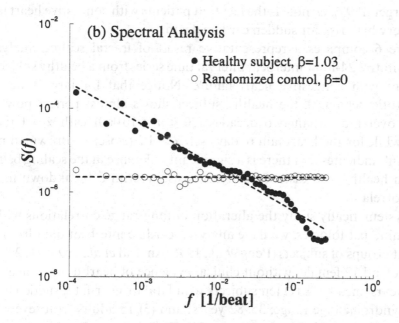

Figure 5. Scaling analyses for two 24-hour interbeat interval time series shown in Figure 3. The solid circles represent data from a healthy subject, while the open circles are for the artificial time series generated by randomizing the sequential order of data points in the original time series. (a) Plot of $\log F(n)$ vs. $\log n$ by the DFA analysis. (b) Fourier power spectrum analysis. The spectra have been smoothed (binned) to reduce scatter.

3.3.2 *Does fractal scaling break down in disease and aging?*

The presence of long-range (fractal) correlations for healthy heartbeat fluctuations has important implications for understanding and modeling neuroautonomic regulation, as discussed below. A corollary question is whether pathological states and aging are associated with distinctive alterations in these scaling properties that could be of practical diagnostic and prognostic use.

Analysis of data from patients with congestive heart failure is likely to be particularly informative in assessing correlations under pathological conditions, since these individuals have abnormalities in both the sympathetic and parasympathetic control mechanisms that regulate beat-to-beat variability (Goldberger *et al.*, 1988). Previous studies have demonstrated marked changes in short-range heart rate dynamics in heart failure compared to healthy function, including the emergence of intermittent relatively low frequency ($\sim$ one cycle/min) heart rate oscillations associated with the well-recognized syndrome of periodic (Cheyne–Stokes) respiration, an abnormal breathing pattern often associated with low cardiac output (Goldberger *et al.*, 1988; Goldberger, 1997). Of note is the fact that patients with congestive heart failure are at very high risk for sudden cardiac death.

Figure 6 compares a representative result of fractal scaling analysis of representative 24-hour interbeat interval time series from a healthy subject and a patient with congestive heart failure. Notice that for large time scales (asymptotic behavior), the healthy subject shows almost perfect power-law scaling over more than two decades ($20 \leq n \leq 10\,000$) with $\alpha = 1$ (i.e., $1/f$ noise), while for the heart failure data set, $\alpha \approx 1.3$ (closer to Brownian noise). This result indicates that there is a significant difference in the scaling behavior between healthy and diseased states, consistent with a breakdown in long-range correlations.

To systematically study the alteration of long-range correlations with life-threatening pathologies, we have analyzed cardiac interbeat data from three different groups of subjects (Peng *et al.*, 1995; Amaral *et al.*, 1998): (1) 29 adults (17 male and 12 female) without clinical evidence of heart disease (age range: 20–64 years, mean 41), (2) ten subjects with fatal or near-fatal sudden cardiac death syndrome (age range: 35–82 years) and (3) 15 adults with severe heart failure (age range: 22–71 years; mean 56). Data from each subject contained approximately 24 hours of electrocardiogram (ECG) recording encompassing $\sim 10^5$ heartbeats.

For the normal control group, we observed $\alpha = 1.0 \pm 0.1$ (mean value $\pm$ SD). These results confirm that healthy heart rate fluctuations exhibit long-range power-law (fractal) correlation behavior over three decades, similar

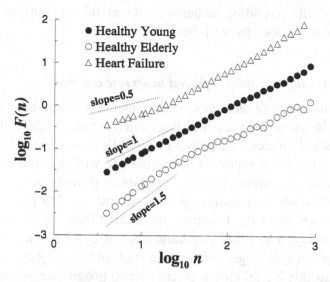

Figure 6. Plot of log $F(n)$ vs. log n for three interbeat interval time series: healthy young subject, elderly subject, and a subject with congestive heart failure. Compared with the healthy young subject, the heart failure and healthy elderly subjects show different patterns of altered scaling behavior (for details, see text).

to that observed in many dynamical systems far from equilibrium (Mallamace and Stanley, 1997; Meakin, 1997). Furthermore, both pathological groups showed significant deviation of the long-range correlations exponent α from the normal value, $\alpha = 1$. For the group of heart failure subjects, we found that $\alpha = 1.24 \pm 0.22$, while for the group of sudden cardiac death syndrome subjects, we found that $\alpha = 1.22 \pm 0.25$. Of particular note, we obtained similar results when we divided the time series into three consecutive subsets (of ~ 8 h each) and repeated the above analysis. Therefore our findings are not simply attributable to different levels of daily activities.[7]

Similar analysis was applied to study the effect of physiological aging. Ten young (21–34 years) and ten elderly (68–81 years) healthy subjects underwent 2 h of continuous supine resting ECG recording (Figure 6). In healthy young subjects, the scaling exponent had an α value close to 1.0. In the group of healthy elderly subjects, the interbeat interval time series showed two scaling regions. Over the short range, interbeat interval fluctuations resembled a random walk process (Brownian noise, $\alpha = 1.5$), whereas over the longer range they resembled white noise $\alpha = 0.5$). Short-range (α_1) and long-range (α_2) exponents were significantly different in the elderly subjects compared with young subjects (Iyengar *et al.*, 1996). Interestingly, the alterations of scaling

---

[7] More recent analysis does indicate subtle but important differences in fractal scaling between sleep and wake periods under healthy as well as diseased conditions (P. C. Ivanov *et al.*, unpublished data).

behavior associated with physiological aging exhibited different patterns compared with the changes associated with heart failure.

3.3.3 *Clinical utility of fractal heart rate analysis*

A relevant question regarding these new measurements is 'does fractal analysis, such as the DFA method, have clinically predictive value, independent of conventional statistical indices?' To answer this question, we have studied the predictive power of the DFA exponent in comparison with multiple conventional measures based on mean, variance and spectral analysis (Ho *et al.*, 1997). We analyzed two-hour ambulatory ECG recordings of 69 participants (mean age 71.7 ± 8 years) in the Framingham Heart Study – a prospective, population-based study. The study population consisted of chronic congestive heart failure patients, and age- and sex-matched control subjects. Importantly, we found that this fractal measurement carried prognostic information about mortality not extractable from traditional methods of heart rate variability analysis (Figure 7). Subsequent studies have confirmed and extended these observations (Mäkikallio *et al.*, 1997, 1998, 1999), suggesting that fractal scaling measures may have a practical use in bedside and ambulatory monitoring.

3.4 Fractal dynamics of human walking

In the previous section, we described the fractal fluctuations in the healthy human heartbeat, as well as alterations of these normal scale-invariant patterns with both aging and disease. In this section, we turn our attention from the dynamics of the involuntary (autonomic) nervous system to the voluntary nervous system.

Our focus here is on the step-to-step fluctuations in walking rhythm; that is, the duration of the gait cycle, also referred to as the *stride interval* (see Figure 8). The stride interval is analogous to the cardiac interbeat interval, and, like the heartbeat, it was traditionally thought to be quite regular under healthy conditions. However, as shown in Figure 8, subtle and complex fluctuations are apparent in the duration of the stride interval. While this 'noise' had been previously observed (Gabell and Nayak, 1984; Yamasaki *et al.*, 1991), until recently these fluctuations had not been characterized and their origin was largely unknown. Our goal is to analyze these step-to-step fluctuations in gait in order to gain insight into the neuronal control of locomotion in health and disease.

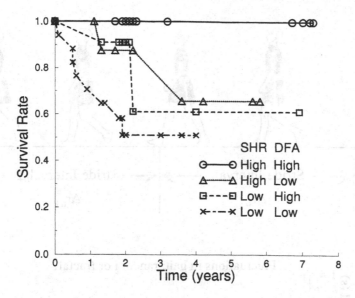

Figure 7. Assessment of patient survival rate by using an index (DFA index) derived from DFA analysis along with the information about the standard deviation of heart rate variability (SHR). In this population-based (Framingham Heart) study, we found, using multivariable analysis, that the DFA and SHR were the two most powerful independent heart rate variability predictors of mortality. Here, high and low DFA indices (or SHR) refer to their median values. (After Ho *et al.*, 1997, Predicting survival in heart failure cases and controls using fully automated methods for deriving non-linear and conventional indices of heart rate dynamics, *Circulation* **96**: 842–848.)

The simplest explanation for these step-to-step variations in walking rhythm is that they trivially represent uncorrelated (white) noise superimposed on a basically regular process – random fluctuations riding on top of the normal, constant walking rhythm. A second possibility is that these fluctuations have short-range correlations ('memory') as one might expect to see in a Markov process or a biological system where there is exponential decay of the system 'memory'. In that case, the current value of the stride interval would be influenced by only the most recent stride intervals, but, over the long term, fluctuations would vary randomly. A third, less intuitive possibility is that the fluctuations in the stride interval could exhibit the type of long-range correlations seen in the healthy human heartbeat (see above), as well as other scale-free, fractal phenomena (Feder, 1988; Bassingthwaighte *et al.*, 1994). If this were the case, the stride interval at any instant would depend (at least in a statistical sense) on the intervals at relatively remote times, and this dependence ('memory effect') would decay in a power-law fashion.

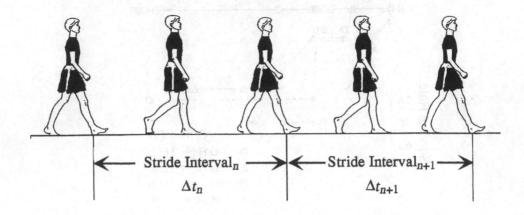

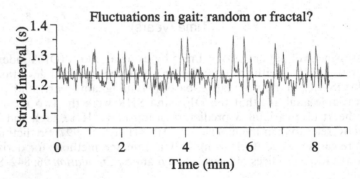

Figure 8. (Top) The gait cycle duration is termed the stride interval and is typically measured as the time between consecutive heel strikes of the same foot. (Bottom) Stride interval time series of a healthy subject while walking under constant environmental conditions. Although the stride interval is fairly stable (varying only between 1.1 and 1.4 s), it fluctuates about its mean (solid line) in an apparently unpredictable manner. A key question is whether these fluctuations represent uncorrelated randomness or whether there is a hidden fractal temporal structure, like that seen for the heartbeat. (Adapted from Hausdorff *et al.*, 1995b.)

3.4.1 Is healthy gait rhythm fractal?

To test these possibilities, we first measured the stride interval in healthy young adult men as they walked continuously on level ground at their self-deter-mined, usual rate for about 9 min (Hausdorff *et al.*, 1995b). To measure the stride interval in health and disease, ultra-thin, force-sensitive switches were placed inside the shoe. We recorded the footswitch force on an ambulatory recorder and then determined heel strike timing (Hausdorff *et al.*, 1995a). This recently devised, inexpensive and portable technique enables, for the first time, continuous and relatively long-term measurement of gait, and is roughly analogous to the use of Holter monitoring for recording continuous heartbeat activity.

A representative stride interval time series from a healthy subject is shown in Figure 9 (top). Of note is the stability of the stride interval; during a 9-min walk, the coefficient of variation is only 4%. Thus, as in Figure 8, a reasonable first approximation of the dynamics of the stride interval would be a constant. Nonetheless, the stride interval, like the healthy heartbeat, does vary irregularly, raising the intriguing possibility of some underlying complex temporal 'structure'. Further, this complicated pattern changes after random shuffling of the data points (Figure 9), demonstrating that the original temporal pattern is related to the *sequential ordering* of the stride intervals, and is not simply a result of the *distribution* of the data points. Figure 9 (bottom left) shows the DFA plots for the original time series and the shuffled time series. The slope of the line relating $\log F(n)$ to $\log n$ is 0.83 for the original time series and 0.50 after random shuffling. Thus, fluctuations in the stride interval scale as $F(n) \approx n^{0.83}$ indicating long-range correlations, while the shuffled data set behaves as uncorrelated (white) noise; $\alpha = 0.5$. Figure 9 (bottom right) displays the power spectrum of the original time series. The spectrum is broad-band and scales as $1/f^{\beta}$ with $\beta \approx 0.92$. The two scaling exponents are consistent with each other within statistical error due to finite data length (Peng *et al.*, 1993a), and both α and β are consistent with long-range (fractal) correlations (compare with Figure 5).

For a group of ten healthy adults, we confirmed that the scaling exponents α and β both indicated the presence of long-range correlations consistent with a fractal gait rhythm. After random shuffling of the original stride interval time series, α approaches the value of a completely uncorrelated process (0.5). The shuffled time series has the same mean and standard deviation as the original time series, indicating that this fractal property of healthy human gait is related to the sequential ordering of the stride interval time series, but not to the first or second moments of the time series.

3.4.2 Stability of healthy fractal rhythm: effects of walking rate

The unexpected observation of fractal variability in human gait raises a number of questions. Does the fractal gait rhythm exist only during walking at one's normal pace, or does it occur at slower and faster walking rates as well? Does the influence of one stride interval on another continue beyond a few hundred strides, or do the long-range correlations eventually break down during an extended walk? To answer these questions, we asked young healthy men to walk for 1 h at their usual rate as well as at slow and fast paces around an outdoor track (Hausdorff *et al.*, 1996). A representative example of the effect

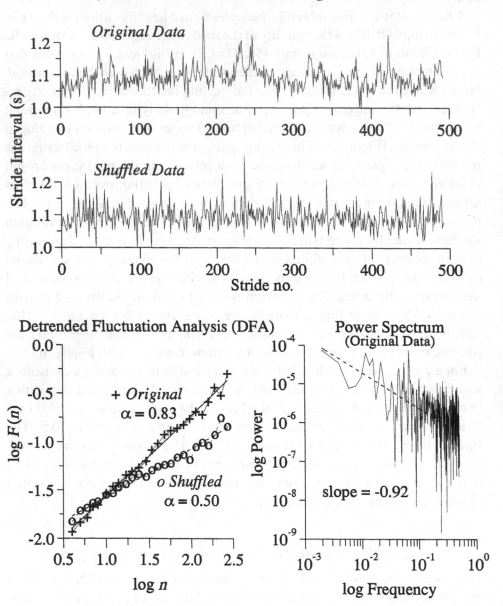

Figure 9. (Top) Representative stride interval time series before and after random shuffling of the data points. (Bottom) The detrended fluctuation analysis (DFA) and power spectrum analysis. The structure in the original time series disappears after random shuffling of the data. DFA indicates that this structure represents a fractal process with long-range correlations ($\alpha = 0.83$). (Adapted from Hausdorff *et al.*, 1995b.)

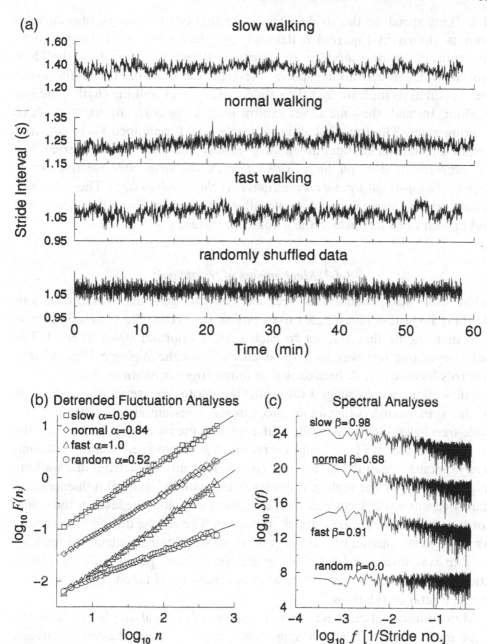

Figure 10. An example of the effects of walking rate on stride interval dynamics. (a) One-hour stride interval time series for slow (1.0 m/s), normal (1.3 m/s), and fast (1.7 m/s) walking rates. Note the breakdown of the temporal structure with random reordering of the fast walking trial data points, even though this shuffled time series has the same mean and standard deviation as the original, fast time series. (b, c) Fluctuation and power spectrum analyses confirm the presence of long-range correlations at all three walking speeds and their absence after random shuffling of the data points. (Adapted from Hausdorff *et al.*, 1996.)

of walking speed on the stride interval fluctuations and long-range correlations is shown in Figure 10. Remarkably, the locomotor control system maintains the stride interval at an almost constant level throughout the 1 h of walking at all three walking speeds. Nevertheless, both the DFA and power spectral analysis indicate that the subtle variations in walking rhythm are not random. Instead, the time series exhibit long-range correlations at all three walking rates. The fractal scaling indices α and β remained fairly constant despite substantial changes in walking velocity and mean stride interval. *For all subjects tested at all three walking rates, the stride interval time series displayed long-range, fractal correlations over thousands of steps.* These findings indicate that the fractal dynamics of walking rhythm are normally quite robust and appear to be intrinsic to the locomotor system.

3.4.3 *Mechanisms of fractal gait*

What biological mechanisms are necessary to generate this fractal gait rhythm? To further investigate this question, we asked subjects to walk in time to a metronome that was set to each subject's normal stride interval. The purpose of this test was to help to characterize the biological 'clock' that controls locomotion. A breakdown of long-range correlations during metronomic walking would suggest that some locomotor pacesetter above the level of the spinal cord (supraspinal mechanism) is essential in generating this scale-free behavior or, at least, that centrally mediated entrainment of the clock can 'overcome' long-range correlations generated peripherally. Alternatively, persistence of the long-range correlations during metronomic walking might imply that the scaling property is unrelated to central influences and that it results either from neuronal circuits at or below the level of the spinal cord, or from peripheral feedback influences. The results during metronomic walking were consistently different from those obtained when the walking rhythm was unconstrained. During metronomically paced walking, fluctuations in the stride interval were always random and failed to exhibit long-range, fractal correlations.

Metronomic walking and normal, unconstrained walking both utilize the same mechanical systems, the same force generators, and the same feedback networks. The breakdown of fractal, long-range correlations during metronomically paced walking demonstrates that influences above the spinal cord (a metronome) can override the normally present long-range correlations. This finding is of interest because it demonstrates that supraspinal nervous system control is critical in generating the robust, fractal pattern in normal human gait.

3.4.4 *Alterations of fractal dynamics with aging and disease*

These findings indicate that fractal gait dynamics depend on central nervous system function. Therefore, we hypothesized that, just as aging and cardiovascular disease may alter the fractal nature of the heartbeat, so too might changes in central nervous system function alter the fractal gait pattern. To test this hypothesis, we have begun to systematically study the effects of advanced age and neurodegenerative disorders on fractal gait rhythm (Hausdorff *et al.*, 1997b).

3.4.4.1 *Effects of aging*

We compared the gait of a group of very healthy elderly adults (ages 76 ± 3 years) to healthy young adults (ages 25 ± 2 years). Interestingly, both groups had identical mean stride intervals (elderly 1.05 s; young 1.05 s), and required almost identical amounts of time to perform a standardized functional test of gait and balance. The magnitude of stride-to-stride variability (i.e., stride interval coefficient of variation) was also very similar in the two groups (elderly 2.0%; young 1.9%). Figure 11 (left) compares the stride interval time series for one young and one elderly subject. Visual inspection suggests a possible subtle difference in the dynamics of the two time series (the data from the young subject appearing more 'patchy'). Fluctuation analysis reveals a marked distinction in how the fluctuations change with time scale for these subjects. The stride interval fluctuations are more random (less correlated) for the elderly subject than for the young subject, a difference *not* detectable by comparing the first and second moments.

Similar results were obtained for other subjects in these groups, indicating a subtle, previously undetected alteration in the fractal scaling of gait with healthy aging. Even among healthy elderly adults who have otherwise normal measures of gait and lower extremity function, the fractal-scaling pattern is significantly altered when compared with young adults. From a practical clinical perspective, the breakdown of long-range correlations of gait with aging is of interest for a number of reasons. An exciting prospect is that quantitative assessment of fractal properties of locomotion may provide a simple, inexpensive way to obtain important information about gait instability among the elderly. Falls are a major cause of disability and death in this age group (Hausdorff *et al.*, 1997a). The ability to identify individuals at greatest risk, as well as to assess interventions designed to restore gait stability (e.g., exercise, footwear), could have major public health implications. From a more basic physiological viewpoint, realistic models of gait dynamics must account not only for the unexpected long-range correlations in stride

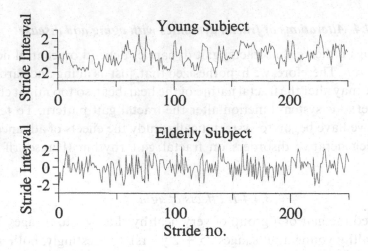

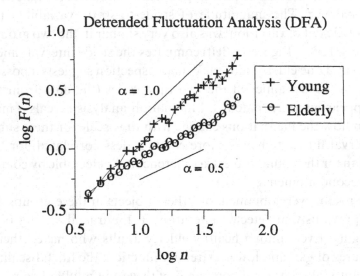

Figure 11. *Left:* Example of the effects of aging. Stride interval time series are shown (above) and DFA (below) for a 71-year-old elderly subject and a 23-year-old young adult. For illustrative purposes, each time series is normalized by subtracting its mean and dividing by its standard deviation. This normalization process highlights any temporal 'structure' in the time series, but does not affect the fluctuation analysis. Therefore, in this figure, stride interval is without units. For the elderly subject, DFA indicates a more random and less correlated time series. Indeed, $\alpha = 0.56$ ($\approx$ white noise) for the elderly subject and 1.04 ($1/f$ noise) for the young adult.

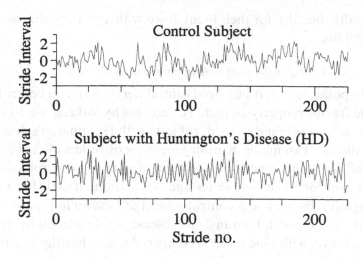

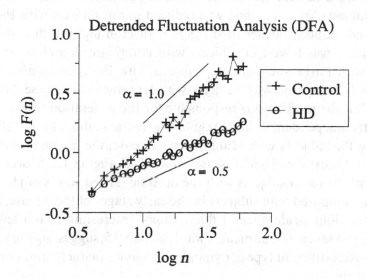

Right: Example of the effects of Huntington's disease (HD). For the subject with Huntington's disease (age: 41-years old), as compared with a healthy control, the stride interval fluctuations, $F(n)$, increase more slowly with time scale, n. This indicates a more random and less correlated time series. Indeed, $\alpha = 0.40$ for this subject with Huntington's disease and 0.92 for this healthy control subject. (Adapted from Hausdorff *et al.*, 1997b.)

interval in health, but also for their breakdown with aging and disease (Hausdorff *et al.*, 1995b).

3.4.4.2 *Effects of neurodegenerative disease*

We further hypothesized that impaired central nervous system control might also alter the fractal property of gait. To test this hypothesis, we have compared the stride interval time series of subjects with Huntington's disease and Parkinson's disease, two major neurodegenerative disorders of the basal ganglia (a part of the brain responsible for regulating motor control), with data from healthy controls. The time series and fluctuation analysis for a subject with Huntington's disease and a control subject are shown in Figure 11 (right panel). For the subject with Huntington's disease, stride interval fluctuations, $F(n)$, increase slowly with time scale, n, compared with a healthy control. This finding indicates increased randomness and reduced stride interval correlations as compared with the control subject. In general, compared with healthy control subjects, fractal scaling was reduced in the subjects with Parkinson's disease and reduced further in subjects with Huntington's disease. Interestingly, while α was lowest in subjects with Huntington's and intermediate in subjects with Parkinson's disease, subjects with Parkinson's disease walked more slowly compared with subjects with Huntington's disease, further confirming that the mechanisms responsible for the generation of gait speed are apparently independent of those regulating fractal scaling (Figure 10a).

Among the subjects with Huntington's disease, the fractal scaling index α was inversely correlated with disease severity (see Figure 12). Moreover, α was significantly lower in subjects with the most advanced stages of Huntington's disease as compared with subjects in the early stages of the disease, indicative of more random stride interval fluctuations. Interestingly, in a few subjects with the most severe impairment, α was less than 0.5, suggesting the presence of a qualitatively different type of dynamical behavior (namely, anti-correlations) in the gait rhythm.

These results indicate that, with both Parkinson's and Huntington's disease, there is a breakdown of the normal fractal, long-range correlations in the stride interval, especially apparent in subjects with advanced Huntington's disease. Step-to-step fluctuations are more random (i.e., more like white noise), suggesting that the fractal property of gait is modulated in part by central nervous system (i.e., basal ganglia) function. Although fractal scaling is altered both with aging and certain diseases, the magnitude of these changes varies in different conditions, and other measures of gait dynamics may also distinguish among different disease states and aging (Hausdorff *et al.*, 1998), adding specificity to these new dynamical measures (compare with Figure 6).

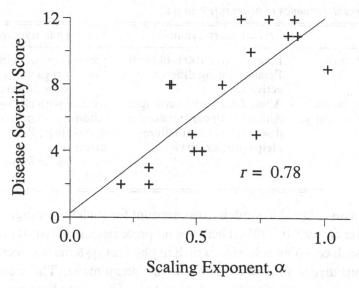

Figure 12. Among subjects with Huntington's disease, disease severity score (0 = most impairment; 13 = no impairment), measured using an index that correlates with positron emission tomography (PET) scan indices of caudate metabolism (Young *et al.*, 1986), is strongly ($p < 0.0005$) associated with fractal scaling of gait. (Adapted from Hausdorff *et al.*, 1997b.)

3.5 Fractal dynamics of heart rate and gait: implications and general conclusions

In this chapter, we have investigated the output of two types of neuro-physiological control systems, one involuntary (heartbeat regulation), and the other voluntary (gait regulation). We find that the time series of both human heart rate and stride interval show 'noisy' fluctuations. According to classical physiological paradigms based on homeostasis, such systems should be designed to damp out noise and settle down to a constant equilibrium-like state (Cannon, 1929). However, analysis of both heartbeat and gait fluctuations under apparently steady-state conditions reveals the presence of long-range correlations (see Table 1). This 'hidden' fractal property is more consistent with a regulatory system driven away from equilibrium, reminiscent of the behavior of dynamical systems near a critical point, or, in the case of physiological systems, perhaps a *critical zone* of parameter values (Ivanov *et al.*, 1998). The discovery of such long-range organization poses a remarkable challenge to contemporary efforts to understand and eventually simulate physiological

Table 1. *Fractal dynamics of heart rate and gait*

	Fractal heart dynamics	Fractal gait dynamics
Features in health	Extends over 1000s of beats Persists during different activities	Extends over 1000s of steps Persists regardless of gait speed (slow, normal, fast)
Potential diagnostic and prognostic utility	Altered with advanced age Altered with cardiovascular disease (e.g., heart failure) Helps predict survival	Altered with advanced age Altered with nervous system disease (e.g., Parkinson's disease) May predict falls among elderly

control systems. Plausible models must account for such long-range 'memory' (Hausdorff *et al.*, 1995b, 1996). There are no precedents in classical physiology to explain such complex behavior, which in physical systems has been connected with turbulence and related multiscale phenomena. The discovery of fractal dynamics as a possibly 'universal' feature of integrated neuronal control networks raises the intriguing possibility that the mechanisms regulating such systems interact as part of coupled cascade of feedback loops in a system driven far from equilibrium (Ivanov *et al.*, 1999).

The long-range power-law correlations in healthy heart rate and gait dynamics may be adaptive for at least two reasons (Peng *et al.*, 1993b): (1) the long-range correlations may serve as a newly described organizing principle for highly complex, nonlinear processes that generate fluctuations on a wide range of time scales, and (2) the lack of a characteristic scale may help to prevent excessive *mode locking* that would restrict the functional responsiveness (plasticity) of the organism. Support for these two related conjectures is provided by the findings described here from severely pathological states, such as heart failure, where the *breakdown* of long-range correlations is often accompanied by the emergence of a dominant frequency mode (e.g., the Cheyne–Stokes frequency; compare Figure 1b). Analogous transitions to highly periodic behavior have been observed in a wide range of other disease states, including certain malignancies, sudden cardiac death, epilepsy, fetal distress syndromes, and with certain drug toxicities (Goldberger, 1996, 1997).

Unanswered questions currently under study include the following. What are the physiological mechanisms underlying such long-range correlations in heartbeat and gait? How do these macroscopic dynamics relate to microscopic fluctuations and self-organization at the cellular and molecular levels (Liebovitch and Toth, 1990)? Are these fluctuations entirely stochastic or do they represent the interplay of deterministic and stochastic mechanisms (Goldberger, 1997; Ivanov *et al.*, 1998)?

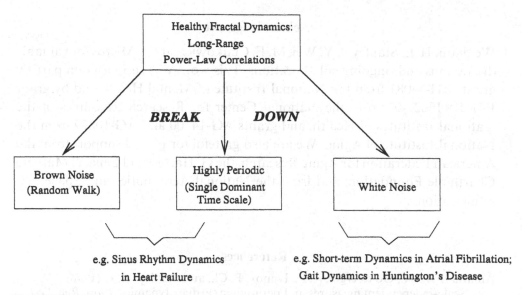

Figure 13. The breakdown of long-range power-law correlations may lead to any of three dynamical states: (1) a random walk ('brown noise') as observed in low-frequency heart rate fluctuations in certain cases of severe heart failure; (2) highly periodic oscillations, as also observed in Cheyne–Stokes pathophysiology in heart failure, as well as with sleep apnea (Figure 1c), and (3) completely uncorrelated behavior (white noise), perhaps exemplified by the short-term heart rate dynamics during atrial fibrillation. (After Peng *et al.*, 1994b.)

From a practical viewpoint, these findings may have implications for physiological monitoring. The breakdown of normal long-range correlations in any physiological system could theoretically lead to three possible dynamical states (Figure 13; Peng *et al.*, 1994b): (1) a random walk (brown noise), (2) highly periodic behavior, or (3) completely uncorrelated behavior (white noise). Cases (1) and (2) both indicate only 'trivial' long-range correlations of the types observed in severe heart failure. Case (3) may correspond to certain cardiac arrhythmias such as fibrillation, or to gait disorders such as Huntington's disease. Such alterations are not detectable with traditional clinical statistics (e.g., those based upon comparison of means and variances). The application of fractal and related analysis techniques is likely to provide an important, complementary set of tools to assess the stability of such systems and their changes with aging and disease (Figures 6 and 11). Perhaps most exciting is the prospect that such new approaches may be the basis for the development of *dynamical assays* designed to assess the efficacy and exclude the toxicity of new interventions, which hopefully will maintain and restore the multiscale complexity and correlated noisiness that appear to be defining features of healthy, adaptive physiology.

Acknowledgments

We thank H. E. Stanley, J. Y. Wei, M. E. Cudkowicz, and J. Mietus for valuable discussions and ongoing collaborations. This work was supported in part by grant MH-54081 from the National Institute of Mental Health and by grant P41 RR13622-01 from the National Center for Research Resources of the National Institutes of Health, and grants AG-14100 and AG-10829 from the National Institute of Aging. We are also grateful for partial support from the American Federation for Aging Research, the G. Harold and Leila Y. Mathers Charitable Foundation, and from the National Aeronautics and Space Administration.

References

Amaral, L. A. N., Goldberger, A. L., Ivanov, P. Ch. and Stanley, H. E. (1998) Scale-independent measures and pathologic cardiac dynamics. *Phys. Rev. Lett.* **81**: 2388–2391.

Bassingthwaighte, J. B., Liebovitch, L. S. and West, B. J. (1994) *Fractal Physiology.* New York: Oxford University Press.

Beran, J. (1994) *Statistics for Long-Memory Processes.* New York: Chapman & Hall.

Buldyrev, S. V., Goldberger, A. L., Havlin, S., Peng, C.-K., Stanley, H. E. and Simons, M. (1993) Fractal landscapes and molecular evolution: modeling the myosin heavy chain gene family. *Biophys. J.* **65**: 2673–2679.

Bunde, A. and Havlin, S. (eds.) (1994) *Fractals in Science.* Berlin: Springer-Verlag.

Cannon, W. B. (1929) Organization for physiological homeostasis. *Physiol. Rev.* **9**: 399–431.

Feder, J. (1988) *Fractals.* New York: Plenum Press.

Gabell, A. and Nayak, U. S. L. (1984) The effect of age on variability in gait. *J. Gerontol.* **39**: 662–666.

Goldberger, A. L. (1996) Non-linear dynamics for clinicians: chaos theory, fractals, and complexity at the bedside. *Lancet* **347**: 1312–1314.

Goldberger, A. L. (1997) Fractal variability versus pathologic periodicity: complexity loss and stereotypy in disease. *Perspect. Biol. Med.* **40**: 543–561.

Goldberger, A. L., Rigney, D. R., Mietus, J., Antman, E. M. and Greenwald, M. (1988) Nonlinear dynamics in sudden cardiac death syndrome: heart rate oscillations and bifurcations. *Experientia* **44**: 983–987.

Goldberger, A. L., Rigney, D. R. and West, B. J. (1990) Chaos and fractals in human physiology. *Sci. Am.* **262**: 42–49.

Hausdorff, J. M., Cudkowicz, M. E., Firtion, R., Wei, J. Y. and Goldberger, A. L. (1998) Gait variability and basal ganglia disorders: stride-to-stride variations of gait cycle timing in Parkinson's and Huntington's disease. *Move Disord.* **13**: 428–437.

Hausdorff, J. M., Edelberg, H. E., Mitchell, S. and Wei, J. Y. (1997a) Increased gait instability in community dwelling elderly fallers. *Arch. Phys. Med. Rehabil.* **78**: 278–283.

Hausdorff, J. M., Ladin, Z. and Wei, J. Y. (1995a) Footswitch system for measurement of the temporal parameters of gait. *J. Biomech.* **28**: 347–351.

Hausdorff, J. M., Mitchell, S. L., Firtion, R., Peng, C.-K., Cudkowicz, M. E., Wei, J. Y. and Goldberger, A. L. (1997b) Altered fractal dynamics of gait: reduced stride interval correlations with aging and Huntington's disease. *J. Appl. Physiol.* **82**: 262–269.

Hausdorff, J. M. and Peng, C.-K. (1996) Multi-scaled randomness: a possible source of 1/f noise in biology. *Phys. Rev. E* **54**: 2154–2157.

Hausdorff, J. M., Peng, C.-K., Ladin, Z., Wei, J. Y. and Goldberger, A. L. (1995b) Is walking a random walk? Evidence for long-range correlations in the stride interval of human gait. *J. Appl. Physiol.* **78**: 349–358.

Hausdorff, J. M., Purdon, P., Peng, C.-K., Ladin, Z., Wei, J. Y. and Goldberger, A. L. (1996) Fractal dynamics of human gait: stability of long-range correlations in stride interval fluctuations. *J. Appl. Physiol.* **80**: 1448–1457.

Ho, K. K. L., Moody, G. B., Peng, C.-K., Mietus, J. E., Larson, M. G., Levy, D. and Goldberger, A. L. (1997) Predicting survival in heart failure cases and controls using fully automated methods for deriving nonlinear and conventional indices of heart rate dynamics. *Circulation* **96**: 842–848.

Hurst, H. E. (1951) Long-term storage capacity of reservoirs. *Trans. Am. Soc. Civil. Engrs.* **116**: 770–799.

Iannaconne, P. and Khokha, M. K. (eds.) (1996) *Fractal Geometry in Biological Systems: An Analytical Approach.* Boca Raton, FL: CRC Press.

Ivanov, P. Ch., Amaral, L. A. N., Goldberger, A. L., Havlin, S., Rosenblum, M. G., Struzik, Z. and Stanley, H. E. (1999) Multifractality in human heartbeat dynamics. *Nature* **399**, 461–465.

Ivanov, P. Ch., Amaral, L. A. N., Goldberger, A. L. and Stanley, H. E. (1998) Stochastic feedback and the regulation of biological rhythms. *Europhys. Lett.* **43**: 363–368.

Iyengar, N., Peng, C.-K., Morin, R., Goldberger, A. L. and Lipsitz, L. A. (1996) Age-related alterations in the fractal scaling of cardiac interbeat interval dynamics. *Am. J. Physiol.* **271**: 1078–1084.

Kitney, R. I. and Rompelman, O. (1980) *The Study of Heart-Rate Variability.* Oxford: Oxford University Press.

Kobayashi, M. and Musha, T. (1982) 1/f fluctuation of heartbeat period. *IEEE Trans. Biomed. Eng.* **29**: 456.

Kolmogorov, A. N. (1961) The local structure of turbulence in incompressible viscous fluid for very large Reynolds number. *Dokl. Akad. Nauk SSSR* **30**: 9–13 (reprinted in *Proc. R. Soc. Lond. A* **434**: 9–13).

Liebovitch, L. S. and Toth, T. I. (1990) Fractal activity in cell membrane ion channels. *Ann. NY Acad. Sci.* **591**: 375–391.

Mäkikallio, T. H., Hoiber, S., Kober, L., Torp-Pedersen, C., Peng, C.-K., Goldberger, A. L. and Huikuri, H. V. (1999) Fractal analysis of heart rate dynamics as a predictor of mortality in patients with depressed left ventricular function after acute myocardial infarction. *Am. J. Cardiol.* **83**: 836–839.

Mäkikallio, T. H., Ristimäe, T., Airaksinen, K. E. J., Peng, C.-K., Goldberger, A. L. and Huikuri, H. V. (1998) Heart rate dynamics in patients with stable angina pectoris and utility of fractal and complexity measures. *Am. J. Cardiol.* **81**: 27–31.

Mäkikallio, T. H., Seppänen, T., Airaksinen, K. E. J., Koistinen, J., Tulppo, M. P., Peng, C.-K., Goldberger, A. L. and Huikuri, H. V. (1997) Dynamic analysis of heart rate may predict subsequent ventricular tachycardia after myocardial infarction. *Am. J. Cardiol.* **80**: 779–783.

Mallamace, F. and Stanley, H. E. (eds.) (1997) *Physics of Complex Systems: Proceedings of Enrico Fermi School on Physics, Course CXXXIV.* Amsterdam: IOS Press.

Meakin, P. (1997) *Fractals, Scaling, and Growth Far from Equilibrium*. Cambridge: Cambridge University Press.

Montroll, E. W. and Shlesinger, M. F. (1984) The wonderful world of random walks. In *Nonequilibrium Phenomena II. From Stochastics to Hydrodynamics* (ed. J. L. Lebowitz and E. W. Montroll), pp. 1–121. Amsterdam: North-Holland.

Ossadnik, S. M., Buldyrev, S. V., Goldberger, A. L., Havlin, S., Mantegna, R. N., Peng, C.-K., Simons, M. and Stanley, H. E. (1994) Correlation approach to identify coding regions in DNA sequences. *Biophys. J.* **67**: 64–70.

Peng, C.-K., Buldyrev, S. V., Goldberger, A. L., Havlin, S., Simons, M. and Stanley, H. E. (1993a) Finite size effects on long-range correlations: implications for analyzing DNA sequences. *Phys. Rev. E* **47**: 3730–3733.

Peng, C.-K., Buldyrev, S. V., Havlin, S., Simons, M., Stanley, H. E. and Goldberger, A. L. (1994a) On the mosaic organization of DNA sequences. *Phys. Rev. E* **49**: 1685–1689.

Peng, C.-K., Buldyrev, S. V., Hausdorff, J. M., Havlin, S., Mietus, J. E., Simons, M., Stanley, H. E. and Goldberger, A. L. (1994b) Fractal landscapes in physiology and medicine: long-range correlations in DNA sequences and heart rate intervals. In *Fractals in Biology and Medicine* (ed. G. A. Losa, T. F. Nonnenmacher and E. R. Weibel), pp. 55–65. Basel, Berlin: Birkhäuser Verlag.

Peng, C.-K., Havlin, S., Stanley, H. E. and Goldberger, A. L. (1995) Quantification of scaling exponents and crossover phenomena in nonstationary heartbeat time series. *Chaos* **5**: 82–87.

Peng, C.-K., Mietus, J., Hausdorff, J. M., Havlin, S., Stanley, H. E. and Goldberger, A. L. (1993b) Long-range anti-correlations and non-Gaussian behavior of the heartbeat. *Phys. Rev. Lett.* **70**: 1343–1346.

Press, W. H. (1978) Flicker noise in astronomy and elsewhere. *Comments Astrophys.* **7**: 103–119.

Yamasaki, M., Sasaki, T. and Torii, M. (1991) Sex difference in the pattern of lower limb movement during treadmill walking. *Eur. J. Appl. Phys.* **62**: 99–103.

Young, A. B., Penney, J. B. and Starosta-Rubenstein, S. (1986) PET scan investigations of Huntington's disease: cerebral metabolic correlates of neurological features and functional decline. *Arch. Phys. Med. Rehabil.* **78**: 278–283.

4

Self-organizing dynamics in human sensorimotor coordination and perception

MINGZHOU DING, YANQING CHEN,
J. A. SCOTT KELSO AND BETTY TULLER

4.1 Introduction

The human brain is composed of 100 billion to a trillion neurons and as many neuroglia. The human-and-environment system is open and complex. Human behavior is adaptive and multifunctional, arising from interactions that occur on many levels among diverse organizational components. How is the vast material complexity of the brain on the one hand and the behavioral complexity that emerges on the other to be understood? In this chapter we describe experiments that illustrate recent research efforts aimed at uncovering the basic principles and mechanisms governing the brain and behavioral function. In particular, we focus on the following specific questions. (1) How do we react to and coordinate with the environment (see Section 4.2), and (2) how do we perceive and categorize the world around us (see Section 4.3)? Our work is based on the joint premises that a more complete understanding of how the brain works will come: (1) when experimental research in the laboratory is combined with new theoretical approaches investigating how the brain functions *as a whole*; and (2) as a result of direct, multidisciplinary collaborations between neuroscientists, experimental psychologists, mathematicians and physicists.

4.2 Evidence for self-organized dynamics from a human sensorimotor coordination experiment

One of the simplest forms of human–environment coordination involves producing motor outputs at a specific timing relationship with regular external events. Many human activities such as music and dance depend on the efficient execution of this sensorimotor task. We approach this problem by carrying out a simple experiment in which a subject taps his finger on a computer keyboard

97

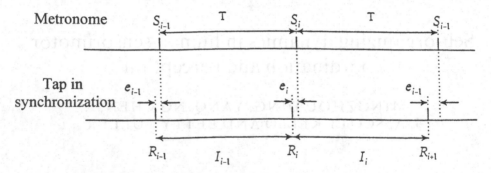

Figure 1. Definition of the synchronization error, e_i, and inter-response interval, I_i. (From Chen *et al.*, 1995, with permission.)

in synchrony with a periodic sequence of metronome beeps (Chen *et al.*, 1997). The variability of his performance is quantified by the *synchronization error, e_i*, defined as the difference between the computer-recorded tapping time and the metronome onset time (see Figure 1). The time course of this variable is erratic, showing clear evidence of an underlying *random* process (Figure 2). It has long been surmised that understanding the nature of this putative random process is an important step towards unraveling the brain's strategy of timing control (Hary and Moore, 1985). Previous work in this area has focused mainly on measuring the mean and the variance of e_i from short trials (< 100 cycles). These averaged quantities ignore the *temporal structure* of the synchronization error time series. Motivated by ideas and concepts from physics and mathematics, we redesigned the experiment by extending the length of experimental trials substantially beyond that employed in traditional experiments, and applied a host of new techniques to analyze the data, including the rescaled range method and the spectral maximum likelihood estimator. This new methodology enabled us to establish that the temporal structure of the synchronization error time series is characterized by $1/f^{\alpha}$ type of long memory (i.e., long-range correlations), and that the underlying stochastic process can be modeled by *fractional* Gaussian noise.

4.2.1 *Experimental design and observations*

Five right-handed male subjects took part in the synchronization experiment. Seated in a sound-attenuated chamber, each subject was instructed to cyclically press his index finger against a computer key in synchrony with a periodic series of auditory beeps, delivered through headphones. Two frequency conditions, $F_1 = 2\,\mathrm{Hz}\,(T_1 = 500\,\mathrm{ms})$ and $F_2 = 1.25\,\mathrm{Hz}\,(T_2 = 800\,\mathrm{ms})$ were studied.

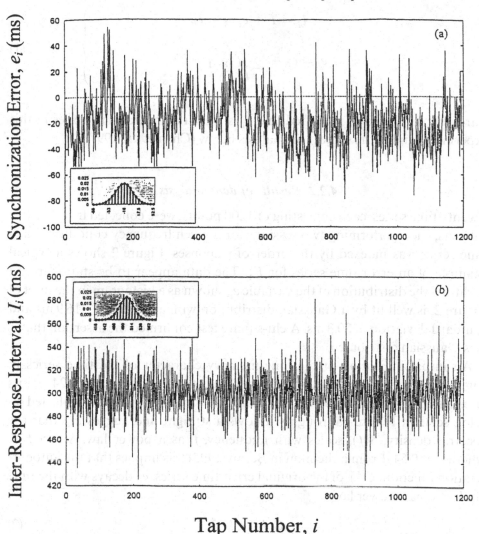

Figure 2. Example of a synchronization error time series. Histogram and its Gaussian fit are shown in the insets. Notice that most synchronization errors and their average are negative, meaning that, on average, the subject tapped before the beep. (From Chen *et al.*, 1995, with permission.)

These frequencies were chosen such that the subject was able to perform the required tapping motion continuously. Each experimental session consisted of the subject performing 1200 continuous taps for a given frequency. A computer program was used to register the time of a specific point in the tapping cycle in microsecond resolution. The data collected were the *inter-response intervals*, I_i, and the *synchronization* or *tapping errors*, e_i. As shown in Figure 1, I_i and e_i relate to each other through

$$I_i = T + e_{i+1} - e_i,$$

$$e_i = e_0 + \sum_{k=1}^{i} (I_{k-1} - T). \tag{1}$$

Careful considerations indicated that e_i is the fundamental time series in this experiment and is the subject of analysis below (Chen *et al.*, 1997).

4.2.2 *Results of data analysis*

Twenty time series, each consisting of 1200 points, were collected from the five subjects, each performing two sessions for a given frequency condition. Each time series was indexed by the order of responses. Figure 2 shows a typical example of an error time series for F_1. The data appear to be stationary. In addition, the distribution of the variable e_i, shown as a histogram in the inset of Figure 2, is well fit by a Gaussian distribution with a mean of -16.9 ms and standard deviation of 20.3 ms. A chi-square test confirmed the assertion that e_i was Gaussian distributed.

An initial indication of the long-memory character of the time series in Figure 2 was provided by computing its spectral density using 1024 points after discarding the first 50 points to eliminate transients. The result, plotted on a log scale in Figure 3a, roughly follows a straight line, suggesting that the spectral density, $S(f)$, scales with frequency, f, as a power law, $S(f) \propto f^{-\alpha}$, where $\alpha \approx 0.54$. From a theorem in Beran (1994) this implies that the autocorrelation function, $C(k)$, of the original error time series, e_i, decays with the time lag k also as a power law,

$$C(k) \propto k^{-\beta}, \tag{2}$$

where $\beta = 1 - \alpha \approx 0.46$. Recall that a long-memory process is mathematically defined as a process whose autocorrelation function, $C(k)$, sums to infinity (Beran, 1994),

$$\sum_{k=0}^{\infty} C(k) = \infty. \tag{3}$$

The autocorrelation function in Equation (2), with $0 < \beta \approx 0.46 < 1$, meets this definition. This establishes the error time series in Figure 2 as coming from a long-memory process, specifically a fractional Gaussian noise process (Mandelbrot and Van Ness, 1968). Similar results were obtained for all 20 error time series from the experiment. Also, the average spectral density for the 10 error

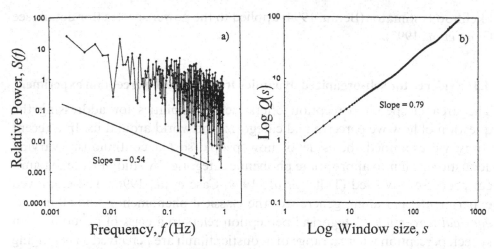

Figure 3. (a) Spectral density of the error time series in Figure 2. We have converted the unit of frequency from 1/beat to Hz. (b) Log–log plot of averaged R/S value, $Q(s)$, against window size, s, for the time series in Figure 2. (From Chen *et al.*, 1995, with permission.)

time series from each frequency condition was observed to obey a power law with slope close to 1/2.

Another index for long-memory processes is the Hurst exponent, H. It relates to α through (Beran, 1994),

$$H = (1 + \alpha)/2. \qquad (4)$$

A direct way to estimate the value of H is the trend-corrected rescaled range analysis originally used by Hurst to analyze yearly minima of the Nile River (Hurst, 1951). Let the trend-corrected range of the random walk be denoted as $R(n,s)$. Let $S^2(n,s)$ denote the sample variance of the data set. If the average rescaled statistic $Q(s) = \langle R(n,s)/S(n,s) \rangle_n$ scales with s as a power law for large s, $Q(s) \propto s^H$, then H is the Hurst exponent. One can show that, if the autocorrelation function, $C(k)$, sums to a finite number, then generally $H = 1/2$, corresponding to the case of short-term memory. If Equation (3) holds, then $1/2 < H < 1$, and the time series is said to have long-persistent memory.

Figure 3b shows the log–log plot of $Q(s)$ versus s for the error time series shown in Figure 2. A straight-line fit to the data gives $H = 0.79$, which is consistent with $H = 0.77$ obtained from Figure 3a and Equation (4). Applying the same rescaled range analysis to all the error time series, we found the average Hurst exponent to be about 0.723 ± 0.071, which is significantly greater than $H = 1/2$. Similar results were obtained using the maximum likelihood estimator (Beran, 1994) applied to the power spectra (for details see Chen *et al.*, 1997).

4.3 Evidence for self-organized dynamics from a speech perception experiment

The area of speech perception offers rich possibilities for addressing the question of how we perceive and categorize the world around us. In a recent study we examined the issue of how people sort a continuously varying acoustic signal into appropriate phonemic categories by studying the dynamical processes involved (Tuller *et al.*, 1994; Case *et al.*, 1996). The employed experimental paradigm generalizes the classical phenomenon known as *categorical perception*. Categorical perception refers to a class of phenomena in speech perception where a range of acoustic stimuli are perceived as belonging to the same phonetic category. For our experiments, the stimuli consisted of a natural 120-ms 's' excised from a male utterance of 'say', followed by a silent gap of variable duration (0 to 76 ms) denoted by τ, which is then followed by a synthetic speech token 'ay'. If τ is small (from 0 to 20 or 30 ms), the stimulus is perceived as 'say'. If τ is large, around 40 to 76 ms, the stimulus is perceived as 'stay'. Thus, if we vary τ systematically as a control parameter, transitions from 'say' to 'stay' or from 'stay' to 'say' take place. This systematic variation of a control parameter is typical in nonlinear dynamics studies, and it allows detailed examinations of important questions such as how and by what mechanism human perception changes from one state to another.

4.3.1 Experimental design and basic findings

In an experimental run, the subject is presented with a sequence of stimuli in which the gap duration is systematically increased from 0 to 76 ms in increments of 4 ms, and then decreased with the same step size back to 0 ms. Between two consecutive stimuli, there is a resting period of 2.5 s, which is called the interstimulus interval. The three observed perceptual patterns are shown in Figure 4. Figure 4a describes a pattern where the switch from one percept to another ('say' to 'stay' or 'stay' to 'say') occurs at the *same* gap duration for both increasing and decreasing τ. The pattern in Figure 4b represents a classic *hysteresis* effect where the overlapping region indicates that a given stimulus can be perceived differently depending on the *direction* of the gap variation. The third pattern, shown in Figure 4c, is a more peculiar one in that the percepts switch from 'say' to 'stay' *earlier* as τ increases, and again from 'stay' to 'say' earlier as τ decreases. We call this phenomenon the *enhanced contrast effect*, which is related to selective adaptation and range effects in

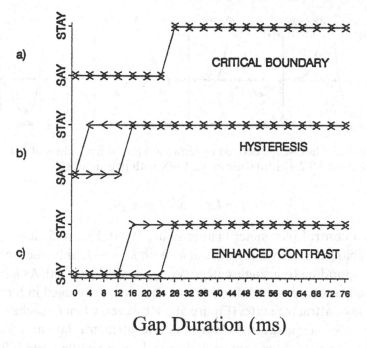

Gap Duration (ms)

Figure 4. (a–c) Individual experimental runs showing three prototypical patterns. For details, see text in Section 4.3.1. (From Tuller *et al.*, 1994, with permission.)

speech perception, implying that boundary shifts act to enhance contrast between perceptual states. The pattern in Figure 4a is rarely observed, while patterns in Figure 4b and c occur about equally often.

The dependence of speech categorization on recent percepts and on the direction of parameter change is a strong indicator of *nonlinearity* and *multistability*. In what follows, we briefly describe a theoretical model proposed to capture the observed patterns of category change within a unified dynamical account. Then, we describe one model prediction that is evaluated by further experiments.

4.3.2 *A dynamical model of categorization*

Speech perception can be regarded as a *pattern formation* process in the brain. For the present experiment, we modeled the coexistence of two distinct patterns ('say' and 'stay') and the spontaneous switch among the patterns by an overdamped oscillator,

$$dx/dt = -\,dV(x)/dx = -\,k + x - x^3, \tag{5}$$

with the following potential,

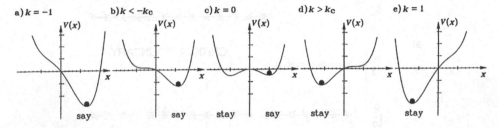

Figure 5. Potential landscape defined by Equation (6) for five values of k. For details, see text in Section 4.3.2. (From Case *et al.*, 1995, with permission.)

$$V(x) = kx - x^2/2 + x^4/4, \tag{6}$$

where k is a control parameter (Tuller *et al.*, 1994). Figure 5 shows how the landscape changes for several values of k. With $k = -1$, only one stable point exists corresponding to a single category (e.g., 'say'; Figure 5a). As k increases, the potential landscape tilts but otherwise remains unchanged in terms of the composition of attractor states (Figure 5b). However, when k reaches a critical point, $k = -k_c$, a *qualitative* change in the attractor layout takes place. Specifically, the particular change at $k = -k_c$ is a saddle-node bifurcation. Thus, where there was once only a single perceptual category, there are now two possible categories. When $-k_c < 0 < k_c$, both 'say' and 'stay' are available categories (Figure 5c). The coexistence of both categories continues until $k = k_c$, where the attractor corresponding to 'say' ceases to exist via a reverse saddle-node bifurcation (where the qualitative change is from two available categories to one), leaving only the stable fixed point corresponding to 'stay' (Figure 5d). Further increases in k serve only to deepen the potential minimum corresponding to 'stay' (Figure 5e). Thus, the model captures the three observed states of the system: at the smallest values of the acoustic parameter only 'say' is reported, for an intermediate range of parameter values either 'say' or 'stay' are reported, and for the largest values of gap duration only 'stay' is reported.

To accommodate the three observed patterns in Figure 4, we assume that k in Equation (6) can be expressed as

$$k(\lambda) = -k_0 + \lambda + \varepsilon/2 + \varepsilon\theta(n - n_c)(\lambda - \lambda_f). \tag{7}$$

The meaning of each term is explained below. First, λ is a variable linearly proportional to the gap duration τ and varies from $\lambda_i = 0$ to $\lambda_f = 3$ (the corresponding experimental gap duration varies from 0 to 76 ms; in Figures 6 and 7, for ease of comparison with the experiment, we have scaled the λ parameter from 0 to 76 ms). Second, $\theta(n - n_c)$ is the step function defined by

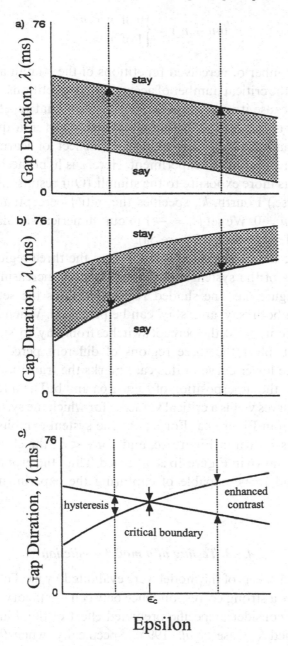

Figure 6. (a) Perceptual states in the ε–λ plane for increasing λ (small n). (b) Perceptual states in the ε–λ plane for decreasing λ (large n). (c) Superposition of transitions between categories in (a) and (b). For details, see text in Section 4.3.2. (From Case *et al.*, 1995, with permission.)

$$\theta(n - n_c) = \begin{cases} 0 & \text{if } n < n_c \\ 1 & \text{if } n \geq n_c, \end{cases} \tag{8}$$

where n is the number of perceived repetitions of the stimuli and $n_c = n_{total/2}$ = 20 represents the critical number of accumulated repetitions. The θ-function is introduced because it is noted in the experiments that repetitive presentations of the stimuli tend to alter the perceptual states in a qualitative way. Third, ε in Equation (7) corresponds to the effect of learning and prior experience with respect to the experiment. Here, ε is hypothesized to increase as the subject gets more exposure to the stimuli. (Our results indicate that this is indeed the case.) Fourth, k_0 specifies the initial perceptual configuration when $\lambda = 0$ and $n = 0$. We set $k_0 = -1$ in our numerical simulations, for only 'say' is perceived at $\lambda = 0$ (see Figure 5a).

By solving the above equations we obtained the three regions of different perceptual states of the system in the ε–λ plane for increasing λ (thus τ) as illustrated in Figure 6a. The shaded region indicates the set of parameter values for which both 'say' and 'stay' can be perceived. When λ is monotonically increased from $\lambda_i = 0$, the percept switches from 'say' to 'stay' at the upper curve. Figure 6b plots the three regions of different perceptual states for decreasing λ. The lower curve in this case marks the transition from 'stay' to 'say'. Figure 6c is the superposition of Figure 6a and b. The intersection of the two transition curves yields a critical value, ε_c, for which the system exhibits the dynamics as seen in Figure 4a. For $\varepsilon > \varepsilon_c$, the system exhibits the enhanced contrast effect as shown in Figure 4c, and, for $\varepsilon < \varepsilon_c$, the classical hysteresis phenomenon as shown in Figure 4b is obtained. Thus, the proposed dynamical model was found to be capable of capturing the experimentally observed patterns.

4.3.3 *Testing of a model prediction*

Several predicted effects of the model were evaluated by us (Tuller *et al.*, 1994). Our results show a strong correspondence between the theory and the experiments. Here we consider a specific predicted effect of the θ-function on $k(\lambda)$, henceforth termed k (Case *et al.*, 1995). Specifically, when $\theta(n - n_c) = 1$ in Equation (7), i.e., when the number of perceived repetitions, n, is larger than the critical value, n_c, each step change in gap duration (λ) entails a larger change in k than when $\theta(n - n_c) = 0$. This is illustrated in Figure 7, which plots k versus λ. Consider what happens for the 'say'–'stay' continuum when gap duration sequentially *increases* (solid line), then *decreases* (dashed line). When λ is at its minimum value (0 ms) at the beginning of the run, k_0 is arbitrarily assigned the

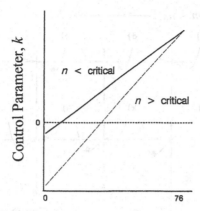

Figure 7. k–λ plot. The solid line represents increasing gap duration, and the dashed line represents decreasing gap duration. There is a dotted reference line at $k = 0$. For details, see text.

value -1, so that k is also negative. As λ increases to its maximum (76 ms in our continuum), k continuously increases and the stimulus with maximum λ is categorized as 'stay'. Now λ begins to decrease, although the stimuli are still identified as 'stay'. When $n > n_c$, the θ-function (the last term in Equation (7)) acts to increase the rate of change of k. Hence, the steeper slope of the dashed line in the k–λ plane. The net result is that, as λ decreases back to 0 ms, the absolute value of k corresponding to a given value of λ is larger for the second portion of the run than for the first portion, for all response patterns.

Figure 8 shows the potential $V(x)$ for expanded k when $n < n_c$ and gap duration is increasing (top row, left to right), and when $n > n_c$ and gap duration is decreasing (bottom row, right to left). Potentials in the top row are for a gap duration identical with those of the corresponding potentials in the bottom row. Nevertheless, the shapes of the minima are different. A comparison of Figure 8a with 8n, or of Figure 8b with 8m reveals that the depth of the potential is greater for a stimulus presented near the end of the sequence than for the identical stimulus presented near the beginning. In contrast, the value of k associated with the largest value of λ (the turnaround point) is the same whether that stimulus is presented as the final stimulus of the first half of the run (Figure 8g), or as the first stimulus of the second half of the run (Figure 8h). To reiterate, although Figure 8 shows an instance of enhanced contrast, the relative difference in the depth of the potential at the beginning and end of a run is not dependent on the response pattern. This observation leads to the hypothesis that the same physical stimulus presented at the end of a sequence is judged a better exemplar of the category than the identical stimulus presented at the beginning of the sequence, as a result of dynamical context effects.

Gap Duration Increasing, $n < n_c$ ———————→

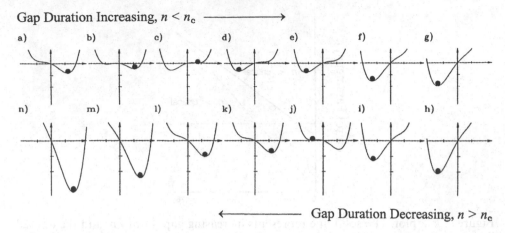

←——————— Gap Duration Decreasing, $n > n_c$

Figure 8. Potential landscape defined by Equation (6) for several values of k as determined by Equation (7). For explanations, see text in Section 4.3.3. This sequence illustrates enhanced contrast with the perceptual shift from 'say' to 'stay' occurring between panels (c) and (d), and the switch back between panels (j) and (k). (From Case *et al.*, 1995, with permission.)

Below we describe the experiment that was designed to test this hypothesis by exploiting the findings that listeners are able (1) to make fine distinctions among members of a given phonetic category and (2) to give reliable judgments about the extent to which a given stimulus constitutes a good exemplar of a category. We used the subjects' judgments of goodness of a stimulus to index perceptual stability, and compared the ratings of physically identical stimuli that occur in different positions in the sequence. If the model predictions hold, sequential changes in an acoustic parameter should result in a stimulus at the end of a run being judged as a better exemplar of the category than the same stimulus at the beginning of a run. On the other hand, nonsequential, *random* changes in the acoustic parameter should not influence judged category goodness.

Sixteen native speakers of American English with normal hearing took part in the experiment. The subjects were divided into two groups of eight each. The *sequential*-presentation group heard the stimulus sequence (0 ms (endpoint), 8,...,72, 76 (turnaround), 76 (turnaround), 72,...,4, 0 ms (endpoint (*condition 1*)), or (76 ms (endpoint), 72,...,4, 0 ms (turnaround), 0 ms (turnaround), 4,...,72, 76 ms (endpoint) (*condition 2*)). The *mixed*-presentation group heard the stimulus sequence, (0 ms (endpoint), 4, 8, 12, 16,...,72, 76 (turnaround), 76 (turnaround), (72 through 4 in *random* order), 0 ms (endpoint) (*condition 1*)), or (76 ms (endpoint), 72, 68, 64,...,4, 0 (turnaround), 0 (turnaround), (72 through 4 in *random* order), 76 ms (endpoint) (*condition 2*)).

The subjects' task was to identify each stimulus as 'say' or 'stay', and then rate from 1 to 7 how good an exemplar of the category the stimulus was. They were given the following instructions for rating the stimuli: 'Choose the number 1 only if you really could not tell whether the stimulus was "say" or "stay". Choose the number 2 if you thought you heard "say" or "stay", but were not completely sure. Choose 3 if you were sure of what you heard, but the stimulus was a very poor example. Choose 4 if you were sure, and it was an "okay" example, 5 if you heard a good example of "say" or "stay", 6 if you heard a very good example, and 7 if the word you heard was the best possible example of "say" or "stay" given the examples you have heard. Feel free to make comparisons between the stimulus words.' The subjects entered their identification and rating responses on an appropriately labeled computer keypad. All stimuli were presented binaurally through headphones at a comfortable listening level.

The obtained experimental results, i.e., the mean differences in judged goodness of stimuli versus position in the sequence, are illustrated in Figure 9. For the sequential-presentation group (filled symbols), there were obvious differences in judged goodness between the first and second presentation of the endpoint stimuli for both condition 1 (squares) and condition 2 (circles) in the direction predicted by the model. That is, the second presentation is judged as a better exemplar than the first, yielding positive mean differences. In contrast,

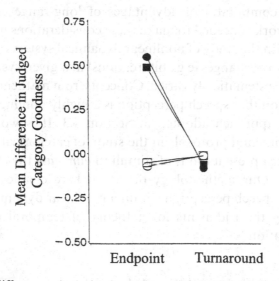

Figure 9. Mean differences in judged goodness (vertical axis) between stimuli with 0-ms gaps and between stimuli with 76-ms gaps. Filled symbols, sequential presentation; open symbols, mixed presentation. Squares, condition 1; circles, condition 2. For explanation, see text in Section 4.3.3. (From Case *et al.*, 1995, with permission.)

no such differences in judged goodness were observed with mixed presentation (open symbols), or between the first and second presentation of a stimulus at the turnaround. Thus, the model prediction was confirmed by the experiment.

4.4 Discussion and outlook

Our approach to the study of the synchronization problem is clearly motivated by statistical physics. The wide occurrence of $1/f^\alpha$ type long-memory processes in electrical systems and solid-state devices has long posed a challenging problem for physics. Increasingly, this type of process is being observed in biological systems (e.g., Bassingthwaighte *et al.*, 1994). Over the years, a number of mechanisms, ranging from the superposition of many independent relaxation processes (Granger, 1980) to self-organized criticality (Bak *et al.*, 1987), were proffered to explain long memory. It remains unclear, however, what the specific mechanism is that could account for the results reported in Section 4.2. In addition to the question of mechanism, another important problem concerns the function of long-memory processes. It is known, for example, that neurons in many brain systems fire spike trains that exhibit long-range correlations (Teich *et al.*, 1997).

It is an intriguing possibility that the long memory seen in the synchronization error time series is a behavioral manifestation of the long-range correlated firing properties of neuronal assemblies. Research has begun to address the question of computational advantages of long-range correlated firing patterns. More work is needed to make these considerations more concrete.

A central issue in the study of nonlinear dynamical systems is the characterization of qualitative changes (e.g., bifurcations) in a given system's dynamics, as a parameter is systematically varied. Concepts from mathematics, combined with the realization that speech perception is a highly nonlinear process, lead naturally to the approach adopted in Section 4.3. It is worth noting that traditional experimental protocols in the study of categorical perception emphasize *randomized* presentations of stimuli to *eliminate* the effects of contextual dependence. Our methodology described here enables us to *examine context effects* in speech perception within a nonlinear dynamical framework, thereby affording the rudiments for a theory of temporal organization of speech categorization.

Acknowledgment

This work was supported by the Office of Naval Research, the National Science Foundation, and the National Institute of Mental Health.

References

Bak, P., Tang, C. and Wiesenfeld, K. (1987) Self-organized criticality: an explanation of 1/f noise. *Phys. Rev. Lett.* **59**: 381–384.

Bassingthwaighte, J. B., Liebovitch, L. S. and West, B. J. (1994) *Fractal Physiology.* New York: Oxford University Press.

Beran, J. (1994) *Statistics for Long-Memory Processes.* New York: Chapman & Hall.

Case, P., Tuller, B., Ding, M. and Kelso, J. A. S. (1995) Evaluation of a dynamical model of speech perception. *Percept. Psychophys.* **57**: 977–988.

Chen, Y., Ding, M. and Kelso, J. A. S. (1997) Long memory processes in human coordination. *Phys. Rev. Lett.* **79**: 4501–4504.

Granger, C. W. L. (1980) Long memory relationships and the aggregation of dynamic models. *J. Econometr.* **14**: 227–238.

Hary, D. and Moore, G. P. (1985) Temporal tracking and synchronization strategies. *Hum. Neurobiol.* **4**: 73–77.

Hurst, H. E. (1951) Long-term storage capacity of reservoirs. *Trans. Am. Soc. Civ. Engrs.* **116**: 770–799.

Mandelbrot, B. B. and Van Ness, J. W. (1968) Fractional Brownian motions, fractional noises, and applications. *SIAM Rev.* **10**: 422–437.

Teich, M. C., Heneghan, C., Lowen, S. B., Ozaki, T. and Kaplan, E. (1997) Fractal character of the neural spike train in the visual system of the cat. *J. Opt. Soc. Am. A* **14**: 529–546.

Tuller, B., Case, P., Ding, M. and Kelso, J. A. S. (1994) The nonlinear dynamics of speech categorization. *J. Exp. Psychol. Hum. Percept. Perform.* **20**: 3–16.

5

Signal processing by biochemical reaction networks

ADAM P. ARKIN

5.1 Introduction

One cannot help but be impressed by the engineering, by evolution, of the cellular machinery. The cellular program that governs cell cycle and cell development does so robustly in the face of a fluctuating environment and energy sources. It integrates numerous signals, chemical and otherwise, each of which contains, perhaps, incomplete information of events that the cell must track in order to determine which biochemical subroutines to bring on- and off-line, or slow down and speed up. These signals, which are derived from internal processes, other cells and changes in the extracellular medium, arrive *asynchronously* and are *multi-valued*; that is, they are not merely 'on' or 'off' but have many values of meaning to the cellular apparatus. The cellular program also has a *memory* of signals that it has received in the past, and of its own particular history as written in the complement and concentrations of chemicals contained in the cell at any instant. These characteristics of robust, integrative, asynchronous, sequential and analog control are the hallmark of cellular control systems. Below, arguments will be made that there is also another characteristic of such control systems: there is often an *irreducible nondeterminism* in their function that, besides leading to differences in timing of cellular events across an otherwise genetically identical (isogenic) cell population, can also lead to profound differences in cell fate. The circuitry that implements these control systems is a network of interconnected chemical reactions. Included in these reactions are the genetic reactions involving: the gene expression reactions such as transcription initiation, transcript elongation, and translation; gene rearrangements such as DNA inversion reactions; and epigenetic control reactions such as DNA methylation. Enzymatic reactions, biosynthetic and mechanochemical interactions, and a host of other chemical reaction types also are central elements of this control system.

112

This chapter focuses on biochemical systems in which spatial concentration distributions and purely mechanical interactions are ignored or are not present. A full specification of cellular function is gained by (1) the determination of all the chemical parts making up the system, (2) the deduction of the mechanisms of interactions between these parts, and (3) the designation of the parameters necessary to describe the physics of all of these mechanisms. However, this specification, which theoretically could produce a computer simulation that exactly predicts the temporal behavior of all cellular constituents, represents a full understanding of the cellular system no more than a fully specified model of a Pentium chip gives us an understanding of the principles of its designs and function. In the case of the Pentium, understanding is best achieved by grouping individual transistors into logic gates, gates into devices such as counters, registers, and amplifiers, and then these devices into large devices such as arithmetic logic units, multiplexers/demultiplexers, clocks and bus controllers. The function of the chip can then be described by a relatively high-level programming language that makes clear the interactions between these composite devices and allows for a vastly simplified mathematical analysis. To achieve progress toward the description of cellular function on a similar basis is the ultimate motive for the work presented here.

5.1.1 Research goals

The work described herein represents efforts whose goal is the deduction of the engineering principles and logic of large biochemical reaction networks (BRNs). Specifically, the capacity of BRNs is explored (1) to sense and respond to multiple time-varying and conflicting signals (often chemical concentrations) in a robust and timely manner as well as (2) to execute internal developmental and behavioral programs. Two complementary types of analysis are presented. First, the utility of the circuit analogy for BRNs is examined and methods for the dissection of large networks into 'functional units' or 'devices' discussed. For this purpose, the device physics for a number of recurrent regulatory architectures is outlined to provide some background. In addition, the role of *thermal noise* in determining chemical reaction outcomes in cells is shown to be significant, especially for reactions involving genetic material. All these analyses assume that the individual components and their interactions have already been identified. However, this is often not the case. Thus, in addition, I briefly describe experimental methods for deducing BRN structures and assigning groups of chemicals into composite devices. The methods are designed to produce these deductions from measurements on the whole chemical reaction system rather than by breaking the system into small pieces.

Methods of network deduction and analysis are of special importance now that many genome projects are completing the inventory of all of the cellular proteins and genetic regulatory systems. If the full promise of these projects, i.e., to uncover the program of cellular life, is to be realized, it is necessary to compose these parts into functioning networks whose temporal behavior we may understand, whose properties we can control, and whose failures we can diagnose and ameliorate. Analytical tools such as the ones described herein, and in other contributions to this volume, lay the groundwork for this endeavor.

5.2 The circuit analogy and network analysis

In biology we are faced with often very complicated networks of interacting components. Ignoring atomic levels of detail, the lowest-level 'devices' in a BRN are often the individual chemicals and the set of reaction channels. Perhaps the best-characterized biochemical network of such devices is intermediary metabolism. Figure 1a is taken from Peter Karp's and Monica Riley's EcoCyc database of *Escherichia coli* (*E. coli*) metabolism, and is a representation of this network wherein every circle in the diagram represents one of the small organic molecules transformed in the course of metabolism, each black line represents a (possibly reversible) chemical reaction that converts one set of small organics into another (Karp *et al.*, 1999). Each grey line indicates that the connected circles are the same species of molecule that appears in multiple pathways. The macromolecules and macromolecular complexes that catalyze these conversions are not shown nor are the regulatory interactions that allow combinatorial control of the rate of one reaction by a set of other chemicals in the network. Were these components to be included in the diagram it would resemble Figure 1b; the diagram would be black with interactions. This highlights the main difficulty in forming a qualitative understanding of how metabolism actually works: biochemical systems are highly nonlinear and interconnected and composed of large numbers of chemical components. It is natural, therefore, to look for other systems that have these properties and ask whether we can apply the tools developed for their analysis to biochemical systems. Because diagrams of BRNs bear some resemblance to *electronic* circuits, it is tempting to ask whether it is possible to map them onto analogous electrical or electronic circuits so that we may apply the well-developed methods of electrical circuit analysis, the theory of computation and Boolean algebra. The subsequent section discusses some current methodology for the analysis of BRNs, and similarities to and differences from electrical engineering analyses.

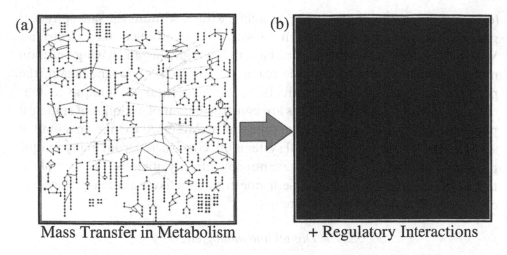

(a) Mass Transfer in Metabolism

(b) + Regulatory Interactions

Figure 1. A biochemical reaction network. (a) The left panel shows all the mass transformation reactions of *E. coli* intermediary metabolism from EcoCyc (Karp *et al.*, 1999). Details of the enzymatic reactions, the enzymes themselves and the regulatory interactions are not shown. If they were to be included, the vast number of interactions would make the diagram appear black as suggested in the right panel (b).

5.2.1 *Comparisons of electrical and chemical networks*

In this chapter, the words 'circuit' and 'network' are used somewhat interchangeably to mean a group of elements that have some property that is affected dynamically by interaction with other elements. In electrical circuits, these elements are, for example, resistors, capacitors, wires, and power sources. A signal in these networks is a voltage (or current) received at some node in the network, such as one lead of a resistor. Different signals are distinguished by the different points at which they impinge within the circuit and sometimes by their temporal pattern. In all cases, the currency of signals in electrical networks is carried by electrons. In chemical circuits, elements are, for example, enzymes, ions, reaction channels, and DNA. A signal in a chemical network is composed, most often, of the appearance of an amount of some chemical species at a point in the network such as at an allosteric regulatory site of an enzyme. The currency of signals in chemical networks is not uniform. It is the individual concentrations and chemical potentials of the particular species interacting at that point in the network. This lack of common currency for information transfer is one of the complications of BRNs: different chemical signals are often distinguished only by their interaction specificities with other members of the network.

This suggests another one of the significant complicating differences between electrical and chemical circuits. In electrical circuits, component types (e.g., resistors) are used over and over again, with or without the same physical parameters, but sharing common substrates and products (i.e., electrons). Whereas, in (nonspatial) chemical circuits, though the underlying reaction motifs of first- and second-order reactions are used again and again, the parameters, substrates and products for each reaction are unique to that reaction. That is, electrical circuits are constructed out of a toolkit of standard parts whose physics are well understood and are designed to take on a limited set of values. In contrast, chemical systems are constructed from many unique pieces whose physical parameters are not immediately obvious. This complication in the analysis is even worse if one tries to make an analogy between chemical and *digital* electronic circuits.

5.2.1.1 Digital and analog circuitry

There are a number of instances when the behavior of a biochemical system is such that it suggests *digital* rather than analog circuitry. The distinguishing feature of a digital circuit is that the signals within a circuit take on discrete values rather than a continuum. Thus, the major benefit of a mapping of a chemical circuit to a digital one is the reduction in the number of states (state-space) of the constituent concentrations and activities that must be considered for analysis. The powerful machinery of Boolean algebra and digital circuit analysis can then be brought to bear on the problem. In addition, it is computationally more efficient to simulate Boolean networks than to simulate differential equations. At some level, all chemical signals might be considered digital, since their values are discrete; that is, their values are measured in numbers of molecules (per unit volume). However, the number of states available to a chemical signal in a kinetic network is usually far greater than two. We are usually concerned with large numbers of molecules so that individual reaction events cause changes in the number of molecules, which are very small, compared to the average, and thus can be approximated as infinitesimal or continuous changes of a concentration variable. (Important exceptions to this are discussed in Section 5.2.3.3). This is no different from disregarding the quantal nature of electrons flowing through an electrical circuit, because their numbers are so large in metal wires. Even neglecting the discrete nature of chemistry, the *dynamics* of a chemical system might be such that it is sometimes feasible to treat the system as a digital circuit.

Cooperative enzyme activity The typical example often used to justify Boolean approximations in biochemistry is the *cooperative* enzyme. For simplicity, consider the cooperativity to be in the action of an allosteric effector on the maximal activity of the enzyme. In this case, for various models of cooperativity, the maximal activity of the enzyme is a sigmoidal function of the effector concentration. If the sigmoid is very steep (high cooperativity), then it looks much like a *threshold* function – zero below a critical value of the effector concentration, and some constant, nonzero value above it. However, for this 'switch' in enzyme activity to be truly Boolean (digital) two criteria must be met: (1) the effector concentration signals must cause the effector to transit from values well below the critical value to values well above it (or vice versa), and (2) the time spent in the transition region of the sigmoidal activity curve should be small compared with the response time of the receptive system. If the first of these criteria is not met, then small variations in effector concentration could cause entry into the transition region resulting in large fluctuations in enzyme activity, thus destroying the two-state property of the system. If changes in the effector concentration fail to cross the transition region, then again the two-state behavior is ill defined. If the second criterion is unfulfilled, then the integrated activity of the enzyme during the transition time might become significant to the receptive process. This means that these intermediate activity values cannot be ignored and so this 'switch' becomes multi-valued at best. (Sometimes this criterion is violated in very fast electronic circuits wherein certain components are fast enough to 'see' the transition time of a transistor. Digital designs that fail to take this into account will fail.)

Multiple signal encoding Finally, another complication of applying digital signal analysis to chemical networks of even 'switch-like' reaction mechanisms is the following: unlike most electronic digital systems, for each signal-receptive mechanism, the value of an 'On'-signal is, in general, different from other systems that also might receive that same signal. For example, one kind of enzyme might become active above a calcium ion concentration, $[Ca^{2+}]$, of $10\,\mu M$, and another one only above $100\,\mu M$. Functionally, then, there are at least three significant values for the $[Ca^{2+}]$ signal: below $10\,\mu M$, between 10 and $100\,\mu M$, and above $100\,\mu M$. The dynamics of the $[Ca^{2+}]$ signal might be such that it is driven very rapidly from 0 to greater than $100\,\mu M$; in which case the early sensitivity of the first enzyme is only functionally important if the difference in activation time between the two enzymes is significant. In chemical systems, where there are many different chemical signals whose value ranges and time scales are all unique, discerning the 'logic' of the network is even more difficult and dependent on the exact parameters of the system.

5.2.1.2 Synchronous and asynchronous design

A final point of contrast between engineering principles of electrical or, more specifically, electronic circuits and biochemical ones is the use of *synchronous* designs. The majority of sequential (as opposed to combinational) digital designs rely on a *system clock* for synchronization of processes. In digital design, clock synchronization is considered desirable for two reasons. First, susceptibility to noise is improved, since the transient dynamics of component devices that occur before the edge of a synchronizing clock pulse do not affect circuit function. Second, the different delays of various signals through a circuit to their respective outputs can be ignored, since the time between clock-pulses in a synchronous circuit is generally set to be longer than the longest delay; thus outputs are not read until all signals have reached their destination.

With very few exceptions, biochemical circuits are unclocked; that is, *asynchronous*. Even when there seems to be a central oscillator, such as that underlying the timing of the cell cycle, microscopic examination of the process reveals a large variability in timing of the oscillations. Progressive dephasing arises both due to noise in the underlying processes (see Section 5.2.3.3) and the fact that the cell cycle oscillator has check points so that the cycle does not proceed until all necessary subprocesses have completed their work. Since cellularly uncontrolled variables such as externally available nutrients control how fast certain of these processes can be executed, the cell cycle is designed to be tolerant to these metabolically induced large changes in timing. Interestingly, electronic asynchronous design (traditionally used for interface circuits) has become increasingly popular as circuit size and complexity has increased. The reasons cited by electrical engineers for asynchronous circuit design are precisely the reasons a biological circuit would be expected to be asynchronous. Five such reasons are stated by Myers (1995):

Average case performance: The clock period for synchronous systems must be set long enough so that the circuit can accommodate the slowest operation possible even though the average delay of an operation is often much shorter. Asynchronous circuit designs allow the speed of the circuit to change dynamically. The speed of the circuit is, therefore, governed by average case delay.

Adaptivity to processing and environmental conditions: Since variables such as temperature change with the environment, circuit up-time and processing rate, and circuit component speeds can be greatly affected by such changes, synchronous designs must be simulated under a wide range of conditions and the clocking set so that the circuit functions under the widest range of variation. Asynchronous designs, in contrast, are adaptive and speed up and slow down as necessary.

Component modularity: In asynchronous systems, components (functional subcircuits) can be interfaced without the difficulties of synchronizing clocks necessary in synchronous systems. Also, when a new version of a component with different timing is developed, the old component can often be replaced without requiring any other changes in the rest of the system. In other words, the system is robust to (some) changes in its component circuitry.

Elimination of clock-skew: In large digital circuits, the time it takes a clock pulse to reach different parts of the chip can be different, leading to loss of synchronization. To minimize this skew in arrival times, a great deal of extra circuitry must be designed in. (Nearly a third of the silicon area is required for clock distribution in a DEC Alpha microprocessor.) Asynchronous circuits are tolerant to signal timing differences among components.

Lower system power requirements: Since they do not require all the extra clock circuitry, asynchronous circuits reduce synchronization power. They can also be easily adjusted to make use of dynamic power supplies.

The advantages of asynchrony have to do, then, with robustness to changes in the circuit environment and in the dynamical state of its various components and efficiency both in speed and energy. The noise-filtering behavior of synchronous design is an advantage only because clocks make rejection of noise and transients relatively easy to design. Asynchronous circuits can be designed to be as stable to spurious signals. It is likely that biochemical circuits have evolved for this robustness, efficiency and adaptability to environmental changes. It is not surprising, therefore, that most biochemical circuits are found to be asynchronous. However, even wholly digital, asynchronous circuits are notoriously difficult to analyze. Thus one can expect similar difficulties for analog biochemical circuits. On the other hand, study of biological circuits may provide unthought of stable electronic asynchronous circuit designs and any analytical tools developed for the biological circuits may have application to the electronic ones and vice versa.

5.2.2 Device function and state

Metabolic charts like the one in Figure 1a are daunting in their complexity, but perhaps no more daunting in topological complexity than the schematic for a modern computer chip. The difference between these two interaction maps is that the device physics for every element on the chip schematic are fairly well characterized. The behavior of the circuit is fully specified by these physics and the functioning and reliability of the chip can be probed efficiently using simulation tools such as the *SPICE* software package (Tuinenga, 1995). Even better, because of the precisely designed physical characteristics of these elements, their function may be partially abstracted into a higher-level language: *digital Boolean logic*. Thus, most analyses do not need to include the detailed differential equations that most completely describe transistor function. Instead, the device details can be abstracted to a higher level, i.e., to perform as *logic gates.* Circuits composed of such gates can be grouped together to form higher-level devices whose input/output behavior can be derived and used without reference to the exact mechanism from which this behavior is derived. This type of grouping of subnetworks into functional components greatly facilitates the analysis of the larger circuit. One challenge, then, for the analysis of biochemical and genetic networks is to dissect complex networks into individually analyzable devices that can be hooked back together to predict the total system behavior.

What, then, constitutes a device? In electrical circuits, elementary devices are objects such as wires, resistors, capacitors and inductors. In digital circuits, elementary devices are parts such as transistors and gates. Perhaps, the defining property of a device is not that it may be separated physically from its network, but rather that the physics of the device may be derived for a general case without reference to the precise dynamics of the rest of the network in which it might be embedded. Resistors, for example, must always obey Ohm's law and Kirchoff's laws, no matter the circuit in which they are used (within broad limits). These laws, along with perhaps some equations for effects of dissipation on the resistivity of the material, fully specify the device function. Practically, this results in the need for only a single parameter to characterize a resistor, the *resistance*. This value is the same no matter how the circuit elements up- and downstream are functioning.

A single chemical reaction step is an elementary device in a chemical network in much the same way as a resistor is an elementary device in an electrical circuit. A single number may characterize the behavior of the reaction: the *rate constant*. Though it is possible to describe the reaction event in much more detail, via quantum mechanics, collision theory, etc., it is generally not necessary. Just as with the resistor, the rate constant for a given elementary reaction does not depend, to first approximation, on the other reactions going on around it. Vast networks of chemical reactions such as in metabolism or during signal transduction, then, are circuits of these elementary devices in which each device accepts chemical concentrations as inputs, and outputs *chemical fluxes*. This is conceptually different from digital devices that accept voltage and output voltage. Electronic devices, on the one hand, simply feed their output voltage to the 'voltage receptor' on the downstream device. Chemical reactions, on the other hand, convert concentration to flux; the output must be reconverted to concentration for input to the next device. In the elementary case, it is both the upstream and downstream elements together that dynamically determine this conversion of the upstream flux to instantaneous chemical concentration and, thence, the downstream flux.

5.2.2.1 *Elementary electronic and chemical devices*

The main advantage of a device description is that composite devices may be constructed out of a 'basis' set of elementary devices. A simple, but informative, example from electronics is the *voltage divider*, the most primitive of power supplies (Figure 2a). The output of the voltage divider is a voltage and current, and the voltage is given by $V_{in} R_2/(R_1 + R_2)$. Next to the divider is represented a *chemical analogy* (Figure 2b) that, here, let us call the 'A' buffer. If we assume that the steady-state concentration value of reaction species A, [A], is the output to this device, then its value is given by $k_1[B]/k_2$ (this circuit can both divide and amplify the signal [B]).

The voltage divider can be considered a device only if the circuits driven from its output do not affect its function. This is only the case if the downstream devices, the load, have very high impedances compared to R_2. Figure 2d shows these devices as a single load with resistance, R_L, connected from the output of the divider to ground. This arrangement puts R_L in parallel with R_2; thus the two resistances can be combined into one (as guaranteed by Thevinin's theorem) that has an effective resistance of $R_2 R_L/(R_2 + R_L)$. Thus, when $R_L \gg R_2$ the effective resistance is equal to R_2. In this case, the voltage divider remains an intact device. However, as the load resistance decreases, the value at the output of the device becomes more and more dependent on the properties of the devices to which it is attached. Similarly, in the 'A' buffer (Figure 2b), if A is consumed by a third reaction then this reaction rate must be very small compared to k_2 in order for the chemical device to remain intact. If many other reactions consume A, then the sum of their rate constants must be much less than k_2. On the other hand, consider the case when A is an allosteric effector of a set of enzymes downstream. Assuming binding to the enzyme is a reversible step, the steady-state [A] is unaffected by the interaction with the downstream enzymes. Seemingly, then, this chemical device remains intact when connected to the rest of the network in this way. However, the time it takes to achieve the steady-state value of [A] after 'turning on' the device (by, for example, adding a catalyst required for B-to-A conversion) increases as the concentrations and binding constants of the downstream enzymes increase.

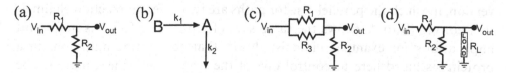

Figure 2. Examples of electrical and chemical 'voltage dividers'. For explanations see text in Section 5.2.2.1.

Thus, some aspects of the device function are changed by connection to the rest of the network. In summary, any analysis of devices dissected from the rest of a large network, therefore, relies on recognizing when these changes are significant and when they are not.

Composite devices like the voltage divider (or the 'A' buffer) serve not only as recognizable units of function but also as a means for *simplifying* circuit analyses. For many analyses of circuits containing the voltage divider, the two voltages (V_{in} and V_{out}), the current (I), and the two resistor parameters (R_1 and R_2) can be replaced by the single parameter, V_{out}. There are no approximations in this simplification, the single parameter is derived directly from the device physics of the underlying components. If we see two resistors and a power supply in the same configuration as shown in Figure 2a, and we can see that the downstream impedances are high, then we know that we need only measure the output voltage in order to determine the central functionality of the circuit. That is, we do not need to determine the particular resistances R_1 and R_2 or the properties of the power supply to obtain the circuit function. This ability to reduce the number of physical measurements that must be performed on the system is of special importance in the biochemical case. Here, it is often at great expense in time and resources that a particular variable can be determined quantitatively and *in vivo*.

Often, reductions in number of parameters or in the dimensionality of a dynamical system are fundamental steps in analyzing the overall function of the circuit. If these reductions are derived directly from the device physics as above, then much of the circuit behavior is retained in the new simplified circuitry. But sometimes, especially in biology, such reductions remove important experimental features of the system. Again, an analogy from electronics provides the simplest explanation. Consider the voltage divider in Figure 2c: here, R_1 has been placed in parallel with R_3. Application of the parallel resistor rule allows us to replace this circuit with one identical in structure with that shown in Figure 2a in which the top resistance is $R_1 R_3/ (R_1 + R_3)$. However, the *reliability* of the reduced circuit is much different from that of the full circuit. Failure of the top resistor is catastrophic for circuit function in the reduced circuit, whereas failure of both R_1 and R_3 is necessary to completely destroy the function of the circuit in Figure 2c. The chemical version, in which the parallel resistor paths are two different reaction channels that convert B to A, shows the same sort of sensitivity. This has important implications: for example, biologists know that debilitating mutations in a protein, assumed here to control one of the reaction channels, may not be lethal to the function of the whole network.

5.2.2.2 *Definition of state in electronic and chemical networks*

Finally, an important concept for circuit analysis of chemical networks is the definition of *state*. We distinguish between the state of a particular input or output, the *local* state, and the state of the *system*, the *global* state. In digital electronics, local states can take on only two values, 0 and 1. The state of the system is a vector of the local states for each distinguishable input and output. In chemical systems, nominally the local state is often the value of a particular concentration that may take on any positive-indefinite number. The global state is the vector of concentrations of all chemically distinguishable species in the system. Each global state is also associated with properties such as dynamical stability and type (stable node, stable focus, limit cycle, etc.) This theoretically infinite state-space of a chemical system makes its analysis extremely difficult compared with digital systems. However, the dynamical and stoichiometric structure of the system may strongly restrict the range of concentrations that can be reached by any particular chemical species.

Switching in an enzymatic futile cycle As an example consider the circuit shown in Figure 3a. This is a standard futile cycle in which a protein, here labeled A, is phosphorylated by another protein, called a kinase (B), and then subsequently dephosphorylated by a phosphatase or by hydrolysis. It is called a futile cycle because it takes energy (usually in the form of adenosine triphosphate, ATP) to achieve the unidirectional phosphorylation step only to have it seemingly wasted when the protein spontaneously dephosphorylates. The total amount of A and A-p (A_{tot}) remains constant. This is the first restriction

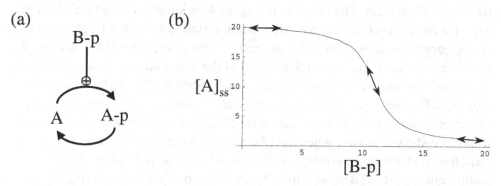

(a)   B-p

A   A-p

(b)   [A]$_{ss}$ [B-p]

Figure 3. (a) A simple futile cycle in an enzyme-based reaction system. This configuration of reactions (i.e., a regulatory architecture or motif) is a ubiquitous control structure found in many prokaryotic and eukaryotic signal transduction circuits. (b) The stationary-state concentration of A, [A]$_{ss}$, as a function of B-p concentration, [B-p]. For details see text in Section 5.2.2.2.

on the state of the system: concentrations of A, [A], and A-p, [A-p], are restricted between 0 and A_{tot}. As shown in Figure 3b, the kinetics of the system represented in Figure 3a are such that there is a sigmoidal transition from high [A] to low [A] as a function of the concentration of B-p, [B-p]. The steepness of this sigmoid is largely dependent on the fraction of the [B-p] range in which the kinase and phosphatase reactions are both saturated (thus, causing the system to enter a state called 'zero-order ultrasensitivity'; ZOU). Thus the smaller the dissociation constants of A and A-p from their respective enzymes, the more the curve in Figure 3b resembles a Boolean step function. In the case of high ZOU, then, it may be reasonable to say that, in the steady state, the variable A takes on two states, low and high, whose physical values are roughly 0 and A_{tot}, respectively. However, the applicability of this simplification depends ultimately on the dynamics of B-p. Even if the changes in [B-p] were slow enough, compared with the dynamics of the futile cycle, such that the cycle was always near the steady state, [A] only *functionally* has two states if two further conditions are met. First, the controlling physiological changes in [B-p] must cross the threshold region of the [A] steady-state curve completely, and, second, some downstream targets of A activity respond differentially to the high and low states of [A] (or [A-p]).

Bistability and hysteresis The chemical switch represented by Figure 3a is a 'soft-switch'. That is, it is not a true *bistable* state. Rather, [B-p] is a control parameter that smoothly transforms the single steady-state solution of the kinetic equations from a high to a low value. However, one change in the circuit topology (the addition of another reaction) and small quantitative changes in the circuit parameters convert this soft-switch to a 'hard-switch', a truly bistable system. The reaction in Figure 4a is identical with that of Figure 3a, with the exception of a positive feedback that allows A-p to catalyze the phosphorylation of A. When the strength of this positive feedback is low, i.e., when the maximal rate is a small fraction of the maximal rate of B-p-catalyzed phosphorylation, the switch behaves nearly identically to the 'soft-switch' (compare Figures 3b and 4b). However, relatively small changes in the strength of the feedback cause a strong qualitative change in the behavior of the switch. Figure 4c shows the case where the feedback strength has been doubled. The switch now exhibits *hysteresis*: the [B-p] at which A switches from high to low is different from the [B-p] at which A switches from low to high. In Figure 4d the feedback strength has been doubled yet again and now the switch is *irreversible*. Once a switch changes from high to low, switching back from low to high is now physically impossible with B-p as the sole control parameter. These qualitative changes in behavior can have profound effects on the func-

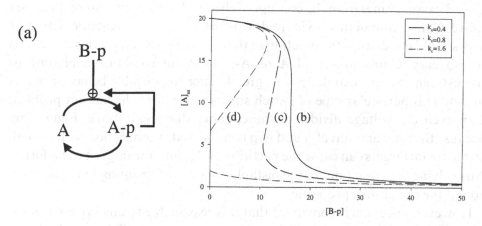

Figure 4. A biochemical 'switch'. (a) The same futile cycle as in Figure 3 is shown; however, here the phosphorylated from of A catalyzes its own production. Depending on the exact value of the feedback strength, this system can behave either exactly like the futile cycle in Figure 3, or generate true bistability with its attendant hysteresis and 'memory' at higher feedback strengths. (b–c) A family of stationary-state concentration curves for A as a function of B-p concentration. For details see text in Section 5.2.2.2.

tion of the rest of the network in which this 'hard-switch' is embedded. These qualitative changes may result from changes in kinetic parameters of well under an order of magnitude. The addition of the positive feedback in Figure 4a may seem to be a large perturbation to the system represented in Figure 3a, but such topological changes in a network structure can be found in control processes that occur in 'real' biological systems. For example, the pp125 focal adhesion kinase (FAK), a cytoplasmic tyrosine kinase-transducing signal initiated by integrin engagement and G-protein-coupled receptors, is alternatively spliced (and more highly expressed) in brain tissue. Some alternative splices that are preferentially expressed in brain tissue have an increased autophosphorylation activity, suggesting that FAK may have properties that are specific to neurons. It has been suggested that these isoforms of FAK may play an increased role in turn over of point contacts in motile or invasive cells.

Limiting assumptions and caveats The analysis of the futile-cycle switches above assumes that they may be treated as *self-contained* subcircuits whose dynamics may be analyzed without reference to the rest of the network in which they are embedded. The first caveat to this assumption arises from the ambiguity of the *functional* definition of the state of A and A-p discussed above. For example, simply because a bifurcation analysis predicts that a system is bistable does not mean that both states are used by the biological system. The second caveat comes from the concept of chemical device impedance discussed above. Connection of this device to downstream targets (by reaction with A or A-p) should not disrupt the function of the device. However, for this device it is not so easy. Consumption of A or A-p, or rapid equilibrium binding to downstream targets, can destroy or greatly alter the bistable behavior of this circuit. This particular type of switch suffers more from the fan-out problem than even the voltage divider in the circuitry discussed above (Figure 2b). Because the total amount of A and A-p is conserved, a connection with enough downstream targets can cause the partition of A_{tot} into the target-bound forms, thus driving the circuit out of the bistable region, and resulting in the destruction of the switch-like properties.

However, once one is convinced that it is reasonable to analyze a chemical subunit as a self-contained device, there are many available methods for predicting the classes of possible circuit behaviors. For example, full solution of the differential equations, bifurcation analysis, and stoichiometric network analysis, all provide means for predicting the range of qualitatively different states that the circuit dynamics may achieve. The sets of kinetic parameters required to switch between each of these states, and control the exact position within, may sometimes be derived as well.

One of the criticisms often leveled at quantitative analysis of biochemical and genetic networks is that one needs measurements of all the mechanisms and of the possibly hundreds of kinetics parameters for those mechanisms, and that obtaining these data is nearly impossible *in vitro*, let alone *in vivo*. The response to this is three-fold: (1) using the analyses just mentioned, it is possible to derive limited classes of behaviors that even relatively roughly measured networks may express; (2) these same analyses can sometimes yield sets of parameter estimates each of which specifies a range of parameters necessary to achieve each qualitatively different behavior; and (3) since most of the BRNs that govern cellular function must be robust to often large fluctuations in the environment and to molecular noise in their own apparatus, chemical circuit behavior should not be overly sensitive to the exact values of each of the kinetics parameters; otherwise there would be a high rate of cell failure. The detailed study of these circuit motifs yields a better qualitative understanding of how a biological pathway is controlled and, as shown below, may point to biologically important physical phenomena that have not yet been fully considered by bench biologists. Further, experimental observation of a particular dynamic behavior may suggest that one or few types of regulatory motifs are responsible for the behavior. Knowledge of how the different motifs achieve a particular behavior can then lead to targeted experiments to differentiate among them.

5.2.3 *Regulatory architecture, motifs, and circuit elements*

In this section, I outline some of the work we have done on identifying common control architectures and elements in biochemical systems. It is an underlying assumption that these elements have evolved to perform one or more specific functions that are useful to an organism. One indicator that these elements are 'functions' is that their architectures recur across organisms and across pathways within a single organism. Not only does identification of the elements simplify the analysis of larger BRNs but they also provide a basis set from which researchers might possibly construct custom networks that perform *novel* functions. The following is far from a complete set of such devices. They are chosen simply to illustrate what sorts of network function can be realized by biochemical systems.

5.2.3.1 *Single enzymes and enzyme networks*

Enzymes and other proteins are examples of how a single molecule or a small molecular complex may be a fairly complicated chemical device. The presence of an enzyme that catalyzes a simple Michaelis–Menten-type reaction is described by an already-composite chemical device composed of three elementary reaction steps. If an enzyme is considered a device in itself, then its inputs are (at least) its substrates and effectors, and its outputs are the rates of product generation. These output rates are usually, but not always, monotonically increasing or decreasing, saturating functions of the various substrates, inhibitors and activators. That is to say, the outputs are often sigmoidal functions of the inputs. In the extreme case, sigmoid functions look like step functions and, therefore, it is tempting to use a Boolean truth-table to describe their function rather than the full enzymological description (Arkin and Ross, 1994). Even if the sigmoid functions are not so steep, they resemble various models of computational 'neurons'. Thus, networks of such enzymes resemble these formal neuronal networks (Bray, 1990). In fact, Hjelmfelt and Ross have demonstrated the equivalence of a particular parameterization of the example given in Figure 3a to a McCulloch–Pitts artificial neuron, and they showed how to make various computational circuits out of interconnected networks of these elements (Hjelmfelt *et al.*, 1991, 1992; Hjelmfelt and Ross, 1992). On the basis of these results, in collaboration with F. Schneider, they experimentally implemented a chemical neuronal network made out of bistable chemical reactions with a dynamical function similar to the futile cycle, as discussed in Section 5.2.2.2 (Hjelmfelt *et al.*, 1993). Furthermore, Bray (1990) has suggested that the neuronal network-like properties of chemical parallel-distributed processes may help to explain, in part, the reliability and evolutionary adaptability of these networks.

Finally, it is worth noting that some protein-based devices have dynamics fundamentally different from the standard enzymological mechanisms. Molecular machines such as polymerases, ribosomes, and kinesin can be very intricate molecular devices. New techniques are allowing the quantitative measurement of the microscopic and mesoscopic dynamics of motion of these molecular complexes on their macromolecular substrates (Guthold *et al.*, 1994; Yin *et al.*, 1995; Bustamante and Rivetti, 1996; Wang *et al.*, 1997). However, consistent models of the *in vivo* dynamics of these machines are still in their infancy. The initial attempts to treat mathematically the dynamics of various of these molecular motors and their input/output behavior have laid a good foundation against which to test future measurements and models (Peccoud and Ycart, 1995; Peskin and Oster, 1995; Astumian and Bier, 1996; Duke and Leibler, 1996; McAdams and Arkin, 1997; Arkin *et al.*, 1998; Goss and Peccoud, 1998). One of the interesting common dynamical phenomena found in these machines is that their operation is fundamentally *stochastic*. A similar observation is made when examining the function of, for example, ion channels (Collins *et al.*, 1995). Noise in the operation of these devices, as is discussed further below, necessitates a consideration of robustness and reliability in the design of cellular signal-processing networks.

5.2.3.2 Biochemical oscillators

Biochemical oscillators are found to play a number of roles in the control of cellular and organismal behavior (Berridge and Rapp, 1979; Rapp, 1979). The most pervasive form of oscillator in biology is the cell cycle oscillations that underlie repeated patterns of cell growth and division (Borisuk and Tyson, 1998; Novak *et al.*, 1998). Though the chemical network that drives a particular cell cycle is usually not a 'free running' oscillator in that it is regulated by checkpoints that can stop, slow or even redirect the cycle (e.g., in order to wait for unsynchronized processes to catch up, to deal with damages and stresses in the cell or to change the chemical pathway responsible for the cell cycle in different cell types), at root is a chemical system capable of repeatedly leaving and very nearly restoring an initial condition. Biochemical oscillations are also found in mitochondrial volume, in yeast glycolytic flux (Jonnalagadda *et al.*, 1982), in GTP/G-protein activity, in cytoplasmic calcium concentrations, in neuronal signaling, in circadian rhythms (Goto *et al.*, 1985; Ouyang *et al.*, 1998) and in certain reconstituted enzyme systems such as horseradish peroxidase (Stemwedel *et al.*, 1994; Hung and Ross, 1995). These oscillations have many different functional roles in the cells in which they are found. Timing and synchronization are the most obvious ones. However, there is some evidence that oscillatory dynamics can reject noise while propagating signals, and that the frequency and amplitude of an oscillation can carry information that can be decoded by chemical frequency filters like the ones discussed further below. Thus, these oscillators are *signal generators* whose output can be modulated in amplitude, frequency and phase by chemical, thermal and/or light inputs.

Ross and co-workers have attempted to classify chemical and biochemical oscillators into a finite set of classes distinguished by their network topology and their responses to various experimental perturbations (Eiswirth *et al.*, 1991a,b). Chemical species within such oscillator devices are classified as essential or nonessential depending on whether or not quenching of their oscillatory behavior destroys the overall ability of the network to support oscillation.

These classification methods demonstrate a number of the advantages of a device analysis. They provide a theoretical framework for understanding the different ways in which chemical systems can provide oscillatory signals. They also provide an ordered set of diagnostic experiments by which a novel, oscillatory chemical species may be classified in a small number of experimental steps. This classification, then, severely restricts the underlying mechanisms and their parameterizations that give rise to the experimental observations.

5.2.3.3 Genetic regulatory circuits

In 1961, at the Cold Spring Harbor Symposium in Quantitative Biology, Jacob and Monod first outlined a circuit theory of genetic control in prokaryotes (Monod and Jacob, 1961). The basic theory describing combinational control of transcription initiation, expression of polycistronic operons and feedback control as a basis for control of metabolism, growth and development remains largely unchanged today. Most of the basic mechanisms proposed are used in prokaryotes and eukaryotes alike, although eukaryotic gene control has a few more levels of complexity to it. The central process is the transcription of DNA to RNA via the multiprotein complex RNA polymerase (RNAP), and then the translation of RNA to protein via transfer RNA and ribosomes. Transcription can be broken up into at least two processes: transcription initiation and transcript elongation. Translation can be broken into three processes: translation initiation, protein elongation and transcript degradation. Each of these processes may, in turn, be regulated by cellular signals.

Transcription initiation The best characterized of these controls is the regulation of transcription initiation. Initiation begins from a region of DNA called the *promoter*, upstream (at the 5′ end of DNA) from the genes of interest. In prokaryotes, this is most often accomplished by the binding of proteins, transcription factors, to sites on the DNA called *operator* sites. The pattern of transcription factors bound to sites can modulate both the strength with which RNAP binds to the promoter and the rate at which it begins transcription. The number of patterns (states) of the promoter can be quite large. For example, the λ phage P_R/P_{RM} control region is composed of two promoters and three operator sites (shown schematically in Figure 5). The operator sites can bind homodimers of two proteins, Cro and CI with different affinities. The region can have 40 different configurations of RNAP, Cro_2 and CI_2 bound, each of which is characterized by its stability (free energy) and its transcriptional activity (Ackers *et al.*, 1982; Shea and Ackers, 1985).

It is tempting to think of these states as 40 different logical states of a complex Boolean switch that transmits a set of RNA signals when particular sets of transcription factors are present or absent. In some cases, this may indeed be a good approximation, but a number of issues need to be addressed before such an abstraction is made. Most important is probably the time that a given configuration of transcription factors and RNAP at a given promoter persists. In many cases, in prokaryotes, these proteins are assumed to be in rapid equilibrium with their respective binding sites. The binding dynamics of RNAP to its promoter, especially, is likely more complicated than this; however, empirically this does not seem to be a bad approximation in most cases. In this approximation, the 40 molecular configurations of P_R/P_{RM} are sampled many times in between transcription initiation events. The total time spent in any configuration is related to its stability. Thus an average transcriptional activity may be calculated for any instantaneous concentration of proteins. The thermodynamic and kinetic parameters for the P_R/P_{RM} promoters have been determined (Ackers *et al.*, 1982; Shea and Ackers, 1985). The graph of P_{RM} activity as a function of CI_2 and Cro_2 concentrations, $[CI_2]$ and $[Cro_2]$,

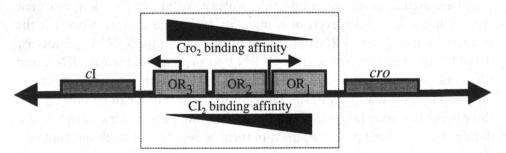

Figure 5. The organization of the P_R and P_{RM} divergent promoters from bacteriophage λ. The gene products of *cI* and *cro* dimerize, then bind to operator sites (OR_1, OR_2 and OR_3) in the promoter region with differential affinities. The pattern of CI_2, Cro_2 and RNA polymerase binding to the operator and promoter sites determines the frequency of transcription initiation from P_{RM} and P_R.

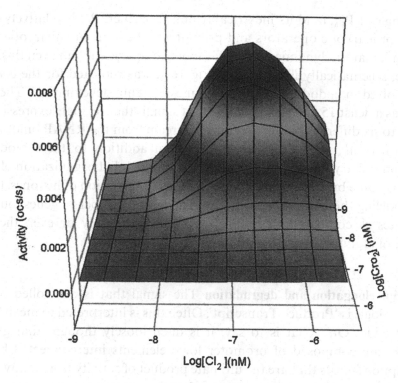

Figure 6. A plot of the activity of the P_{RM} promoter as function of the concentration of its two transcription factors, CI_2 and Cro_2. Activity is expressed in units per time of open-complex formation (ocs/s). For details see Section 5.2.3.3.

respectively, is shown in Figure 6 for a constant (available) RNAP concentration of 30 nM. This plot summarizes the 'control logic' for transcription initiation at P_{RM}. The 40 'logical' states of the promoter region are not directly visible in this activity curve. The smoothness of this curve arises because of the rapid equilibrium assumption. Were the transcription factors 'sticky' (i.e., were their off-rates from DNA comparable with or slower than the rate of transcription initiation), then the timing (order) of CI and Cro binding would become important and a time-independent control surface could not be plotted. Further, the activity of P_{RM} as a function of CI_2 is not a monotonically saturating function. Instead, as $[CI_2]$ increases, the activity of P_{RM} at first increases, then decreases. There are, then, at least three functionally different regions of $[CI_2]$ and a Boolean abstraction of this 'switch' could not necessarily map $[CI_2]$ into a single binary variable. In addition, as discussed above, the appropriateness of a Boolean approximation to this curve is dependent on the time it takes the effector molecules to traverse from their initial to final values and vice versa.

The logic of P_{RM} (and its inextricably linked partner, P_R) is relatively complex for prokaryotic operators and promoters. The logic of eukaryotic transcription initiation dynamics can be far more complicated than even this. One example, schematically illustrated in Figure 7, was reported for the *endo-16* gene involved in endodermal formation in sea urchin development. The promoter has at least 15 different protein input signals that regulate expression by binding to six different binding regions upstream from the RNAP binding site (Davidson *et al.*, 1998; Yuh *et al.*, 1998). In addition to these modes of regulation, eukaryotes also can regulate the more global organization of their nuclear genome by, for example, controlling the acetylation of histones, thereby remodeling the chromatin structure. The larger number of genes found in eukaryotes as compared to prokaryotes is not the only or even the best measure of organismal complexity.

Transcript elongation and degradation The signal that is controlled by the promoter logic is 'Produce Transcript'. Often this is interpreted to mean 'Turn Gene Product On'. That is to say, it is often loosely thought that genetic networks are composed of promoter logic elements interconnected by the transcription factors that are the ultimate product of activity from many of the constituent promoters. But there are other factors that must be considered before such an abstraction can be made. After transcription initiation, there are numerous mechanisms of elongation control, including terminators and antiterminators (regions of DNA at which a transcribing RNAP can fall off the template, or at which RNAP can be modified to be resistant to such termination, respectively), downstream binding sites for proteins that block a processive RNAP, and RNAP pause sites. In polycistronic operons, mostly found in prokaryotes, these can lead to strong polarity effects in which there is higher expression of transcripts proximal to the promoter than for those that are distal. Further, each gene in the transcript may or may not have its own

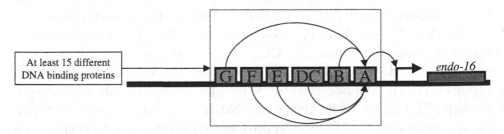

Figure 7. Diagram of the promoter control of the sea urchin *endo-16* gene involved in early embryogenesis. Derived from Yuh *et al.* (1998).

ribosome binding site and degradation rate. Thus each gene can express different numbers of proteins per transcript. Further, in prokaryotes, translation of a protein product from a transcript is often tightly coupled to transcription. Production of protein rapidly follows production of transcript. In eukaryotes, this coupling is much weaker, since many processes can act on a given transcript before and after it is exported from the nucleus to be translated. The correlation between the concentration of transcript and its protein product is lower in eukaryotic systems. In all cases, protein products may or may not be actively degraded. All these processes have to be taken into account before realistic models can be constructed.

Stochastic processes in gene regulation There is one further level of complexity to the genetic machinery that must be carefully considered. Many of the molecules that control gene expression are often present in small numbers inside the cell. The genes themselves are usually present in only one or a few copies. Further, the genetic reaction rates are often rather slow compared to the other biochemical reactions inside the cell. These facts indicate that a deterministic chemical kinetic treatment of these reactions may not always be possible and that the discrete molecular nature of the expression machinery and their thermal fluctuation must be taken into account as well (Kampen, 1981; Ko, 1991, 1992; Peccoud and Ycart, 1995; McAdams and Arkin, 1997; Arkin *et al.*, 1998; Goss and Peccoud, 1998). Consider, for example, the rate of transcription initiation from P_{RM}. The maximum activity of P_{RM} shown in Figure 6 is about 0.007 open-complexes/s; that is, one transcript initiation about every 2.5 min on average. This occurs at $[Cl_2]$ of the order of 200 nM, and at $[Cro_2]$ of zero. However, during early λ phage development, $[Cro_2]$ and $[CI_2]$ are generally less than 100 nM. In *E. coli*, which has a cell volume of approximately one femtoliter, 1 nM corresponds to about one molecule only. Given that a well-fed *E. coli* cell has a cell cycle time of about 20 min, there are, on average, fewer than ten transcription initiations from a fully activated P_{RM} per cell division. However, the actual number of transcription initiations from P_{RM} is most likely a stochastic process. Further, the number of proteins produced per transcript is also probably a random process. The 'back-of-the-envelope' explanation for this is as follows. We can roughly divide the gene expression process into four stages: (1) RNAP binding to the promoter, (2) transcription initiation, (3) RNAP arrival at the end of a gene, and (4) competitive binding of ribosomes and RNA degrading enzymes to the RNA transcript. The probability of RNAP being bound at its promoter is determined by the partition function of operator/promoter states that enters into the calculation of the curve in Figure 6. The probability of transcription initiation is very roughly a first-order rate process whose rate constant can be read off the figure. The time until transcription initiation, after RNAP has bound to its promoter in a given state, is distributed approximately exponentially (McAdams and Arkin, 1997). Elongation then proceeds by a series of independent steps each exponentially distributed in time. The arrival time at the end of a gene is, therefore, the sum of a set of independent exponential distributions (one for each nucleotide in the sequence) which has the form of a Γ-distribution.

Finally, the number of proteins per transcript is determined by how many ribosomes can bind to the ribosome binding site on the transcript before the transcript is degraded by an RNase protein. These two processes are often competitive, so ribosome binding temporarily protects the transcript from degradation. Thus, the question arises of how many ribosomes can bind before degradation by RNase occurs. This is analogous to asking how many heads does one get before one gets a tail when flipping a biased coin. Such processes are described by a geometric distribution. Each of these distributions can be rather broad and skewed. Consequently, the pattern of protein production from a single promoter can be expected to be burst-like and erratic.

When all of the above arguments are put forth in a chemically more rigorous fashion, the dynamics of gene expression may be described by a chemical master equation (McAdams and Arkin, 1997). Figure 8 shows the pattern of protein production from a model of the λ phage P_R promoter (McAdams and Arkin, 1997). Each curve is one realization of the stochastic gene expression process started from the *same* initial conditions in *identical* cells. Our theoretical model indicates that individual cells can have quite different expression patterns. There is ample experimental evidence that this is indeed the case in cell populations (Novick and Weiner, 1957; Ko, 1992; Ross *et al.*, 1994; Siegele and Hu, 1997). The implications of this noise for the control of cellular behavior and development, and for the engineering of reliable genetic circuitry, has been discussed in detail (McAdams and Arkin, 1997, 1998, 1999; Arkin *et al.*, 1998).

5.2.3.4 *Electrical and chemical frequency filters*

Since biological signals can be *periodic*, as described in Section 5.2.3.2, and *noisy*, as described in Section 5.2.3.3, a consideration of the frequency-dependent responses of chemical reactions to time-varying chemical signals is in order. A frequency dependence can be considered to be a type of *filtering*. In an electrical context, frequency filters are devices that accept a time-dependent input and differentially pass on some frequencies in the signal, while suppressing others to different degrees.

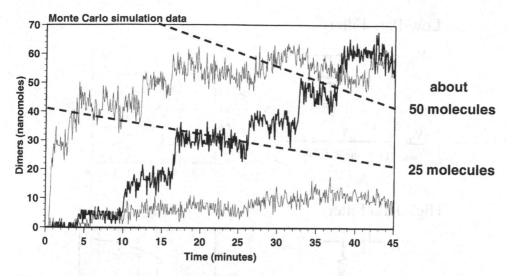

Figure 8. Master equation simulation of stochastic gene expression from a P_R-like promoter in three initially identical cells. For details see Section 5.2.3.3. (From McAdams and Arkin, 1997.)

The two basic, passive electrical filters, the *low*-pass filter and the *high*-pass filter, are represented in Figure 9. The amplitude of each frequency component of the output signal is always less than, or equal to, the corresponding amplitude in the input signal. These small circuits are composed of 'linear' elements (resistors and capacitors) and thus are noise filters as well as frequency filters. The spectrum of the output signal is the superposition of the filtered amplitudes of each frequency component. The filter causes no interference among the different components of the input signal to arise in the output signal.

Chemical low-pass filter A chemical version of the low-pass filter is shown in Figure 10 (top). The input is the amplitude of time-varying (positive) input of a chemical species; the output is the amplitude of the concentration of species A. The frequency response function has the same fall-off as that of the low-pass filter. However, this 'filter' can (at low frequencies) *amplify* a signal as well, since it has a factor of '*k*' in the numerator. In fact, any network of such linear chemical reactions (with one input) is a low-pass filter. The additional reactions between the input and output change the phase retardation of the signal as well as the exact shape of the monotonically decreasing filtering profile.

Chemical band-pass filter As soon as a *nonlinear* chemical reaction is considered, there exists the possibility of *band*-pass filtering (all chemical systems, like all electronic systems, are low-pass at high enough frequencies). As an

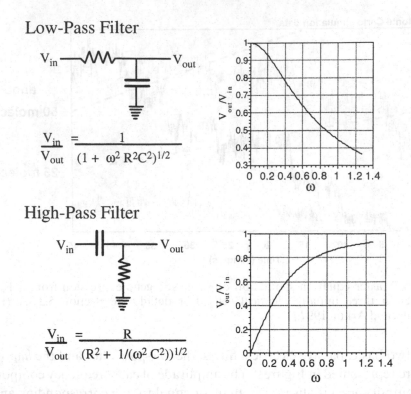

Low-Pass Filter

$$\frac{V_{in}}{V_{out}} = \frac{1}{(1 + \omega^2 R^2 C^2)^{1/2}}$$

High-Pass Filter

$$\frac{V_{in}}{V_{out}} = \frac{R}{(R^2 + 1/(\omega^2 C^2))^{1/2}}$$

Figure 9. Electrical frequency filters. (Top) A filter that passes only low frequencies (a low-pass filter). (Bottom) A filter that passes only high frequencies (a high-pass filter). For details see text in Section 5.2.3.4.

example, consider the bimolecular reaction shown in Figure 10 (bottom). In this reaction scheme, γ must be less than or equal to P, which, in turn, must be equal to RB. Solving the Riccati equation that describes this system analytically is at best difficult, but very good asymptotic solutions may be found. The filtering profile shown in Figure 10 (bottom) shows a pronounced band-pass region between the critical frequency, ω_c, and k_2. This particular chemical circuit is admittedly artificial, since the constraint that 'P match RB' is not likely to be met in biological systems. The example demonstrates, however, that chemical systems can be both band-pass and low-pass filters. Note, however, that this does not describe a *linear* filter, and if the driving signal has many frequency components, then there are in fact (very small) interference bands that appear at frequencies not contained in the original signal.

More complex chemical filters The above types of mechanism can be easily linked in serial or parallel fashion to form very complicated filtering profiles

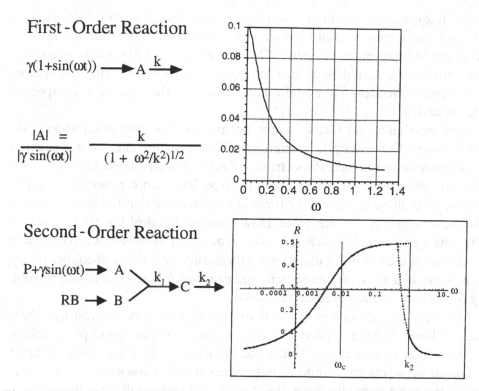

First-Order Reaction

$$\gamma(1+\sin(\omega t)) \longrightarrow A \xrightarrow{k}$$

$$\frac{|A|}{|\gamma \sin(\omega t)|} = \frac{k}{(1+\omega^2/k^2)^{1/2}}$$

Second-Order Reaction

$$P+\gamma\sin(\omega t) \longrightarrow A$$
$$RB \longrightarrow B$$
$$\searrow_{k_1} C \xrightarrow{k_2}$$

Figure 10. The frequency filtering properties of two different chemical mechanisms. *Top*: A reaction pathway with linear (first-order) kinetics acts as a low-pass filter. Here, the concentration of the species A is driven by a sinusoidal influx of material (amplitude = γ). The amplitude of the oscillation in A is plotted as a function of the frequency of the influx species. The analytical equation describing this curve is shown below the reaction mechanism. *Bottom*: As soon as a simple nonlinearity is found in the reaction (second-order kinetics) then the system can behave as a band-pass filter. Here, there is influx of material from independent sources into species A and B. Species A is driven by a sinusoidal signal that has an amplitude equal to γ around an average influx value of P (the pedestal). The amplitude of the oscillation in C is plotted as a function of the driving frequency of the A influx. The equation for this curve is too complicated to show (see Samoilov, 1997).

indeed, such as *notch* filters that do not pass some intermediate band of frequencies. The ability to easily construct complex filters out of relatively simple chemical reactions suggests that such filtering could be used by biological systems to respond differentially to the types of oscillatory input discussed above. This is particularly interesting when considering, for example, that in T lymphocytes some expression of some transcription factors are stimulated by much lower frequencies of cytosolic Ca^{2+} oscillations than others (Dolmetsch *et al.*, 1997, 1998). Also, although this has not yet been directly observed, chemical frequencies can allow the demultiplexing of multiple signals carried in the frequency spectrum of a signal chemical concentration. For example, perhaps one hormone could induce one frequency of Ca^{2+} spiking and another could superimpose another. Thus the presence or absence of two external signal species could be carried throughout the cell using only one species; downstream chemical filters could then decompose the signal at the respective sites of action.

Nonlinear chemical filters do not behave, as was mentioned above, like linear filters. For example, interference effects can introduce new frequency components into the output spectrum. Thus these macroscopic kinetic circuits are not really noise filters, but they may perform more general transformations. Also, these treatments of these circuits assume that the fluctuations in chemical reactions are negligible. How a given chemical kinetic mechanism filters its own internal noise is still an open question. However, it is certain that mechanisms such as molecular dimerization, and other forms of cooperativity, filter molecular noise to some extent and can therefore lend increased reliability to the genetic circuits described above.

More complex chemistry can have ever more exotic and interesting behaviors. These include analytical delays and strong band-pass effects, quasiperiodicity and chaos. When the input signal is noisy, some complex chemical networks can exhibit a phenomenon called *stochastic resonance* in which the noise signal improves the detection of very small periodic signals in the system over some region of power in the noise spectrum (Collins *et al.*, 1995, 1996; Braun *et al.*, 1997; Astumian and Moss, 1998; Jung *et al.*, 1998). These examples only underscore the point that biological systems are nonstationary dynamical systems whose signal-processing machinery may be far more sophisticated than is generally understood.

5.3 Comments on the parameterization of models, nonlinear systems and cellular reliability

One of the concerns often voiced about the business of biological systems modeling concerns the fact that most mathematical descriptions of biological processes contain a good many parameters most of which cannot easily be experimentally measured. Further, these descriptions are most often sets of coupled nonlinear differential equations that may, in general, show an extreme sensitivity to parameters (and initial conditions). Since it is currently difficult, at best, to measure all of the kinetic parameters for even isolated network components, such as enzymes, and since such measurements of *in vitro* kinetics are not guaranteed to be the same as those that are obtained *in vivo*, it might seem a hopeless task to construct confirmatory and predictive models of complex BRNs. This fear is valid; however, there are a number of phenomenological observations and mathematical facts that argue that the situation is not so bleak as might be initially thought.

The first trivial observation is that all that is nonlinear is not necessarily sensitive to parameter changes. Extreme sensitivity, such as that found in deterministically chaotic systems, at least so far, has proven to be a relatively rare phenomenon in biological and biochemical systems. Though chaos has certainly been detected at the tissue level such as in heart and brain dynamics, and on the chemical level in reconstituted and forced peroxidase–oxidase enzyme systems, it is certainly not the rule in even very complex, nonlinear biochemical systems. The reason for this must reside in the engineering specifications for good biological function. Most cells operate in fluctuating environments (wherein variables such as temperature, pressure, volume and ionic strength can change unpredictably) and must both detect and use chemical components that are at very low concentrations and whose kinetics, therefore, is likely to exhibit large fluctuations in reaction rates. In addition, there is a finite chance that a given component of a regulatory network may fail, due either to this noise in its components or to more extreme processes such as mutations. In order for a cell to survive under such conditions, the function of its regulatory networks cannot be so sensitive to their parameters (which are sensitive to these fluctuations) and must be reliable in the face of individual component failure and mutation. In order to achieve this robustness, cells use *functional redundancy* and *feedback stabilization* among other design strategies to obtain reliable operation.

Barkai and Liebler have suggested that this very insensitivity to parameters might be one criterion for judging whether or not a particular biochemical model is reasonable (Barkai and Leibler, 1997). As an example, they have analyzed various models of exact adaptation in bacterial chemotaxis. This is a phenomenon in which the ratio of clockwise to counter-clockwise rotation of the flagellum initially decreases upon cellular exposure to a step of chemo-attractant, but then returns exactly to the initial basal value, even under continued (constant) exposure to the attractant. Since this behavior is judged to be important to the cell's fitness for survival, Barkai and Leibler (1997) have argued that the regulatory network that controls this behavior should be *insensitive* to changes in its parameters. They propose a schematic model of adaptation that maintains 'exactness' despite order-of-magnitude changes in one or more parameters, whereas some previous models of adaptation seemed to require a highly tuned parameter set. However, although the model by Barkai and Liebler exhibits high reliability in exact adaptation, the time to recovery is less robust. Experimental measurements on the dispersion of times to exact recovery in different mutants of chemotactic *E. coli* should further constrain the class of models that can explain the chemotactic behavior (Alon *et al.*, 1999). An important side-note is that the argument that cells should not be too sensitive to small perturbations assumes that the perturbations are natural (i.e. commonly occurring ones). Specific toxins and pharmaceuticals, hard radiation, and other such 'artificial' perturbations are rarely encountered during the normal course of a particular organism's evolution. Its BRNs may therefore be sensitive to very low exposures to these types of attacks.

Biochemical models, therefore, are often subject to a number of restrictive global functional constraints such as robustness and, in some cases, homeostasis as well as an often large list of experimental data that greatly restricts the class of models that can explain and predict organismal behavior. If the basic stoichiometric network of reactions is known, this provides a further restriction on the class of behaviors and the feasible sets of parameters that can reproduce experimental observations (Clarke, 1981).

5.4 Summary and outlook

The challenges of understanding how these incredibly complicated biological systems function to the point where we can predict their behavior, control them and rationally design modifications into them are clear. The chemical and physical systems that underlie their function operate in regimes of which we do not yet have a full theoretical facility. They operate *asynchronously*, *asymmetrically* and *nonlinearly* in fluctuating environments with less than fully reliable parts. In addition, the systems are rather large and highly interconnected networks that operate over a large range of time and space scales. Practically, it is not feasible to derive the equations for each microscopic event that occurs within and among cells. Some higher levels of abstraction will be necessary to make useful and rapid analyses. The work presented here has considered a bottom-up approach that starts with the detailed kinetics of networks of chemical reactions and attempts to derive when and where such networks may be dissected into self-contained 'devices'. It was a hypothesis of this chapter, and one that is confirmed in part by the literature, that these devices are *regulatory motifs* that recur within different pathways of the same organism and across organisms. The motifs may be realized using different (or related) biochemical species, but their functions may be the same. The level of abstraction from the detailed molecular kinetics of each device will be different, and it is a central challenge to develop methods for making models that can combine such heterogeneous submodels in a physically consistent way. Also, in analogy to finding the basis set of protein folds in order to understand the principles of protein folding, it would be advantageous to identify a basis set of such devices, and their restricted class of functions, from which many BRNs may be constructed.

Meeting this challenge is especially crucial in light of the accelerated onslaught of essentially raw data that has fallen out of new high-throughput biological measurement devices and their resultant 'projects'. Genome projects provide partial parts lists for the cellular machinery. Information technologies are providing large numbers of hypotheses for predicting protein activity, structure/function predictions and even network hypotheses. Gene chip technology and two-dimensional protein gel/mass spectrometry methods are beginning to provide quantitative measurements of the condition- and time-dependent variations in concentrations of mRNA-transcript proteins. Advanced microscopy and other cell measurement devices are beginning to create large databases of spatial information, cell motion and cellular interaction data that can be related to changes in ion concentrations and gene expression. In addition to these relatively new stores of data, there are all the data generated from the standard biochemical and genetic research communities as well as massive amounts of clinical and medical diagnostic data. It is one central challenge to deduce from these data the responsible regulatory networks. Such reverse engineering methods are in their very early days (Arkin and Ross, 1995; Arkin *et al.*, 1997; Liang *et al.*, 1998; Thieffry and Thomas, 1998).

Theoretical and computational tools developed to dissect and analyze complex biological systems are essentially tools to make more rigorous the process of hypothesis formation that every biologist must conduct before and after performing such experiments. These tools provide a central structure for organizing the data generated by the above techniques. They help to yield new insights and new biological principles, some of which are discussed above. Several of these insights have profound implications for biological processes such as development, facultative infection and other diseases. Finally, these tools are beginning to aid in the design of novel functions into cells. Since the engineering principles by which such circuitry is constructed in cells comprise a super-set of that used in electrical engineering, it is, in turn, possible that we will learn more about how to design asynchronous, robust electronic circuitry as well.

Acknowledgments

I thank Dr Harley McAdams and Professor John Ross for many helpful discussions on these topics over a number of years. I also acknowledge the support by grants from the Office of Naval Research and Department of Energy.

References

Ackers, G. K., Johnson, A. D. and Shea, M. A. (1982) Quantitative model for gene regulation by phage-1 repressor. *Proc. Natl. Acad. Sci. USA* **79**: 1129–1133.

Alon, U., Surette, M. G., Barkai, N. and Leibler, S. (1999) Robustness in bacterial chemotaxis. *Nature* **397**: 168–171.

Arkin, A. P. and Ross, J. (1994) Computational functions in biochemical reaction networks. *Biophys. J.* **67**: 560–578.

Arkin, A. P. and Ross, J. (1995) Statistical construction of chemical mechanisms from measured time series. *J. Phys. Chem.* **99**: 970–979.

Arkin, A. P., Ross, J. and McAdams, H. H. (1998) Stochastic kinetic analysis of developmental pathway bifurcation in phage 1-infected *E. coli* cells. *Genetics* **149**: 1633–1648.

Arkin, A. P., Shen, P.-D. and Ross, J. (1997) A test case of correlation metric construction of a reaction pathway from measurements. *Science* **277**: 1275–1279.

Astumian, R. D. and Bier, M. (1996) Mechanochemical coupling of the motion of molecular motors to ATP hydrolysis. *Biophys. J.* **70**: 637–653.

Astumian, R. D. and Moss, F. (1998) Overview: the constructive role of noise in fluctuation driven transport and stochastic resonance. *Chaos* **8**: 533–538.

Barkai, N. and Leibler, S. (1997) Robustness in simple biochemical networks. *Nature* **387**: 913–917.

Berridge, M. J. and Rapp, P. E. (1979) A comparative survey of the function, mechanism and control of cellular oscillators. *J. Exp. Biol.* **81**: 217–279.

Borisuk, M. T. and Tyson, J. J. (1998) Bifurcation analysis of a model of mitotic control in frog eggs. *J. Theor. Biol.* **195**: 69–85.

Braun, H. A., Schafer, K., Voigt, K., Peters, R., Bretschneider, F., Xing, P., Wilkens, L. and Moss, F. (1997) Low-dimensional dynamics in sensory biology. I. Thermally sensitive electroreceptors of the catfish. *J. Comput. Neurosci.* **4**: 335–347.

Bray, D. (1990) Intracellular signalling as a parallel distributed process. *J. Theor. Biol.* **143**: 215–231.

Bustamante, C. and Rivetti, C. (1996) Visualizing protein–nucleic acid interactions on a large scale with the scanning force microscope. *Annu. Rev. Biophys. Biomol. Struct.* **25**: 395–429.

Clarke, B. L. (1981) Complete set of steady states for the general stoichiometric dynamical system. *J. Chem. Phys.* **75**: 4970–4979.

Collins, J. J., Chow, C. C. and Imhoff, T. T. (1995) Stochastic resonance without tuning. *Nature* **376**: 236–238.

Collins, J. J., Imhoff, T. T. and Grigg, P. (1996) Noise-enhanced information transmission in rat SA1 cutaneous mechanoreceptors via aperiodic stochastic resonance. *J. Neurophysiol.* **76**: 642–645.

Davidson, E. H., Cameron, R. A. and Ransick, A. (1998) Specification of cell fate in the sea urchin embryo: summary and some proposed mechanisms. *Development* **125**: 3269–3290.

Dolmetsch, R. E., Lewis, R. S., Goodnow, C. C. and Healy, J. I. (1997) Differential activation of transcription factors induced by Ca^{2+} response amplitude and duration. *Nature* **386**: 855–858.

Dolmetsch, R. E., Xu, K. and Lewis, R. S. (1998) Calcium oscillations increase the efficiency and specificity of gene expression. *Nature* **392**: 933–936.

Duke, T. and Leibler, S. (1996) Motor protein mechanics: a stochastic model with minimal mechanochemical coupling. *Biophys. J.* **71**: 1235–1247.

Eiswirth, M., Freund, A. and Ross, J. (1991a) Mechanistic classification of chemical oscillators and the role of species. *Adv. Chem. Phys.* **LXXX**: 127–199.

Eiswirth, M., Freund, A. and Ross, J. (1991b) Operational procedure toward the classification of chemical oscillators. *J. Phys. Chem.* **95**: 1294–1299.

Goss, P. J. and Peccoud, J. (1998) Quantitative modeling of stochastic systems in molecular biology by using stochastic Petri nets. *Proc. Natl. Acad. Sci. USA* **95**: 6750–6755.

Goto, K., Laval-Martin, D. L. and Edmunds, Jr, L. N. (1985) Biochemical modeling of an autonomously oscillatory circadian clock in *Euglena*. *Science* **228**: 1284–1288.

Guthold, M., Bezanilla, M., Erie, D. A., Jenkins, B., Hansma, H. G. and Bustamante, C. (1994) Following the assembly of RNA polymerase–DNA complexes in aqueous solutions with the scanning force microscope. *Proc. Natl. Acad. Sci. USA* **91**: 12927–12931.

Hjelmfelt, A. and Ross, J. (1992) Chemical implementation and thermodynamics of collective neural networks. *Proc. Natl. Acad. Sci. USA* **89**: 388–391.

Hjelmfelt, A., Schneider, F. W. and Ross, J. (1993) Pattern recognition in coupled chemical kinetic systems. *Science* **260**: 335–337.

Hjelmfelt, A., Weinberger, E. D. and Ross, J. (1991) Chemical implementation of neural networks and Turing machines. *Proc. Natl. Acad. Sci. USA* **88**: 10983–10987.

Hjelmfelt, A., Weinberger, E. D. and Ross, J. (1992) Chemical implementation of finite state machines. *Proc. Natl. Acad. Sci. USA* **89**: 383–387.

Hung, Y.-F. and Ross, J. (1995) New experimental methods towards the deduction of the mechanism of the oscillatory peroxidase–oxidase reaction. *J. Phys. Chem.* **99**: 1974–1979.

Jonnalagadda, S. B., Becker, J. U., Sel'kov, E. E. and Betz, A. (1982) Flux regulation in glycogen-induced oscillatory glycolysis in cell-free extracts of *Saccharomyces carlsbergensis*. *Biosystems* **15**: 49–58.

Jung, P., Cornell-Bell, A., Moss, F., Kadar, S., Wang, J. and Showalter, K. (1998) Noise sustained waves in subexcitable media: from chemical waves to brain waves. *Chaos* **8**: 567–575.

Kampen, N. G. (1981) *Stochastic Processes in Physics and Chemistry*. New York: North-Holland.

Karp, P. D., Riley, M., Paley, S. M., Pellegrini-Toole, A. and Krummenacker, M. (1999) Eco Cyc: encyclopedia of *Escherichia coli* genes and metabolism. *Nucl. Acids Res.* **27**: 55–58.

Ko, M. S. H. (1991) A stochastic model for gene induction. *J. Theor. Biol.* **153**: 181–194.

Ko, M. S. H. (1992) Induction mechanism of a single gene molecule: stochastic or deterministic? *BioEssays* **14**: 341–346.

Liang, S., Fuhrman, S. and Somogyi, R. (1998) Reveal, a general reverse engineering algorithm for inference of genetic network architectures. *Pac. Symp. Biocomput.* **4**: 18–29.

McAdams, H. H. and Arkin, A. P. (1997) Stochastic mechanisms in gene expression. *Proc. Natl. Acad. Sci. USA* **94**: 814–819.

McAdams, H. H. and Arkin, A. P. (1998) Simulation of prokaryotic genetic networks. *Annu. Rev. Biophys. Biomol. Struct.* **27**: 199–224.

McAdams, H. H. and Arkin, A. P. (1999) Genetic regulation at the nanomolar scale: it's a noisy business! *Trends Genet.* **15**(2): 65–69.

Monod, J. and Jacob, F. (1961) General conclusions: teleonomic mechanisms in cellular metabolism, growth and differentiation. *Cellular Regulatory Mechanisms*, vol. XXVI, pp. 389–401. Cold Spring Harbor, NY: Cold Spring Harbor Laboratory Press.

Myers, C. J. (1995) Computer-aided synthesis and verification of gate-level timed circuits. Ph.D. thesis, Stanford University.

Novak, B., Csikasz-Nagy, A., Gyorffy, B., Chen, K. and Tyson, J. J. (1998) Mathematical model of the fission yeast cell cycle with checkpoint controls at the G1/S, G2/M and metaphase/anaphase transitions. *Biophys. Chem.* **72**: 185–200.

Novick, A. and Weiner, M. (1957) Mixed population response to induction of *lac* promoter. *Proc. Natl. Acad. Sci. USA* **43**: 553–566.

Ouyang, Y., Andersson, C. R., Kondo, T., Golden, S. S. and Johnson, C. H. (1998) Resonating circadian clocks enhance fitness in cyanobacteria. *Proc. Natl. Acad. Sci. USA* **95**: 8660–8664.

Peccoud, J. and Ycart, B. (1995) Markovian modelling of gene product synthesis. *Theoret. Population Biol.* **48**: 222–234.

Peskin, C. S. and Oster, G. F. (1995) Force production by depolymerizing microtubules: load–velocity curves and run–pause statistics. *Biophys. J.* **69**: 2268–2276.

Rapp, P. E. (1979) An atlas of cellular oscillators. *J. Exp. Biol.* **81**: 281–306.

Ross, I. L., Browne, C. M. and Hume, D. A. (1994) Transcription of individual genes in eukaryotic cells occurs randomly and infrequently. *Immunol. Cell Biol.* **72**: 177–185.

Samoilov, M. S. (1997) Reconstruction and functional analysis of general chemical reactions and reaction networks. Ph.D. thesis, Stanford University.

Shea, M. A. and Ackers, G. K. (1985) The O_R control system of bacteriophage lambda: a physical-chemical model for gene regulation. *J. Mol. Biol.* **181**: 211–230.

Siegele, D. A. and Hu, J. C. (1997) Gene expression from plasmids containing the *araBAD* promoter at subsaturating inducer concentrations represents mixed populations. *Proc. Natl. Acad. Sci. USA* **94**: 8168–8172.

Stemwedel, J. D., Schreiber, I. and Ross, J. (1994) Formulation of oscillatory reaction mechanisms by deduction from experiments. *Adv. Chem. Phys.* **LXXXIX**: 327–387.

Thieffry, D. and Thomas, R. (1998) Qualitative analysis of gene networks. *Pac. Symp. Biocomput.* **4**: 77–88.

Tuinenga, P. W. (1995) *Spice: A Guide to Circuit Simulation and Analysis Using Pspice.* Upper Saddle River, NJ: Prentice Hall.

Wang, M. D., Yin, H., Landick, R., Gelles, J. and Block, S. M. (1997) Stretching DNA with optical tweezers. *Biophys. J.* **72**: 1335–1346.

Yin, H., Wang, M. D., Svoboda, K., Landick, R., Block, S. M. and Gelles, J. (1995) Transcription against an applied force. *Science* **270**: 1653–1657.

Yuh, C. H., Bolouri, H. and Davidson, E. H. (1998) Genomic *cis*-regulatory logic: experimental and computational analysis of a sea urchin gene. *Science* **279**: 1896–1902.

Part II
Nonlinear sensitivity of biological systems to electromagnetic stimuli

Electromagnetic stimuli represent a special class of external perturbations that are discussed in almost all of the remaining book chapters. Part II therefore provides important information in regard to the biophysical foundations of interactions between biological processes and electric or magnetic fields. In addition, experimental examples are described that demonstrate the nonlinear sensitivity to electromagnetic stimuli of enzymes, single cells and tissues. The principles of *electric field* interactions and the functional role of bioelectric fields are reviewed in Chapter 6 by Paul Gailey. He discusses the remarkable electrosensitivity of selected biological systems and how oscillating electric fields may be detected and amplified by biological structures. The chapter concludes with the description of a model based on the concept of long-range coherence, which may explain how relatively weak electric fields may effectively interact with excitable cellular assemblies in the presence of noise. An electric field-sensitive cellular oscillator in cells of the immune system is the subject of Chapter 7 by Howard Petty. He discusses experiments that have led to the discovery of coherent metabolic oscillations in human neutrophils, and describes the response of these cells to time-varying chemical and electric fields. His work shows the critical importance of the phase relationship between internal cellular oscillations and the externally applied field oscillations in the induction of cellular responses. Direct interactions between *magnetic field* stimuli and biological activity is the main theme of Chapter 8 by Jan Walleczek and Clemens Eichwald. Their chapter presents a brief history of research in this area and then describes work showing that enzyme activity, including oscillatory enzyme dynamics, may serve as an effective magnetic field coupling target. Further, they present results from nonlinear modeling studies that propose mechanisms by which biological processes may become sensitive to the frequency of oscillating magnetic field perturbations. Finally, Chapter 9, contributed by Stefan Engström and collaborators, outlines an experimental approach that has revealed the magnetic field sensitivity of hippocampal brain tissue. Their work provides evidence that neuronal tissues may be able to discriminate between different frequencies of magnetic field oscillations.

6

Electrical signal detection and noise in systems with long-range coherence

PAUL C. GAILEY

6.1 Introduction

The long-running controversy over the possibility of health effects from weak, environmental electric and magnetic fields (e.g., 60-Hz power line fields) has both advanced and obscured the study of field interactions with biological systems. While a substantial number of publications focus on theoretical limits of field detection, the efforts by some to disprove the possibility of health effects has drawn attention away from the broad and fascinating range of well-established field interaction phenomena. In this chapter, I review some of the fundamentals of *electric field interactions with biological systems*, extend these concepts to systems with *long-range coherence*, and discuss implications for research and therapy. Those readers interested in the detailed mathematical analysis of these processes are directed to appropriate treatments in the literature.

It should come as no surprise that applied electric fields can affect biological systems. Organisms at all levels of complexity both generate and use electric fields in development and function. Early work by L. Jaffe showed that a number of organisms generate electric fields during development (Jaffe and Nuccitelli, 1977). More recent work by Shi and Borgens (1995), Hotary and Robinson (1994) and others revealed that such fields are ubiquitous and may play a key role in tissue organization. Electric fields produced by embryos appear to *direct* the placement and differentiation of certain cells into structural and functional components of the developing organism. As investigators attempt to tease apart this intricate, self-directing symphony, we can gain insights by looking at a more comprehensible and immediately relevant process – *wound healing*. When our skin is abraded, the natural electric potential maintained between the outer and inner layers of skin is short circuited. The wound provides a low-resistance return path, and the resulting electric field directs the migration of keratinocytes toward the injured area (Sheridan *et al.*, 1996). Many of the details of this process are well understood, but, taken as a whole, it is a stunning example of self-directed organization that is globally mediated by an *endogenous* electric field.

Other examples of electric field effects in biology include phenomena such as ephaptic signaling, or electric field coupling between neurons (Dudek *et al.*, 1986). But we focus instead on externally applied electric fields to discuss the physics of these interactions and *theoretical detection limits*. From this perspective, there is no better starting point than the elasmobranch fish. Pioneering work by Kalmijn (1982) and others revealed the extraordinary sensitivity of these marine animals. Their well-established behavioral responses to 500-nV/m electric fields are best understood by analogy. If wires connected to either end of a single 1.5-V flashlight battery were placed 2000 miles (about 3000 km) apart in the ocean, electrically sensitive sharks and rays would be able to clearly detect the electric field produced in the water by the battery. These animals need such extreme sensitivity in order to detect the weak electric fields produced in seawater by the physiological processes of their prey.

There is much to be learned about the methods and limitations of electric field detection from this remarkable sensory system. First, the fish use highly conductive 'canals' to amplify the field *internally* at the site of their sensory organ. Second, they average the response of large populations of cells to reduce the effects of noise. Finally, they are clearly able to correct for the effects of temperature on the detection process – effects that can be significantly larger than the effect of the electric field itself. These are the same issues relevant to any discussion of electric field detection processes: the electric field must be *amplified* to levels of physiological significance, and the detector must be able to distinguish between the signal and the noise or random fluctuations inherent in the system.

The latter point is currently the subject of intense discussions in the neuroscience, biophysics, and physics literature (Barnes, 1988, 1996; Adair, 1991; Astumian *et al.*, 1995, 1997; Rieke *et al.*, 1997; Valberg *et al.*, 1997). Any physical system operating at biological temperatures will exhibit *thermal noise*, and specific structures introduce additional classes of noise that must also be considered. *Voltage-gated ion channels*, for example, are not deterministic open or shut devices. They are probabilistic devices that switch randomly between different states. Therefore, a potential difference applied to the cell membrane will affect only the probability of channels being open or closed, not their exact states. The random switching produces a type of noise specific to these channels. Release of acetylcholine in a synapse is another example of a random process. These and other sources of noise result in fluctuations in the timing of the spike trains from nerves that encode sensory input and other neuronal signals. The question of how organisms extract the information of interest from the background of natural noisy processes is very much open.

We are faced with similar questions when inquiring into the use of electric fields to perturb nonsensory biological systems except that the potentially affected components are not known *a priori*. Fortunately, we can begin an investigation at the cellular level for which some very detailed analyses of theoretical models are available (Schwan, 1983; Trachina and Nicholson, 1986; Polk and Postow, 1996). More recent work includes numerical simulations for groups of electrically coupled cells (Stuchly and Xi, 1994; Gailey, 1996). Following a review of these concepts, we will push forward the frontier of electric field detection by considering populations of *synchronized oscillatory cells*. Such populations occur widely in nonsensory settings and are fundamental to neuronal processing. Here we will find that the presence of noise can enhance signal detection in a manner reminiscent of *stochastic resonance* (e.g., see Moss, Chapter 10, this volume). In these coupled populations, noise and coupling strength between elements work counter to each other in preventing or permitting synchronous oscillations. As either noise is increased or coupling strength is decreased, the system undergoes a *phase transition* from the synchronous or long-range coherent state to the nonsynchronous state. Near this boundary, the system is very sensitive to *external perturbations*. The possibility of using electric fields to influence this transition may be important because, as described by Ding *et al.*, Chapter 4, this volume, such phase transitions appear to be fundamental to brain function. More generally, we can expect to discover various forms of long-range coherence, occurring widely in biological function, as it is an essential feature of self-organization – the hallmark of living systems.

6.2 Principles of electric field detection in biological systems

Electric fields are defined in terms of the force they exert on electric charge. In biological systems, these charges can occur as electrons, ions and macromolecules containing excess charge or nonsymmetrical charge distributions. A spherically symmetric, charged object in a uniform electric field experiences a force, F, of magnitude $F = qE$, where q is the net charge and E represents the electric field strength. This force is generally very small in the case of ions and charged molecules because the unit of elementary charge is so small. As an example, consider what is called the fair-weather, static electric field. Ordinarily, when no thunderstorm is underway, the electric field near the earth's surface is about 100 V/m. A singly charged ion such as sodium (Na^+) has a net charge of one electron, or about 1.6×10^{-19} coulombs. From the force equation above, one can estimate that the force on Na^+ due to the fair-weather electric field is only about 10^{-17} newtons. Such a small force is hard to imagine, being roughly equivalent to the pull of gravity on an object that weighs one one-thousandth of a picogram.

Electric fields also exert a force on objects that have an inhomogeneous distribution of charge, but an overall net charge of zero. The simplest example is an electric dipole, which one can imagine as a short rod with positive charge on one end and negative charge on the other. Because the net charge is zero, a uniform electric field will not pull the dipole in one direction; it will instead exert a torque on the dipole that tends to align it in the direction of the electric field vector. In summary, while free ions are set in motion by uniform electric fields to produce electric currents, dipoles are only rotated.

Electric dipoles are common in biological systems and can be either free (e.g., water molecules) or bound to a structure, such as some receptors and other proteins bound to the cell membrane. Electric field interactions with dipoles on numerous scales of organization lead to the property known as the *electric permittivity* of a substance. Detailed analyses and measurements of the electrical permittivity of biological materials have been conducted (Foster and Schwan, 1996). The charge distributions on biological macromolecules are complex, and are often described in terms of higher-order multipole moments. For general discussions of the physics of electric field interaction, many good texts are available, including those of Feynman *et al.* (1964) and Iskander (1992).

When one is attempting to analyze or predict electric field interactions, the strength of the electric field must, of course, be known. Determining the field strength experienced by the cells and tissues of a biological organism can be quite difficult, however, because of the complex relationship between the *externally* applied field and the fields induced *internally*. At low frequencies, biological tissues strongly shield electric fields in a way similar to the shielding properties of a metal Faraday cage. The ratio between external and internal electric fields can be very large. A 1000-V/m, 60-Hz external electric field, for example, will produce an internal electric field of only about 0.001 V/m inside the human body. This attenuation by six orders of magnitude is substantial. Actual fields inside the body will vary considerably depending on the orientation of the subject, body size and position within the body. Biological tissues have enormously complex electrical properties because of their organization at the molecular, cellular, tissue and organ levels. A number of researchers have tackled the problem of predicting the internal electric fields that will result from a known externally applied field (Durney *et al.*, 1975; Gandhi and Chen, 1992; Dawson *et al.*, 1997). However, the reliability of these results is limited by the resolution of the anatomical models used in the simulations. Even at millimeter resolution, important electrical features such as thin membranes surrounding organs cannot be accurately modeled.

Internal electric fields can also be induced by external, oscillating magnetic fields via a process known as *Faraday induction*. Low-frequency magnetic fields pass through biological tissues largely unattenuated. Inside the tissues, they induce an electric field with a magnitude and direction that depends on the geometry and electrical properties of the tissues, along with the frequency and polarization of the magnetic field. Therapeutic devices for accelerating bone healing, for example, depend on this principle for inducing electric fields in the tissues of interest. The prediction of internal electric fields induced by external magnetic fields is also difficult and requires detailed numerical models. To provide a rough idea of the strength of field induction, we note that a uniform 60-Hz magnetic field of 0.1 millitesla (mT), which is also 1 gauss (G), will induce an electric field of the order of 1 mV/m inside an average-size human male adult. The internal electric fields induced by either external electric or magnetic fields depend very much on frequency. Higher frequencies induce higher internal electric fields up to the frequency at which the whole body resonates in the 30–100 MHz range (Durney *et al.*, 1975). Internal electric fields may have magnitude similar to that of external electric fields at whole-body resonance. At higher frequencies still (microwave to millimeter wave range), electromagnetic fields are unable to penetrate the body efficiently, and the magnitude of induced internal fields decreases rapidly beneath the surface.

6.2.1 *The cell membrane as a target for electric field coupling*

Exogenously produced internal electric fields can interact with a variety of biological substructures, and summaries of these interaction mechanisms can be found in Polk and Postow (1996). Much attention in recent years has focused on electric field interactions with the cell membrane and membrane-bound proteins. The remainder of this chapter addresses such interactions at several levels of organization.

Interest in the cell membrane as a locus for electric field interactions arises for two reasons. First, the plasma membrane is a primary sensory organ for the cell. Cell membrane processes mediate responses to biochemical signals as well as physiological electrical activity. Second, the extraordinary electrical properties of these lipid-based membranes result in the *amplification* of any electric field in the cell's environment for the following reasons. Electrical current flowing through biological tissues cannot easily pass through the membranes of cells in the tissue. An electrical potential that would drop gradually in a homogeneous medium along a length comparable to the cell dimension instead drops abruptly across the membrane. For purposes of illustration, consider a 10-μm diameter cell in the presence of a 1-V/m electric field. With no cell present, the potential drop over the 10-μm distance along the field direction would only be 0.01 mV. However, because the cell membrane is a poor conductor, most of the potential now drops across the membrane surfaces perpendicular to the direction of the field. The lipid bilayer structure of these membranes is only about 5 nm thick. Thus, the induced electric field across the membrane is roughly 0.005 mV divided by 5 nm, or 1000 V/m. Although this 1000 : 1 amplification is dramatic, some perspective is gained by noting that the cell's resting potential of about 50 mV produces a transmembrane electric field of 10^6 V/m.

Methods for predicting the transmembrane potentials produced by internal electric fields have been reported by several investigators (Schwan, 1983; Trachina and Nicholson, 1986; Gailey, 1996). Such potentials can be of the order of the electric field strength times the cell's dimension in the direction of the field (as in the example above). The maximum potential obtainable in large cells, however, is limited by the electrical properties of the cell membrane, cytoplasm and extracellular media. These properties can be summarized in terms of a characteristic length of the cell, λ, such that the maximum membrane potential inducible by a low-frequency electric field external to the cell is approximately $E\lambda$ (Gaylor et al., 1988). The characteristic length reflects the 'leakiness' of the cell membrane relative to the conductivity of the intracellular and extracellular media.

Cell membranes exhibit a high capacitance per unit area because of the atomic dimensions of the lipid bilayer. This capacitance of about 1 μF/cm^2 is a significant factor in determining the frequency dependence of induced membrane potentials. As noted earlier, because the membrane represents a barrier to current flow at low frequencies, the interior of the cell is shielded from external fields. However, as the capacitive reactance of the membrane is decreasing at the higher frequencies, the external currents now may flow *through* the cell, and produce small perturbations in the membrane potential. A notable exception is the observation that very intense electric fields, even of short duration and high-frequency content, are able to puncture cell membranes (Weaver, 1993; Gailey and Easterly, 1994). This process, known as *electroporation*, is widely utilized by molecular biologists for transfecting cells with genes. It is also thought to underlie the tissue damage occurring in severe electrical trauma (Gaylor et al., 1988). Although there has been some investigation of high-frequency, resonant interactions at the cell membrane level (Fröhlich, 1988), the present discussion will be confined to extremely low-frequency (ELF) interactions ($f < 100\,\text{Hz}$) that are frequently on a time scale comparable to physiological processes associated with electrically excitable membranes.

6.2.2 *Electric field effects on action potentials and cell electrical oscillations*

Most cells maintain a relatively steady resting potential in the neighborhood of $60\,\text{mV}$, but excitable tissues such as neurons and myocytes can transiently or periodically exhibit large excursions known as *action potentials*. As discussed earlier, low-frequency electric fields in biological tissues induce membrane potentials in the cells of the tissues. These induced potentials are generally small in comparison to a cell's resting potential, although they may be more significant for active excitable tissues as will be illustrated next.

The physiology of cellular excitability is complex, but results largely from the *nonlinear* properties of voltage-gated ion channels (Hille, 1992; Aidley, 1998). Individual cells contain thousands to millions of these channels categorized by the types of ion they transport or allow to pass through the membrane. Excitable cells typically contain large populations of Na^+ and K^+ channels that play a major role in the generation of action potentials.

Ion channels gate *stochastically*. That is, they open and close randomly such that only their average properties can be determined under a specific set of conditions. Voltage-gated ion channels in a lipid membrane respond to the membrane potential through changes in these average properties. For example, a channel gate with a mean open time of $10\,\text{ms}$ at a given membrane potential may express a mean open time of $5\,\text{ms}$ at another membrane potential. The actual time the gate stays open will vary, but after many openings will be seen to obey an exponential probability distribution with a rate constant equal to the inverse of the mean open time. A separate rate constant governs the closed-time statistics.

An action potential is initiated when the membrane potential exceeds a certain *threshold* determined by the properties of the ion channels. This threshold may be tens of millivolts above the resting potential of the cell, and is difficult to induce by means of an externally applied field. For example, a person must come into contact with a relatively high-voltage source before muscle contractions are induced (e.g., a defibrillator). The voltage source generates intense electric fields inside the muscle tissues, and the induced membrane potentials exceed the action potential threshold for the muscle cells (Reilly, 1992). Externally applied electric fields, where there is no direct contact with a voltage source, generally cannot induce sufficient membrane potentials to initiate action potentials. Instead, they produce only a small perturbation of the membrane potential. In the case of excitable tissues, such as muscle, these perturbations produce no immediately apparent effect.

More interesting is the case of neurons that are in the process of firing. Some insight into this process can be gained by modeling the neurons as *integrate-and-fire devices*. In this model, a steady endogenous current is assumed to flow into the neuron and continuously raise its potential until a threshold is reached and an action potential is generated. After firing, the potential is reset to some lower level and the process begins again. This repetitive firing or oscillatory behavior is common in the central nervous system and some sensory systems. Such oscillatory systems are far more sensitive to small perturbations than are excitable tissues that require induced membrane potentials of 20 mV or more to initiate an action potential. A detailed analysis reveals that these neurons represent nonlinear, *limit-cycle oscillators*, and can be influenced by extremely weak perturbations (Fohlmeister *et al.*, 1980). The discovery of the sensitivity of limit-cycle oscillators to perturbations was made in the 1600s by Huygens: he reported that two pendulum clocks would self-synchronize through the tiny vibrations of the pendula movements transmitted through the wall on which they were hanging (Bak, 1986). For the theory of perturbation responses of limit-cycle oscillators, see Kaiser, Chapter 1, this volume.

In the case of an integrate-and-fire neuron, we can visualize its sensitivity by considering the point in time when the potential is ramping up and is just about to reach the firing threshold. At this point, a small perturbation in one direction could push it over the threshold ahead of schedule. A small perturbation in the other direction would produce a delay in reaching the threshold. In either case, we expect the membrane potential perturbation to affect the *timing* of the action potential. Such effects have been clearly demonstrated *in vitro*. For example, a constant (DC) electric field has been used to suppress epileptiform activity in hippocampal slices (Gluckman *et al.*, 1996), while oscillating fields have been used to entrain or phase lock cardiac myocytes (Lazrak *et al.*, 1997). In the latter case, the applied electric fields alternated with a frequency near the natural beating frequency of the myocytes. Earlier studies by Wachtel (1979) demonstrated a similar effect on spontaneously firing *Aplysia* neurons.

Extensive studies of periodic stimulation of a nonlinear biological oscillator have been reported (Guevara *et al.*, 1981). These investigators injected current pulses into beating chick cardiac myocytes and studied a range of complex nonlinear interactions including phase locking. However, there is a distinct difference between these current injection experiments and the electric field exposure experiments described above. When electric *current* is injected directly into a cell through a microelectrode, the injected charge causes a *unipolar* modification of the membrane potential; that is, the membrane potential of the entire cell changes in the same direction and by the same amount. In contrast, electric *field* exposure *hyperpolarizes* one side of a cell and *depolarizes* the opposite side. The effects of this *bipolar* stimulation require slightly more complex models for analysis (Gailey *et al.*, 1996).

Nonlinear interactions as described above can be studied with numerical models, if the details of the biological process in question are understood. In the case of chick cardiac myocytes, detailed models of the ionic channels producing the spontaneous activity have been developed (Kowtha *et al.*, 1994). By incorporating this model into an electric field-exposure model, some of the expected interactions can be examined. Figure 1 shows the interbeat interval (IBI) for chick cardiac myocytes exposed to an electric field that induces a 10-mV perturbation of the membrane potential at different frequencies. For frequencies far below the natural frequency of the myocytes, the IBI changes gradually at twice the frequency of the stimulation. Phase locking occurs when the electric field frequency is near the natural frequency, and a different behavior occurs when the electric field frequency is three times higher than the natural frequency. On the basis of mathematical models of this type, there is no fundamental lower limit with respect to the magnitude of perturbation that is still capable of influencing a nonlinear oscillator. Any 'real-world' system, however, must contend with noise, an issue that will be addressed in Section 6.3.

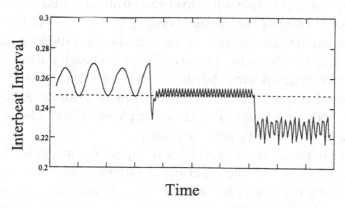

Time

Figure 1. Predicted interbeat interval (IBI) of chick cardiac myocytes exposed to an electric field that induces a 10-mV peak membrane potential perturbation. In the first third of the graph, the electric field has a frequency 20 times less than the natural beating frequency of the myocyte. The myocyte nearly phase locks in the middle third of the graph when the electric field frequency is changed to a value close to the myocyte's natural beating frequency. In the final third, the electric field frequency is increased by a factor of 3. The dashed line indicates the IBI when no electric field is applied.

6.3 Competing with noise

Extracting weak signals from the background of noisy processes is an endeavor of fundamental importance in many different disciplines. Noise is an inescapable reality that occurs in all physical systems, and the detection of a weak signal requires not only a device that will respond to the signal, but the ability to distinguish the signal from all relevant sources of noise. An analogy will be used to establish some common terminology. Assume a person is listening to their favorite radio station as they leave town on an automobile trip. At first the program is received clearly, but as the car travels farther from the radio transmitter, the radio signal becomes weaker, and the quality of the sound from the radio degrades. We will assume that the problem arises due to electronic noise within the radio. All electronic processes produce noise and, when the radio signal becomes weak enough, the magnitude of the noise is comparable to that of the signal. In engineering terms, the signal-to-noise ratio (SNR) is approaching unity. For good reception, we want to have an SNR as high as possible.

Next, assume the driver is a physicist who happens to have a container of liquid helium in the car. Because noise processes are generally reduced at lower temperatures, he immerses his radio in the liquid helium. At temperatures near absolute zero, the *internal noise* of the radio is greatly reduced and his SNR once again improves to acceptable levels. The sound quality soon degrades again, however, as he continues to travel away from the radio transmitter. The problem now is noise in the antenna system, so his next tactic is to install several antennas on the car and to average the outputs from each before routing the signal to the radio. Because the noise in each antenna is random and independent (uncorrelated), he can reduce the antenna noise by the square root of the number of antennas he installs. A total of 16 antennas, for example, improve the SNR by a factor of 4. Thus, our physicist has reduced noise and improved signal detection by *array averaging*.

Continuing to travel, he is again plagued by poor signal quality. With all the antennas on his car's roof, the radio signal, although weaker, is still measurable, and the other measures taken have reduced the radio's *internal* noise as much as possible. The problem now is that his signal is so weak that it is comparable to that of distant radio stations broadcasting at the same frequency and to that of noise in the environment. The noise may arise from weather-related electrical phenomena and emissions from electronic devices in the area. Faced with this *external* noise, our physicist is now out of luck. No further improvements to his system will allow him to improve the SNR for his radio signal.

Biological systems face these same limits in attempts to detect weak signals. Aside from our physicist's liquid helium trick, biological systems have employed every means discussed, and others, to improve signal detection. These principles are clearly demonstrated in the system used by the shark to detect the incredibly weak electric fields produced by the electrophysiological activity of their prey (Kalmijn, 1982). Ultimately, the shark's detector is based on ion channels in specialized cells. But as we will discuss, ion channels are very noisy devices. The shark overcomes this internal noise by using many cells closely coupled in an array (the ampulla of Lorenzini). Signal strength is improved by using conductive canals that open on the shark's body surface at positions up to 10 cm apart. An array of these organs and canals further improves sensitivity, and averaging or other signal processing of these separate signals in the central nervous system may accomplish additional noise reduction.

Sources of noise in biological systems are numerous, but evolution has apparently produced sensory systems that operate close to theoretical limits (Bialek, 1987; Lowe and Gold, 1995). Since humans have no known sensory system for detecting weak electric fields, we now ask how these fields might be detected incidentally by neurons. In keeping with our focus on cell membrane interactions and ion channels in particular, we will address ion channel noise in some detail and comment only briefly on other sources of noise occurring at the cell membrane. Reviews of these sources of membrane noise, including thermal voltage and current noise, shot noise, and $1/f$ noise are available (DeFelice, 1981; Bezrukov and Vodyanoy, 1994).

Thermal electrical noise, in particular, has received attention relative to minimum detection limits. All materials, including biological tissues and membranes, produce thermal electrical noise (Johnson, 1927). A number of early analyses of minimum detection limits for cells were based on thermal voltage noise occurring at the cell membrane (Barnes and Seyed-Madani, 1987; Weaver and Astumian, 1990; Adair, 1991). These analyses addressed the question of the SNR by assigning the induced membrane potential to be the signal of interest, but they did not specify a particular sensor. If membrane ion channels are the sensors of interest, however, a direct comparison of induced membrane potential to thermal voltage noise may not be appropriate. A more detailed model revealed that the thermal voltage noise occurring in different ion channels is poorly correlated when many channels are present (Gailey, 1999a). Thus, array averaging of thermal voltage noise is predicted and its effects are expected to be insignificant in comparison with the effects of ion channel noise.

6.3.1 The role of ion channel noise in electric field detection

Channel noise results from the random opening and closing of membrane-bound ion channels. Consider, for example, a cell with a potential difference between the inside and the outside and only two ion channels in the membrane. If the probability of each channel being open is one half, then, on average, one of the two channels will be open. Channel gating is stochastic, however, thus it is also possible for both channels or neither channel to be open. The current through the membrane will fluctuate from its expected value (with one channel open), to twice this value when both channels are open, or to zero when neither channel is open. A similar process occurs when large numbers of channels are present, and the characteristics of the fluctuations are predictable as long as the statistical properties and rate constants of the channel gates are known. Returning to the integrate-and-fire model of a neuron, we now imagine that the charging of the membrane, or ramping up of the voltage, does not occur smoothly. It fluctuates due to the channel fluctuations that cause the charging current to fluctuate. In Section 6.2.2, we discussed how a small perturbation could advance or delay the firing time of an integrate-and-fire neuron. Now it is clear that the time between nerve firings (action potentials) will vary randomly due to channel noise. This relationship was first described mathematically by Gerstein and Mandelbrot (1964).

In order for a perturbing signal, for example a membrane potential induced by an electric field, to influence a spontaneously firing neuron, it must produce an effect comparable with or exceeding the fluctuations in firing time caused by channel noise. This is the basic signal-to-noise limit discussed earlier. We can tackle this problem mathematically by making some simplifying approximations (for details of the mathematical analysis, see Gailey, 1999b). First, ion channels will be treated as two-state devices that exist in either an opened or a closed state. Although channel behavior is well predicted by a model that includes four separate gates in series (Hodgkin and Huxley, 1952), many of the essential features can be observed using a two-state model (DeFelice, 1981). We will also restrict the analysis to constant, rather than alternating, perturbations. However, the results will also apply to low-frequency oscillating perturbations under the condition that any effects must occur in a time period of less than half the period of the perturbing signal. A 10-Hz signal, for example, undergoes a complete oscillation in 100 ms. The analysis below will apply for time periods of less than 50 ms, during which the perturbing signal remains consistently either positive or negative. If the SNR is 1 or greater during this time period, then longer time scale effects such as phase locking are also possible. These effects occur because of the repetitive influence of the (alternating) perturbing signal inter-acting with the natural oscillations of the neuron.

A membrane potential perturbation will slightly alter the opening and closing rate constants of a voltage-gated ion channel, as discussed earlier. Interestingly, this effect is a continuous function of the membrane potential, so there is no lower limit of the magnitude of perturbation capable of changing these rate constants. The question is whether or not the change is significant compared to the channel noise. The change in the expected fraction of open channels due to a perturbing potential can be approximated using a Boltz-mann function (Hille, 1992). Assuming that the membrane potential is constant over the short period of interest, the Boltzmann function also predicts the change in expected current flow through the membrane due to the perturbing potential. Integrating this effect over time yields the induced change in total charge transferred. This change in total charge is considered to be the signal of interest. In the integrate-and-fire model, it could be the change that either pushes the neuron over its firing threshold early or delays its firing.

The next step is to find the magnitude of the fluctuations in total charge transferred due to stochastic channel gating. This quantity is the 'noise' in our signal-to-noise analysis and can be obtained from a random walk analysis of the channel fluctuations. By setting the signal (derived from the Boltzmann function above) equal to the noise from the random walk analysis, we obtain a mathematical condition that must be met in order for a perturbing signal to meet the SNR = 1 criterion (Gailey, 1999b). This condition is a function not only of the magnitude of the signal but of the channel rate constants, the number of channels and the time period involved. The number of channels is a factor because the relative magnitude of the channel noise decreases as more channels are added. The duration of the perturbing signal is important because the current fluctuations tend to average out over time. Choosing biologically reasonable rate constants and setting the induced membrane potential to 1 μV, one can examine the number of channels and the time required for detecting the signal. Figure 2 illustrates the results of this analysis for some specific conditions approximating an array of Na^+ channels.

It is apparent from Figure 2 that the number of channels and time required to detect a weak membrane perturbation are inversely related. With more channels, less time is required. Although the calculations were performed for a 1-μV signal, the results scale with signal strength. For example, while an array of 10^6 channels requires about 0.1 s to detect a 1-μV signal, only 0.01 s would be required to detect a 10-μV signal. Another result, not shown in Figure 2, is that detection limits are sensitive to the gating rates of the ion channels. Faster gating rates provide more noise averaging during a time interval so that less

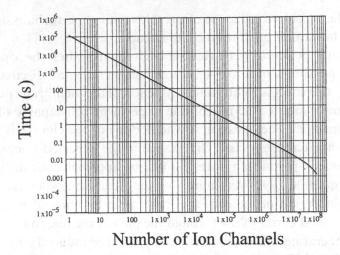

Number of Ion Channels

Figure 2. Number of ion channels and duration of applied signal required for detecting a membrane potential perturbation of 1 µV. Using a Boltzmann approximation for sodium (Na^+) channels ($z = 6$), the change in total charge transferred is calculated for the induced potential of 1 µV. This value is compared with expected fluctuations in total charge transferred as a result of random channel gating. The graph shows the number of channels and time required for the induced effect to equal the expected fluctuations (signal-to-noise ratio, SNR = 1). Note the change in requirements for SNR = 1 when the channel gating time is varied. (For details, see text in Section 6.3.1.)

time and/or number of channels are required to detect a given signal. Although the perturbing signal in this analysis is treated as a constant, the results are applicable to oscillating signals, as discussed earlier. In order to detect a 10-Hz signal, the SNR > 1 criterion must be met in less than 50 ms. In Figure 2 one can observe that the detection of a 10-Hz, 1-µV signal requires about 10^7 ion channels, while the detection of a 60-Hz, 1-µV signal would require 10^8 or more channels.

It is interesting to compare these results with well-established findings in the biophysical literature. Some 30 years ago, Verveen et al. (1967) devised a measure called the relative spread of threshold (RS) to describe the variation in the probability of the firing of a nerve relative to the magnitude of stimulus needed to have it fire 50% of the time. Later models by Clay and DeFelice (1983) and Rubenstein (1995) demonstrated that experimental measurements of this variability can be predicted quite well by numerical studies of channel noise. According to Rubenstein (1995), the RS for an ensemble of 10^7 ion channels is about 0.0009. Assuming a threshold stimulation of about 20 mV, this implies that the probability of nerve firing changes by one standard deviation for a 20-µV change in the magnitude of the stimulus. If the results presented in Figure 2 are scaled to 20-µV signal strength, the time required to detect the signal is about 1 ms – in good agreement with experimental findings and Rubenstein's model. Although the RS measurements and calculations were not designed to address detection limits per se, the general agreement between them and the analysis presented here support the validity of this approach.

The next question that must be addressed concerns the biological relevance of these results. While we have shown that the detection of very small perturbations is theoretically possible, and supported by *in vitro* nerve stimulation experiments, their possible impact on living systems is not clear. Is a tiny alteration in the timing of a nerve impulse important? This question, for reasons quite apart from those addressed here, is currently the subject of intense investigations. For example, researchers in the neuroscience community are attempting to understand how information is encoded in neuronal signals (Rieke *et al.*, 1997). The basic problem is that nerve signals are typically very noisy, and it is not clear whether this variability represents information encoded in a way that we do not understand, or is just an unavoidable nuisance with which organisms must contend. Some interesting results have been reported in this area recently and we will return to this question in the next section.

Another question relates to the number of ion channels needed for detecting a weak signal. Is it possible to have 10^8 or more ion channels in an array? Most cells do not have this many channels, and because of the way electric fields interact with cells, not all the channels in a cell will experience the same membrane potential perturbation. The only way to reach the numbers of channels required to detect very weak signals is for the channels from many cells to work together. Such group activity would simply represent a larger-scale example of the array averaging discussed above. The difficult question is *how* the cells could work together. One possibility is for many cells to be connected together by *gap junctions*. These junctions are tiny pores that provide an ionic pathway directly from the cytoplasm of one cell to another and, thus, connect them *electrically*. A well-studied example of this behavior is the aggregates of chick cardiac myocytes mentioned earlier (DeHaan and DeFelice, 1978; Kowtha *et al.*, 1994; Lazrak *et al.*, 1997). Myocytes form gap junctions when they come into contact with each other, and connect so well that the aggregate behaves electrically as a giant cell with a *single* beating frequency. Cells that initially beat at their own frequencies synchronize once they are brought into contact and form gap junctions. Clay and DeHaan (1979) studied the rhythmic properties of these aggregates and found that the variability or noise in beating rate was inversely proportional to the square root of the number of cells. Larger aggregates, with more cells and more ion channels, beat more regularly. Their findings also support the analysis of detection limits presented above, and demonstrate further that the principles can be extended to groups of cells.

Perhaps more intriguing is the possibility of groups of neurons working together. With some notable exceptions, however, gap junctions generally do not connect neurons. If they do not *directly* share ionic currents (as in the case of the myocytes), is it possible that detection limits will still be reduced for groups of neurons? The answer to this question is the focus of the next section.

6.4 Signal detection in systems with long-range coherence

Synchrony among groups of neurons in the central nervous system is a well-established phenomenon (see, e.g., Traub and Miles, 1991). In a larger context, such synchronous behavior is an example of *long-range coherence* – the organized, cooperative activity of large populations of interacting units. Although we have tread a long path from the specifics of ion channels and noise to electric field interactions, our questions can now assume a somewhat more general nature. Instead of asking only whether interacting neurons can detect weaker signals than single neurons, we can ask, in general, whether systems with long-range coherence are more sensitive to weak perturbations. The relevance of this question to neuronal systems will be developed, but its possible importance for biological systems as a whole should first be noted. A moment's reflection reveals that biological systems should be defined in terms of coherence. With some 10^{18} cells in the human body, most of them about five orders of magnitude smaller than the body as a whole, the unified functionality of the body suggests that long-range coherence is one of our most defining characteristics. If systems with long-range coherence exhibit a peculiar sensitivity to weak perturbations, the implications for carefully planned therapeutic interventions are intriguing.

Interesting examples of long-range coherence have received attention in recent years. Particularly surprising are oscillatory chemical reactions such as the horseradish peroxidase system (see Larter *et al.*, Chapter 2, and Walleczek and Eichwald, Chapter 8, this volume). Traditionally, chemical reactions have been thought of as perfect examples of statistically random processes; that is, thermal energy causes molecules to move rapidly and erratically, and they collide with each other in a completely random fashion. Thus, the time course of a chemical reaction follows a statistically predictable path. In contrast, the molecules in chemical systems with long-range coherence undergo *synchronized transitions*. Considering the enormous numbers of molecules involved and the large relative distances between molecules, such synchrony is remarkable. A more familiar example of long-range coherence is the laser. Here, electrons undergo transitions between energy levels when photons are absorbed or emitted. Under appropriate conditions, the emitted photons provide coupling between electrons and synchronize the transitions. The result is the coherent, highly monochromatic light output of the laser. In both examples, some form of coupling between elements forces the random, independent fluctuations of the individual *microscopic* elements to give way to *macroscopic* synchrony. Order emerges from disorder as coupling overcomes noise.

Mathematical descriptions of long-range coherence can be quite complex. The difficulty arises from complexities in the behavior of the individual elements, the nature of their intrinsic fluctuations, and the form of the coupling. Modeling coupled neurons in detail, for example, would require a representation of the ionic currents, the ion channel noise, and the complex network of excitatory and inhibitory synapses between them. All this complexity can easily obscure the fundamental issue of weak signal detection. In order to avoid such problems, we turn to a highly simplified model of coupled, oscillatory neurons described by Shinomoto and Kuramoto (1986). The neurons in this model are simply active rotators that can be thought of as runners circling a track. Each moves on their own with some random fluctuation in speed, but they experience an attractive or repulsive force with each other depending on their proximity. The coupling in this model is considered *global* because each element is equally coupled to all other elements, thus eliminating all spatial and geometrical considerations associated with structure. Although a highly idealized model of coupled neurons, it captures enough of the essential features to address the importance of long-range coherence in weak signal detection.

By varying either the coupling between elements or the magnitude of noise in each element, the system passes through a phase transition. When the noise magnitude is high and coupling is weak, the system is incoherent and the mean phase of all the elements is constant in time. At lower noise magnitude or stronger coupling, the system becomes coherent and a large fraction of the elements are synchronized so that the mean phase of the system varies *periodically* in time. This emergence of macroscopic, global synchrony after the phase transition is similar to the synchrony observed in large populations of neurons in the central nervous system (Kelso, 1995). Using this model, we have investigated the influence of a weak signal on the behavior of a system of globally coupled neurons (Gailey *et al.*, 1997a). In our model, the signal is a weak perturbation of the coupling between elements. Figure 3 shows a simulation in which the perturbation is applied with too much noise in each element for synchrony to occur. Very little change in this incoherent system is observed

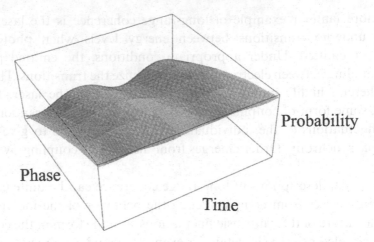

Figure 3. Numerical simulation of a system of globally coupled, noisy phase oscillators illustrating the fraction (probability) of oscillators at various phases as a
function of time. A small, low-frequency signal is applied to the coupling coefficient
between oscillators, but the noise in the system is too large to allow the oscillators to
synchronize at their natural frequency. The relatively smooth distribution indicates
that the phases of the oscillators are largely uncorrelated in time and are not responding to the applied signal.

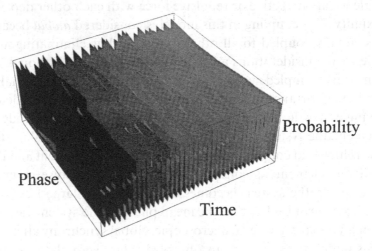

Figure 4. Same system of oscillators shown in Figure 3, but with a *low* value of
intrinsic noise. Because the noise is low, the oscillators are strongly synchronized at
their natural frequency as can be observed along the front edge of the plot. The applied
signal again produces little effect on the system.

due to the perturbing signal. In Figure 4, the noise is low enough for the system
to synchronize, and the perturbing signal again has little effect on the overall

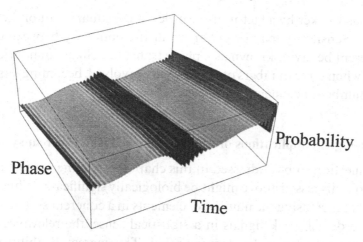

Figure 5. The system of oscillators in Figures 3 and 4 is now modeled with an *intermediate* noise level and the same low-frequency signal applied to the coupling coefficient. During the negative-going part of the applied signal, the system is desynchronized (smooth portions of the plot). Synchrony occurs only during the positive-going portion of the applied signal, where the overall phase of the system is seen to oscillate at its natural frequency. This effect is similar to stochastic resonance in that a weak signal is efficiently transmitted through the system only above some optimum noise level.

behavior of the system. This situation changes, however, in Figure 5 where the noise magnitude poises the system very close to the phase transition between synchronized and nonsynchronized states. The small perturbing signal now produces the very large effect of moving the system in and out of synchrony at a regular rate.

The effect observed in Figures 3 to 5 is reminiscent of *stochastic resonance* in that signal detection is enhanced when noise is varied to some optimum magnitude (see Moss, Chapter 10, this volume). Here, however, the noise we are interested in is *internal* noise rather than externally applied noise. Much of the relevant noise in biological systems is internal noise due to random processes such as channel gating, and can play a constructive role in signal processing (Bezrukov and Vodyanoy, 1997; Gailey *et al.*, 1997b). The results presented above are based on an infinite number of interacting elements, but we have also studied finite populations and observed similar results (Gailey *et al.*, 1997a). As expected, the sensitivity of the system increases in proportion to the square root of the number of elements, similar to the results for arrays of ion channels discussed earlier. Thus, the answer to the question about detection limits posed earlier is affirmative: systems with long-range coherence can detect much weaker signals than can individual elements within the system. Compared with detection limits for a single element, coherent systems can detect signals weaker by a factor of the order of the square root of the number of elements. Sensitivity may be enhanced at the boundary between coherent and incoherent behavior known as a phase transition. Such boundaries can be very sharp when large numbers of elements are involved, becoming less distinct when the number of elements is reduced.

6.5 Biological implications of small perturbations to coherent systems

The final question to be addressed in this chapter is whether or not the small perturbations discussed above might be biologically significant. While we have observed that increasing the number of elements in a coherent system increases its ability to detect weak signals in a statistical sense, the relevance of small perturbations remains to be demonstrated. The increased ability of large systems to detect small signals is a result of the averaging of the noise associated with each element through the coupling between elements. This noise averaging reduces the magnitude of fluctuations in the global behavior of the system. For example, the firing regularity of a large, coupled system of identical oscillating neurons should be much better than that of a similar small system. Smaller noise-driven fluctuations in the oscillation period suggest that smaller perturbations compared to the timing of oscillations can theoretically be detected. Therefore, one way to pose the question of relevance is in terms of the importance of the precision in firing times of neurons during information processing.

How critical is the exact timing of a nerve impulse? This question is hotly debated at present, but recent results have suggested that information may be encoded in the precise timing of neuronal spikes (de Ruyter van Steveninck *et al.*, 1997). Traditionally, neuroscientists have assumed that information was encoded only in the *firing rate* of neurons (Ferster and Spruston, 1995), but recognition of the improved information-carrying ability of *spike timing* has focused attention away from simple rate encoding. For example, Berry *et al.* (1997) have observed timing jitter in salamander retinal spike trains of less than 1 ms, noting that the timing of the spikes conveys several times more information than the spike count. Similar precision was observed in the rat cerebral cortex in response to time-varying stimulation (Mainen and Sejnowski, 1995). Another current dilemma is how the central nervous system decodes important signals in the presence of so much *external* noise. What we interpret as noise in this case, however, is actually ongoing patterns of electrical activity occurring simultaneously in neighboring or co-located neuronal networks (Ferster, 1996). In contrast to internal noise, external noise may be highly correlated among neurons in a network, thus, this noise cannot be minimized by averaging. Pointing out that such signals are actually not noise at all, Arieli *et al.* (1996) suggested that neuronal processing is specifically designed to process the combination of evoked and ongoing signals. What investigators have so far called noise may turn out to be the contextual background necessary for the complete interpretation of incoming signals. If this observation is indeed accurate, then only internal noise must be considered when determining minimum detection limits. As described above, coherent function among many neurons in a network reduces such noise substantially.

The study of long-range coherence in neuronal function is gaining momentum. Alonso *et al.* (1996) reported phase locking among large populations of neurons in a cat visual cortex, and Arieli *et al.* (1995) have measured coherence between sites up to 6 mm apart in the same system. Such networks can oscillate at high frequencies, with attention focused recently on the 40–70 Hz frequency band (Jefferys *et al.*, 1996). Such oscillations can involve billions of synaptically coupled neurons, and have been observed during rapid eye movement (REM) sleep in humans (Llinas and Ribary, 1993; Steriade *et al.*, 1993). Kelso (1995) has used magnetoencephalography (MEG) to study coherent modes of activity in the human brain during various types of stimulation. He emphasizes the shift between modal patterns as a possible key to shifting cognitive processes (see also Ding *et al.*, Chapter 4, this volume). These changes in modal patterns are phase transitions – a real-world version of the transitions described earlier in terms of the coupled rotator model. As has been pointed out, the greatest sensitivity to perturbation occurs at the phase transition boundary. This observation is significant because of the evidence now suggesting that phase transitions underlie many neuronal processes. The rapid onset and offset of synchrony among populations of interacting neurons has been characterized in the cat visual cortex (Gray *et al.*, 1992). Thus, it is possible that long-range coherence may be the physical correlate of unified consciousness (Kahn *et al.*, 1997).

This chapter has outlined a preliminary but scientifically sound basis for the possibility of using relatively weak electric fields to perturb and perhaps control neuronal function. Indeed, investigation of electric field control of epilepsy is already underway (Jerger and Schiff, 1995; Gluckman *et al.*, 1996; see also Ditto and Spano, Chapter 15, this volume). The primary limiting factor is noise, but we have shown that internal noise yields to coherence in coupled systems. Sensitivity is enhanced at phase transition boundaries, and the rapid crossing of such boundaries may form the physical basis of cognition. What appears to be external noise may instead be an intrinsic component of unified information processing. Considerable work has been performed to model the interactions of electric fields with cells, but it is apparent that a more detailed picture of how fields interact with coupled networks is needed. As understanding of the mechanisms involved in rapidly establishing neuronal coherence develops, the ability to influence such transitions will also grow. One might envision electric field therapies for a variety of neurological disorders, along with a scientific basis for evaluating the possibility of undesired interactions due to environmental fields. A comprehensive view of living organisms as multilevel, dynamically changing, coherent systems has yet to be developed. Such an understanding promises to yield deep insights and the possibility of interventions cast in the body's own, unifying language.

Acknowledgment

This work was supported in part by the US Department of Energy, Office of Energy Management.

References

Adair, R. K. (1991) Constraints on biological effects of weak extremely-low-frequency electromagnetic fields. *Phys. Rev. A* **43**: 1039–1048.

Aidley, D. J. (1998) *The Physiology of Excitable Cells*, fourth edition. Cambridge: Cambridge University Press.

Alonso, J.-M., Usrey, W. M. and Reid, R. C. (1996) Precisely correlated firing in cells of the lateral geniculate nucleus. *Nature* **383**: 815–819.

Arieli, A., Shoham, D., Hildesheim, R. and Grinvald. A. (1995) Coherent spatiotemporal patterns of ongoing activity revealed by real-time optical imaging coupled with single-unit recording in the cat visual cortex. *J. Neurophysiol.* **73**: 2072–2093.

Arieli, A., Sterkin, A., Grinvald, A. and Aertsen, A. (1996) Dynamics of ongoing activity: explanation of the large variability in evoked cortical responses. *Science* **273**: 1868–1871.

Astumian, R. D., Adair, R. K. and Weaver, J. C. (1997) Stochastic resonance at the single-cell level [letter]. *Nature* **388**: 632–633.

Astumian, R. D., Weaver, J. C. and Adair, R. K. (1995) Rectification and signal averaging of weak electric fields by biological cells. *Proc. Natl. Acad. Sci. USA* **92**: 3740–3743.

Bak, P. (1986) The devil's staircase. *Physics Today* **39**: 38–45.

Barnes, F. S. (1996) Extremely low frequency (ELF) and very low frequency (VLF) electric fields: rectification, frequency sensitivity, noise, and related phenomena. In *Handbook of Biological Effects of Electromagnetic Fields* (ed. C. Polk and E. Postow), pp. 121–138. Boca Raton, FL: CRC Press.

Barnes, F. S. (1988) Mechanism of interaction of magnetic fields with biological systems. *IEEE Trans. Magnetics* **24**: 2101–2104.

Barnes, F. S. and Seyed-Madani, M. (1987) Some possible limits on the minimum electrical signals of biological significance. In *Mechanistic Approaches to Interactions of Electric and Electromagnetic Fields with Living Systems* (ed. M. Blank and F. Findl), pp. 339–347. New York: Plenum Press.

Berry, M. J., Warland, D. K. and Meister, M. (1997) The structure and precision of retinal spike trains. *Proc. Natl. Acad. Sci. USA* **94**: 5411–5416.

Bezrukov, S. M. and Vodyanoy, I. (1994) Noise in biological membranes and relevant ionic systems. In *Biomembrane Electrochemistry* (ed. M. Blank and I. Vodyanoy), pp. 375–399. Washington, DC: American Chemical Society.

Bezrukov, S. M. and Vodyanoy, I. (1997) Stochastic resonance in non-dynamical systems without response thresholds. *Nature* **385**: 319–321.

Bialek, W. (1987) Physical limits to sensation and perception. *Annu. Rev. Biophys. Biophys. Chem.* **16**: 455–478.

Clay, J. R. and DeFelice, L. J. (1983) Relationship between membrane excitability and single channel open-close kinetics. *Biophys. J.* **42**: 151–157.

Clay, J. R. and DeHaan, R. L. (1979) Fluctuations in interbeat interval in rhythmic heart-cell clusters. *Biophys. J.* **28**: 377–389.

Dawson, T. W., Caputa, K. and Stuchly, M. A. (1997) A comparison of 60 Hz uniform magnetic and electric induction in the human body. *Phys. Med. Biol.* **42**: 2319–2329.

de Ruyter van Steveninck, R. R., Lewen, G. D., Strong, S. P., Koberle, R. and Bialek, W. (1997) Reproducibility and variability in neural spike trains. *Science* **275**: 1805–1808.

DeFelice, L. J. (1981) *Introduction to Membrane Noise*. New York: Plenum Press.

DeHaan, R. L. and DeFelice, L. J. (1978) Electrical noise and rhythmic properties of embryonic heart cell aggregates. *Fed. Proc.* **37**: 2132–2138.

Dudek, F. E., Snow, R. W. and Taylor, C. P. (1986) Role of electrical interactions in synchronization of epileptiform bursts. *Adv. Neurol.* **44**: 593–617.

Durney, C. H., Johnson, C. J. and Massoudi, H. (1975) Long-wavelength analysis of plane wave irradiation of a prolate spheroid model of man. *IEEE Trans. Microwave Theory Tech.* **23**: 246–253.

Ferster, D. (1996) Is neural noise just a nuisance? *Science* **273**: 1812.

Ferster, D. and Spruston, N. (1995) Cracking the neuronal code. *Science* **270**: 756–757.

Feynman, R. P., Leighton, R. B. and Sands, M. (1964) *The Feynman Lectures on Physics*. Reading, MA: Addison-Wesley.

Fohlmeister, J. F., Adelman, W. J. and Poppele, R. E. (1980) Excitation properties of the squid axon membrane and model systems with current stimulation. *Biophys. J.* **30**: 79–97.

Foster, K. R. and Schwan, H. P. (1996) Dielectric properties of tissues. In *Handbook of Biological Effects of Electromagnetic Fields* (ed. C. Polk and E. Postow), pp. 25–102. Boca Raton, FL: CRC Press.

Fröhlich, H. (ed.) (1988) *Biological Coherence and Response to External Stimuli*. New York: Springer-Verlag.

Gailey, P. C. (1996) Comparison of voltage signals induced by power frequency fields to thermal electrical noise at the cell membrane. Ph.D. thesis, University of Michigan (University Microfilms No. 9621968).

Gailey, P. C. (1999a) Correlation of thermal electrical noise in ion channel arrays. In

Electricity and Magnetism in Medicine and Biology (ed. F. Bersani), pp. 227–230. New York: Plenum Press.

Gailey, P. C. (1999b) Membrane potential and time requirements for detection of weak signals by voltage-gated ion channels. *Bioelectromagnetics* **20** (Suppl. 4): 102–109.

Gailey, P. C. and Easterly, C. E. (1994) Cell-membrane potentials induced during exposure to EMP fields. *Electr. Magnetobiol.* **13**: 159–165.

Gailey, P. C., Imhoff, T., Chow, C. C. and Collins, J. J. (1997a) Noise-enhanced information transmission in a network of globally coupled oscillators. In *4th SIAM Conference on Applications of Dynamical Systems*, Snowbird, UT, USA.

Gailey, P. C., Lazrak, A., Wachtel, H. and Griffin, G. D. (1996) Nonlinear dynamics and phase locking of cardiac myocytes exposed to weak ELF electric fields. In *Proc. 18th Annu. Meet. Bioelectromagnetics Soc.*, pp. 136–137, 9–14 June, Victoria, Canada.

Gailey, P. C., Neiman, A., Collins, J. J. and Moss, F. (1997b) Stochastic resonance in ensembles of non-dynamical elements: the role of internal noise. *Phys. Rev. Lett.* **79**: 4701–4704.

Gandhi, O. P. and Chen, J.-Y. (1992) Numerical dosimetry at power-line frequencies using anatomically based models. *Bioelectromagnetics Suppl.* **1**: 43–60.

Gaylor, D. C., Prakah-Asante, K. and Lee, R. C. (1988) Significance of cell size and tissue structure in electrical trauma. *J. Theor. Biol.* **133**: 223–237.

Gerstein, G. L. and Mandelbrot, B. (1964) Random walk models for the spike activity of a single neuron. *Biophys. J.* **4**: 41–68.

Gluckman, B. J., Neel, E. J., Netoff, T. I., Ditto, W. L., Spano, M. L. and Schiff, S. J. (1996) Electric field suppression of epileptiform activity in hippocampal slices. *J. Neurophysiol.* **76**: 4202–4205.

Gray, C. M., Engel, A. K., Konig, P. and Singer, W. (1992) Synchronization of oscillatory neuronal responses in cat striate cortex: temporal properties. *Vis. Neurosci.* **8**: 337–347.

Guevara, M. R., Glass, L. and Shrier, A. (1981) Phase locking, period-doubling bifurcations, and irregular dynamics in periodically stimulated cardiac cells. *Science* **214**: 1350–1353.

Hille, B. (1992) *Ionic Channels of Excitable Membranes*. Sunderland, MA: Sinauer.

Hodgkin, A. L. and Huxley, A. F. (1952) A quantitative description of membrane current and its application to conduction and excitation in nerve. *J. Physiol. (Lond.)* **117**: 500–544.

Hotary, K. B. and Robinson, K. R. (1994) Endogenous electrical currents and voltage gradients in *Xenopus* embryos and the consequences of their disruption. *Dev. Biol.* **166**: 789–800.

Iskander, M. F. (1992) *Electromagnetic Fields and Waves*. Englewood Cliffs, NJ: Prentice-Hall.

Jaffe, L. F. and Nuccitelli, R. (1977) Electrical controls of development. *Annu. Rev. Biophys. Bioeng.* **6**: 445–476.

Jefferys, J. G., Traub, R. D. and Whittington, M. A. (1996) Neuronal networks for induced '40-Hz' rhythms. *Trends Neurosci.* **19**: 202–208.

Jerger, K. and Schiff, S. J. (1995) Periodic pacing of an in vitro epileptic focus. *J. Neurophysiol.* **73**: 876–879.

Johnson, J. B. (1927) Thermal agitation of electricity in conductors. *Phys. Rev.* **29**: 367–368.

Kahn, D., Pace-Schott, E. F. and Hobson, J. A. (1997) Consciousness in waking and dreaming: the roles of neuronal oscillation and neuromodulation in determining the similarities and differences. *Neuroscience* **78**: 13–38.

Kalmijn, A. J. (1982) Electric and magnetic field detection in elasmobranch fishes. *Science* **218**: 916–917.

Kelso, J. A. S. (1995) *Dynamic Patterns*. Cambridge, MA: MIT Press.

Kowtha, V. C., Kunysz, A., Clay, J. R., Glass, L. and Shrier, A. (1994) Ionic mechanisms and nonlinear dynamics of embryonic chick heart cell aggregates. *Progr. Biophys. Mol. Biol.* **61**: 255–281.

Lazrak, A., Griffin, G. and Gailey, P. C. (1997) Studying electric field effects in cardiac myocytes. *BioTechniques* **23**: 736–741.

Llinas, R. and Ribary, U. (1993) Coherent 40-Hz oscillation characterizes dream state in humans. *Proc. Natl. Acad. Sci. USA* **90**: 2078–2081.

Lowe, G. and Gold, G. H. (1995) Olfactory transduction is intrinsically noisy. *Proc. Natl. Acad. Sci. USA* **92**: 7864–7868.

Mainen, Z. F. and Sejnowski, T. J. (1995) Reliability of spike timing in neocortical neurons. *Science* **268**: 1503–1506.

Polk, C. and Postow, E. (eds.) (1996) *Handbook of Biological Effects of Electromagnetic Fields*. Boca Raton, FL: CRC Press.

Reilly, J. P. (1992) *Electrical Stimulation and Electropathology*. New York: Cambridge University Press.

Rieke, R., Warland, D., de Ruyter van Steveninck, R. and Bialek, W. (1997) *Spikes: Exploring the Neural Code*. Cambridge, MA: MIT Press.

Rubenstein, J. T. (1995) Threshold fluctuations in an N sodium channel model of the node of Ranvier. *Biophys. J.* **68**: 779–785.

Schwan, H. P. (1983) Biophysics of the interaction of electromagnetic energy with cells and membranes. In *Biological Effects and Dosimetry of Nonionizing Radiation* (ed. M. Grandolfo, S. M. Michaelson and A. Rindi), pp. 213–231. New York: Plenum Press.

Sheridan, D. M., Isscroff, R. R. and Nuccitelli, R. (1996) Imposition of a physiologic DC electric field alters the migratory response of human keratinocytes on extracellular matrix molecules. *J. Invest. Dermatol.* **106**: 642–646.

Shi, R. and Borgens, R. B. (1995) Three-dimensional gradients of voltage during development of the nervous system's invisible coordinates for the establishment of embryonic pattern. *Dev. Dyn.* **202**: 101–114.

Shinomoto, S. and Kuramoto, Y. (1986) Phase transitions in active rotator systems. *Progr. Theor. Phys.* **75**: 1105–1110.

Steriade, M., McCormick, D. A. and Sejnowski, T. J. (1993) Thalamocortical oscillations in the sleeping and aroused brain. *Science* **262**: 679–685.

Stuchly, M. A. and Xi, W. (1994) Modelling induced currents in biological cells exposed to low-frequency magnetic fields. *Phys. Med. Biol.* **39**: 1319–1330.

Trachina, D. and Nicholson, C. (1986) A model for the polarization of neurons by extrinsically applied electric fields. *Biophys. J.* **50**: 1139–1156.

Traub, R. D. and Miles, R. (1991) Some collective phenomena in the hippocampus in vitro. In *Self Organization, Emerging Properties, and Learning* (ed. A. Babloyantz), pp. 97–112. New York: Plenum Press.

Valberg, P. A., Kavet, R. and Rafferty, C. N. (1997) Can low-level 50/60 Hz electric and magnetic fields cause biological effects? *Radiat. Res.* **148**: 2–21.

Verveen, A. A., Derksen, H. E. and Schick, K. L. (1967) Voltage fluctuations of neural membrane. *Nature* **216**: 588–589.

Wachtel, H. (1979) Firing-pattern changes and transmembrane currents produced by extremely low frequency fields in pacemaker neurons. In *Proc. 18th Annu. Hanford Life Sci. Symp.*, pp. 132–146. NTIS No. CONF-781016.

Weaver, J. C. (1993) Electroporation: a general phenomenon for manipulating cells and tissues. *J. Cell Biol.* **51**: 426–435.

Weaver, J. C. and Astumian, R. D. (1990) The response of living cells to very weak electric fields: the thermal noise limit. *Science* **247**: 459–462.

7

Oscillatory signals in migrating neutrophils: effects of time-varying chemical and electric fields

HOWARD R. PETTY

7.1 Introduction

Cellular oscillators have been found in a great variety of biological settings, from bacteria to humans (Rapp, 1979). They have been phenomenologically associated with a broad array of biological processes including metabolism, intercellular and intracellular signaling, cell division, hormone secretion and muscle contraction (Goldbeter, 1996). One cell system exhibiting oscillatory biochemical and physiological properties is the human *neutrophil*. Studies of neutrophil migration or activation have reported oscillations in actin assembly (Wymann *et al.*, 1989; Omann *et al.*, 1995), shape change (Wymann *et al.*, 1989, 1990; Ehrengruber *et al.*, 1995), velocity change (Hartman *et al.*, 1994), respiratory burst (Wymann *et al.*, 1990; Kindzelskii *et al.*, 1998), membrane potential (Jager *et al.*, 1988), intracellular calcium (Ca^{2+}) (Kruskal and Maxfield, 1987), nicotinamide adenine dinucleotide phosphate (NAD(P)H) autofluorescence (Kindzelskii *et al.*, 1997, 1998) and receptor interactions (Kindzelskii *et al.*, 1997). Interestingly, these neutrophil oscillations generally exhibit periods of about 10 or 20 s. Although oscillators can be correlated with certain cell activities, their biochemical mechanisms and broad physiological relevance are not established. Moreover, the relationships among these oscillators are not clear.

In addition to the well-known 3-min nicotinamide adenine dinucleotide (NADH) oscillations of eukaryotic glycolysis (Hess and Boiteux, 1971; Chance *et al.*, 1973), we have recently discovered more rapid 10- and 20-s NAD(P)H oscillations (Kindzelskii *et al.*, 1997, 1998). Other biochemical (e.g., Ca^{2+}) and physiological (e.g., $O_2^{\cdot -}$ release) oscillators operate at these periods with consistent phase relationships to NAD(P)H oscillations. These and other findings detailed below have led to several suggestions regarding the regulation of neutrophil function. (1) Cell metabolism may entrain cell functions;

(2) metabolic oscillations may act as a transmembrane signaling mechanism; (3) neutrophils respond to time-varying chemical and electric fields in a frequency-dependent fashion; and (4) certain clinical deficiencies in neutrophil function can be traced to aberrant metabolic oscillations and can be ameliorated by perturbation of these oscillators. Thus, the aim of this chapter is briefly to review the potential ability of intracellular oscillators to explain cell properties and clinical disorders.

7.2 Oscillatory receptor interactions and inside-out signaling

Several years ago we reported that cell surface integrins physically and functionally interact with other cell surface molecules (Petty and Todd, 1996). The leukocyte integrins CR3 (complement receptor type 3) and CR4 interact with the pro-inflammatory glycosylphosphatidylinositol (GPI)-linked membrane proteins FcγRIIIB, uPAR (urokinase receptors) and CD14. Although GPI-linked proteins lack transmembrane domains, they elicit transmembrane signals, at least in part, by interacting with integrins. Recent studies have shown that these interactions can be highly dynamic. Using resonance energy transfer (RET) as a qualitative measure of receptor proximity, receptor association/dissociation events have been followed in real time (Kindzelskii et al., 1996; Zarewych et al., 1996).

As neutrophils polarized for migration, CR3 trafficked to the uropod while CR4 and uPAR moved to the lamellipodium. RET experiments indicated that CR4 and uPAR were in close proximity; however, the signal oscillated instead of reaching a stable plateau (see Figure 1a; Kindzelskii et al., 1997). Receptor specificity was suggested by the inability of other labels to exhibit oscillatory RET signals. Physical controls confirmed the signal's RET origin (Kindzelskii et al., 1996, 1997). Indistinguishable frequencies were obtained by monitoring the emission of the acceptor or the quenching of donor fluorescence. This suggested that many CR4 molecules were binding and releasing uPAR simultaneously to create a sinusoidal waveform, i.e., the system was phase locked. As these oscillations were observed in the absence of an extracellular signal, we inferred that they were a manifestation of inside-out signaling via integrins.

To test the role of the signaling apparatus in CR4–uPAR oscillations, we examined the effects of signal transduction inhibitors (Figure 1c and d). Since typical doses of these inhibitors blocked CR4–uPAR oscillations, we titered these drugs to obtain suboptimal doses that crippled, but did not destroy, cell polarization. Importantly, the addition of staurosporine, a kinase inhibitor, led to a flyback sawtooth CR4–uPAR waveform; this suggested that reducing kinase activity slows the formation of CR4–uPAR proximity complexes. On

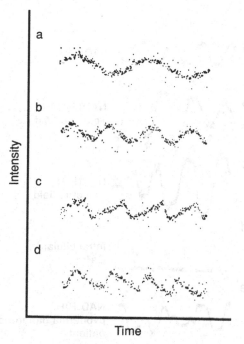

Figure 1. Resonance energy transfer (RET) signal oscillations of normal migrating neutrophils labeled with fluorochrome-conjugated anti-uPAR F(ab′)$_2$ and anti-CR4 Fab fragments. Cells were prepared from peripheral blood using Ficoll–Hypaque density gradient centrifugation. Quantitative microfluorometry was performed as previously described (Kindzelskii *et al.*, 1996). Signal intensities (ordinate) are plotted versus time (abscissa). All cells were morphologically polarized for migration; oscillations decay to background levels when migration ends. (a) Sinewave RET signal of cells ($\tau \cong 20$ s). (b) When cells are stimulated with 10^{-7} M *N*-formylmethionylleucyl-phenylalanine (FMLP), sinewaves of higher frequency are found ($\tau \cong 10$ s). These oscillations are phenomenologically linked to the signal transduction apparatus. (c) When cells are exposed to a suboptimal dose (0.05 μM) of the kinase inhibitor staurosporine, a flyback sawtooth waveform ($\tau \cong 10$ s) was found. (d) On the other hand, suboptimal doses (50 μM) of the phosphatase inhibitor pervanadate leads to a reverse sawtooth waveform ($\tau \cong 10$ s). The RET amplitude oscillations are ∼40% of the peak rate-shuttered count rate.

the other hand, suboptimal doses of pervanadate, a phosphatase inhibitor, led to a reverse sawtooth waveform. Thus, we suggest that the signaling apparatus participates in RET oscillations.

Since kinases utilize ATP as a substrate, we tested the hypothesis that the neutrophil's metabolism participates in signaling oscillations. As measuring ATP levels in living cells would perturb its concentration, we chose to study NAD(P)H, which oscillates 180° out of phase with adenosine triphosphate (ATP). Furthermore, NAD(P)H emission was linear within the cellularly relevant concentration range (Liang and Petty, 1992). We therefore studied

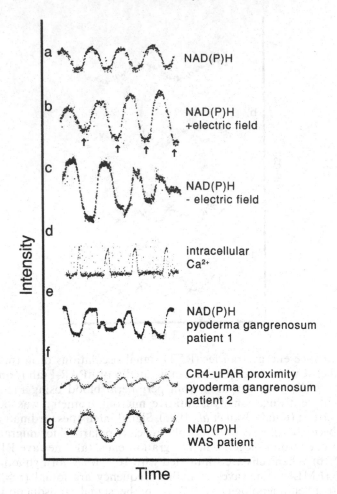

Figure 2. Gallery of oscilloscope recordings of oscillatory physico-chemical properties in migrating neutrophils. Data from normal ((a)–(d)) and clinical ((e)–(g)) blood samples are shown. (a) Sinusoidal NAD(P)H autofluorescence oscillations of normal migrating neutrophils are shown ($\tau \cong 20\,\mathrm{s}$). (b) When an electric field is applied at NAD(P)H troughs, the oscillations grow in amplitude. Electric field strengths were calculated from current readings. (c) When the field is terminated, the amplitude declines. (d) Neutrophils labeled with the fluorescent Ca^{2+} indicator INDO-1 show regular Ca^{2+} spikes ($\tau \cong 20\,\mathrm{s}$), as previously reported by Kruskal and Maxfield (1987). To examine the clinical relevance of these oscillations, neutrophils with reported deficiencies in chemotaxis and cell shape change were studied. (e) Pyoderma gangrenosum patient 1 (KN) displayed random or chaotic NAD(P)H oscillations. (f) Pyoderma gangrenosum patient 2 (SC) exhibited sinusoidal 10-s NAD(P)H oscillations, but a flyback sawtooth CR4–uPAR waveform ($\tau \cong 10\,\mathrm{s}$). (g) Wiskott–Aldrich syndrome (WAS) neutrophils exhibited a shoulder on the rising side of the NAD(P)H oscillation ($\tau \cong 10\,\mathrm{s}$). Data were collected at sweep rates of 5 or 10 s per division depending upon the cells under study.

NAD(P)H oscillations in single migrating neutrophils (Figure 2a). We found that NAD(P)H oscillations were identical in period with the CR4–uPAR oscillations under all experimental conditions. Furthermore, the NAD(P)H oscillations could not be accounted for by variations in neutrophil thickness; in fact, the completion of cell shape changes preceded NAD(P)H peaks by 90° (Figure 3). By rapidly switching between optical set-ups, we found that NAD(P)H oscillations were 180° out of phase with RET oscillations, thus suggesting that RET oscillations were *in phase* with putative ATP oscillations.

To explore the extracellular expression of cellular oscillators, we monitored the pericellular release of reactive oxygen metabolites (ROMs) and extracellular proteolysis during cell migration. Cells were placed in semi-solid collagen

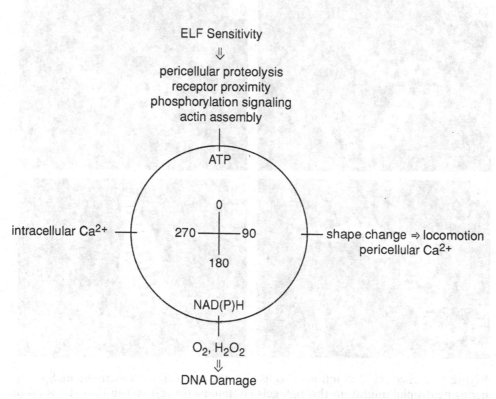

Figure 3. Relative phase angles in cellular chemistry and function. The sketch summarizes the results of several published and unpublished studies from our group. The NAD(P)H peak was chosen to be 180°. Relative phases of these parameters were determined by rapidly switching between their optical set-ups on the microscope's filter slider. In certain cases (e.g., INDO-1 and NAD(P)H), substantial overlap in spectral properties necessitated triangulation with a third parameter relative to these two. ELF, extremely-low-frequency electric field.

gel matrices, doped with various reporters. These matrices resembled base-
ment membranes and impeded the diffusion of reporters, thus permitting
photomicroscopy. Gels were doped with dihydrotetramethylrosamine (H_2-
TMRos) or hydroethidine, which become fluorescent upon exposure to
ROMs, or Bodipy-BSA, which, in turn, becomes fluorescent upon proteolytic
disruption. As unstimulated neutrophils migrated through this environment,
stripes of proteolysis were observed microscopically (Figure 4). Similarly,
ROM production was observed to oscillate. However, ROM production and
proteolysis oscillated 180° apart while ROM production was in-phase with
NAD(P)H oscillations.

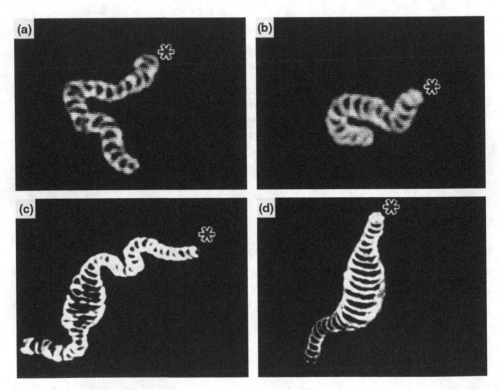

Figure 4. Oscillatory interactions of cells with the extracellular environment. Spontaneous neutrophil migration through gels containing ((a),(c)) 100 ng H_2TMRos/ml or
((b),(d)) 25 µg Bodipy-BSA/ml was observed. The release of oxidative molecules such as
hydrogen peroxide oxidizes H_2TMRos to TMRos, yielding fluorescence emission.
Proteolytic attack on Bodipy-BSA yields highly fluorescent peptides. (a) When cells
migrate through these matrices, a repetitive series of oxidant-induced stripes of
TMRos are observed. (b) Similarly, cell migration through Bodipy-BSA-containing
gels yields a series of fluorescent stripes indicating periodic extracellular proteolysis.
((c),(d)) Phase-matched electric field application enhances the size of the stripes.

7.3 Outside-in signaling in cellular function

If metabolic oscillators are a key component of the signaling apparatus, it should be possible to detect outside-in signaling. Indeed, a variety of factors that activate or inhibit neutrophil function alter the *frequency* and/or *amplitude* of NAD(P)H oscillations. As mentioned above, migrating cells exhibited metabolic oscillations of ~ 10 or $20\,s$. In the absence of exogenous agents, 20-s oscillations were observed (Figure 1a). When neutrophils migrated on fibrinogen-coated surfaces, which ligate CR3 and CR4, 10-s oscillations were also found (Kindzelskii *et al.*, 1997). Similarly, incubation of neutrophils with yeast particles triggered 10-s NAD(P)H oscillations (Petty *et al.*, 1996). 10-s oscillations were also observed when neutrophils migrated on vascular endothelial cell monolayers, which ligated neutrophil selectins and integrins. When neutrophils migrated in gel matrices containing immune complexes composed of bovine serum albumin (BSA) and anti-BSA antibody, these complexes were taken up by the cells, which then displayed 10-s oscillations. Ligation of formyl peptide receptors with *N*-formylmethionylleucylphenylalanine (FMLP) also triggered 10-s oscillations. Thus, multiple pro-inflammatory extracellular mediators lead to higher frequency NAD(P)H oscillations. The increased NAD(P)H frequency in conjunction with a constant ROM step size (Kindzelskii *et al.*, 1998), doubles the ROM production rate, thereby potentiating target destruction and cell migration.

In addition to extracellular signals that double NAD(P)H frequencies, there are factors that block formation of 10-s NAD(P)H oscillations. When immune complexes were opsonized with complement, 20-s NAD(P)H oscillations were found. Thus, complement deposition blocks acquisition of the activated 10-s oscillation. At present, we do not know whether these differences were due to structural differences in the complexes or due to different receptor ligation patterns. Nonetheless, the reduced rate of ROM production may contribute to the diminished phlogistic potential of complement-treated targets. Another exogenous factor found to affect NAD(P)H oscillations was the widely prescribed anti-inflammatory drug indomethacin. When neutrophils were attached to endothelial cells, a 10-s oscillation was observed. However, in the presence of $100\,\mu M$ indomethacin, neutrophils on endothelial cell monolayers exhibited a reduced NAD(P)H oscillation periodicity ($\sim 27\,s$) and a reduction in amplitude ($\sim 30\%$). This is consistent with the ability of indomethacin to reduce ROM production by neutrophils (Smolen and Weissmann, 1980). Thus, pro-inflammatory factors increased NAD(P)H oscillation frequency and potentiated ROM production whereas anti-inflammatory conditions did the opposite. These findings support the concept that NAD(P)H oscillations are important in regulating the NADPH oxidase's activity. Furthermore, it may be possible that NADPH oscillations could function as a signal switching the NADPH oxidase on and off.

7.4 Cell metabolism as a message for activation and polarization

We suggest that metabolic oscillations, perhaps due to the glycolytic oscillator, account for receptor, signaling and functional oscillations during cell migration. This is not unreasonable, since the equilibrium constants of several oscillatory cell properties are within an order of magnitude of approximate substrate concentrations. During cell migration, most CR4 molecules bind and release (or undergo conformational changes with) uPAR molecules in synchrony. A role for metabolic oscillations is plausible, because the concentration of free ATP as measured by the luciferase assay is $\sim 10^{-5}$ M. This is, for example, comparable to the equilibrium constant (K_d) for ATP binding to protein kinase C ($\sim 10^{-6}$ M; Petty, 1993), although the relevance of bulk ATP levels and equilibrium constants is uncertain. This hypothesis is also consistent with the observation that kinase and phosphatase inhibitors retard the rise time and declination of CR4–uPAR oscillations, respectively. Hunter (1987) has speculated regarding such an oscillatory phosphorylation apparatus and suggested an analogy between cell signaling and electronic circuit theory. Our work suggests that this analogy may apply to neutrophils. Furthermore, a panel of neutrophil-activating substances increased metabolic oscillation frequency whereas anti-inflammatory agents blocked these frequency changes. We conjecture that signaling molecules such as kinases are not necessarily signaling molecules per se, but rather represent the *conduit* through which signals pass, just as telephone wires carry signals, but are not the signals. Another derivative of this concept is that some postulated, but unknown, 'second' messengers may not exist; metabolite usage could feedback through the metabolic apparatus creating oscillations that transiently drive metabolite levels to concentrations required for certain cell functions to appear.

Cell functions, in turn, may be driven by metabolic and/or signaling oscillations. A central concept in cell motility refers to *oscillatory adhesive events* (Springer, 1990). Oscillatory integrin behavior may be required for oscillatory cell adhesion. Moreover, leukocyte integrins and uPAR molecules interact with and affect the function of one another (Simon *et al.*, 1996; Sitrin *et al.*, 1996). We speculate that when CR4 is phosphorylated, uPAR and its associated protease are more active, thus contributing to oscillatory proteolysis. Pericellular proteolysis is required for cell migration in complex environments. Oscillatory metabolism may also participate in the oscillatory production of

ROMs. An oscillatory respiratory burst was first reported by Wymann *et al.* (1989). Moreover, we have found that the frequency of NAD(P)H oscillations match that of ROM production. Since the NAD(P)H oscillation peak occurs simultaneously with the release of ROMs and, since ROMs are produced by the NADPH oxidase, we suggest that substrate oscillations *entrain* product oscillations. Although the concept has not been proven, it is not unrealistic because the magnitude of NADPH oscillations (~ 180 to $350 \, \mu M$) are within a factor of 10 of the enzyme's apparent equilibrium constant.

In addition to participating in target destruction, ROMs also inactivate protease inactivators (Weiss, 1989), thereby allowing proteases to function. Thus, as neutrophils migrate through complex environments, they alternatively inactivate protease inactivators, then activate protease action. The metabolic *phase difference* is translated into a repetitive sequence of biochemical reactions in the extracellular matrix to promote locomotion.

Our studies reveal cellular rhythms within the extracellular environment, the plasma membrane and the cytoplasm of a cell, which are all related by consistent phase differences (Figure 3). Figure 5 shows a hypothetical model that seeks to explain the roles of the numerous cellular and extracellular neutrophil oscillators in simple biochemical terms. During neutrophil migration CR4 and/or associated proteins are simultaneously phosphorylated by an oscillatory signaling machinery. This phosphorylation wave leads to CR4–uPAR coupling. CR4–uPAR coupling focuses pericellular proteolysis next to the integrin adherence site where cell extension is to occur. Actin assembly, and consequently, cell extension (shape change) then proceed through this adherence and proteolysis site when ATP is at its peak. The mean free cytoplasmic ATP level ($\sim 10^{-5} \, M$) approximates to the equilibrium constant for ATP-actin binding to microfilaments. In summary, we therefore suggest that the cadence of metabolic oscillations entrains and coordinates downstream oscillatory behaviors.

7.5 Neutrophil response to pulsed chemical fields

We speculate that the neutrophil's signal transduction apparatus *periodically* senses the extracellular environment via a metabolic oscillation-coupled mechanism (e.g., the system could be 'ready to receive' when a substrate concentration is near its peak). Previous theoretical (Schienbein and Gruler, 1995) and experimental (Vicker, 1994) studies have predicted the existence of a signal at the above frequency. To provide a preliminary test of this idea, we exposed neutrophils to spatially uniform, but temporally varying, concentrations of FMLP using a stopped-flow microscope chamber (for details, see Albrecht and

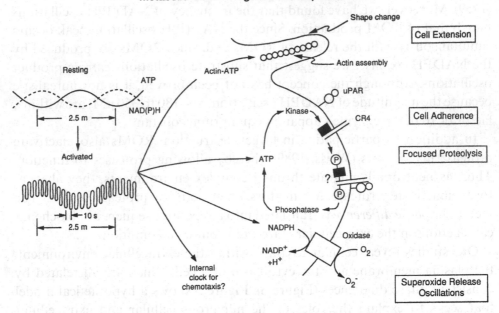

Figure 5. A hypothetical model of the role of intracellular oscillators in cell motility. For details see text in Section 7.4.

Petty, 1998). Briefly, neutrophils were exposed to temporally decreasing FMLP concentrations at various time intervals. When cells were treated at 10-s intervals, the neutrophils apparently perceived that they were migrating in the wrong direction (a temporally decreasing signal), and they reversed their direction of polarization. Most cells chose their new direction at $180 \pm 15°$ relative to their initial direction of polarization. Cells treated with buffer injections or increasing concentrations of FMLP did not change direction. Thus, cells were able to *compare* current FMLP levels with those previously encountered. Importantly, the *frequency* of ligand input was a crucial determinant of the physiological output. Moreover, the 10-s input period was consistent with signal response lags previously noted for chemical and electrical stimulation (Gerisch and Keller, 1981; Franke and Gruler, 1994). Although cells are exposed to chemotactic factor gradients *in vivo*, these findings strongly suggest that temporal events participate in neutrophil signal processing.

7.6 Pulsed electric fields: the signal contribution of metabolic resonance

To test the effect of another class of extracellular signals, we applied pulsed DC electric fields of various intensities to cells. When electric fields were repeatedly applied to cells at NAD(P)H oscillation troughs, the NAD(P)H autofluorescence intensity continued to increase until it saturated at 200% of the normal oscillatory amplitude for both the 10- and 20-s oscillations (see Figure 2b). When the field was terminated, the oscillations slowly returned to their original amplitude (compare Figure 2c). We call this repetitive increase in NAD(P)H amplitude *metabolic resonance* (see also Adachi *et al.*, 1999).

The unique aspect of metabolic resonance is that it accompanies exaggerated normal migratory functions. During metabolic resonance, neutrophils extended from their normal length of $\sim 10\,\mu m$ to ~ 40–$50\,\mu m$. Figure 6 shows a typical example of exaggerated cell extension, but it occurred only when the field was applied at NAD(P)H troughs. This correspondence between field application, phase and cell extension was observed in every trial (amounting to several hundreds) over the past three years. These extraordinary cell shape changes cannot be accounted for by galvanotaxis, electrophoretic motion or dielectric forces (Petty and Kindzelskii, 1997; A. L. Kindzelskii and H. R. Petty, unpublished results). The observed shape change was not a direct (passive) physical effect, but rather was probably due to (active) assembly of cytoplasmic microfilaments. Figure 6d shows a fluorescence micrograph of microfilaments after 10 min of metabolic resonance followed by fixation and labeling (A. L.

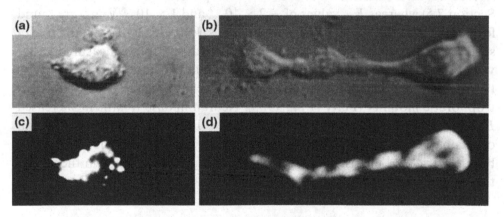

Figure 6. Phase-matched electric fields lead to exaggerated neutrophil shape changes and cytoskeletal polymerization. ((a),(b)) Differential interference contrast (DIC) and ((c),(d)) fluorescence micrographs are shown. (a) Normal migrating neutrophils are triangular in shape and about 10 μm in length. (b) However, when a phase-matched electric field is applied to the sample, cells extend to >40 μm in length. These morphological changes are accompanied by changes in microfilaments. Control or electric field-exposed cells were rapidly fixed with dithiobis(succinimidyl propionate), which provides superior retention of labile filaments, followed by extraction and staining with fluorescein-phalloidin. As previously reported by other groups, a patchy appearance of microfilaments is observed in control cells (c) whereas long microfilaments are observed in cells exposed to phase-matched electric fields (d).

Kindzelskii and H. R. Petty, unpublished results). We suggest, therefore, that microfilament assembly drives cell shape changes. Other changes that accompanied metabolic resonance were the amount of ROMs produced, DNA damage and the extent of pericellular proteolysis taking place during locomotion (Figure 4c and d; Kindzelskii *et al.*, 1998; A. L. Kindzelskii and H. R. Petty, unpublished results). Heightened NAD(P)H amplitudes may have led to exaggerated ROM production. In summary, the amplitudes of multiple cell functions and cell metabolism were influenced by phase-matched electric field application.

To better define the physical parameters affecting metabolic resonance, we examined the effects of electric field intensity (~ 1–10^{-6} V/m) and cell surface charge on metabolic resonance in normal neutrophils. The minimal electric field intensity (E_{mr}) that could support metabolic resonance was 10^{-4} V/m. The same value for E_{mr} was found for both Pt and Ag/AgCl electrodes. Precautions were taken to guard against sample contamination by electrode reaction products (vycro frits, agar and gelatin barriers), ground loops and stray magnetic fields. Interestingly, metabolic resonance was constant above E_{mr}, i.e., the effects of 1 and 10^{-4} V/m were indistinguishable, whereas intensities $< 10^{-4}$ V/m were without effect. In contrast, metabolic resonance was not observed for *Escherichia coli* or neutrophils from WAS patients (see Section 7.5) whereas E_{mr} values of $\sim 2 \times 10^{-2}$ and 1×10^{-5} V/m were found for human lymphocytes and fibrosarcoma cells, respectively. To test the role of surface charges in the metabolic resonance of neutrophils, we measured E_{mr} at various cell surface charge densities. Cell surface charge was reduced by treatment with neuraminidase, which dramatically reduces neutrophil electrophoretic mobility (Petty *et al.*, 1980). Neuraminidase treatment increased E_{mr} from 0.80 ($\pm$ 0.05) $\times 10^{-4}$ V/m to 10.6 ($\pm$ 1.0) $\times 10^{-4}$ V/m after 100 min of incubation (A. L. Kindzelskii and H. R. Petty, unpublished results). Thus, both electric field intensity and cell surface charges participated in the induction of metabolic resonance.

7.6.1 Electromechanical coupling hypothesis

An electric field simultaneously acts on $\sim 10^{10}$ net negative surface charges per cell. This exerts a small force (F_e) on cell membranes relative to their immobile cytoskeleton. The magnitude of this force ($F_e = QE$) at E_{mr} is 0.02 μdyne, which is a significant fraction of the force of gravity on a neutrophil (~ 0.5 μdyne). Cell shape change is an exercise in force balancing, where cytoskeletal polymerization forces (F_p) support cell extension whereas opposing forces derived from surface tension (F_s) promote cell rounding. F_p has been estimated to be

~1 to 6 µdyne per filament (Albrecht-Buehler, 1990), which is consistent with actin load forces (Mogilner and Oster, 1996). The surface-cortical tension (s_w) of neutrophils is 0.036 dyne/cm (Evans and Yeung, 1989). Using the equation $F_s = 2R\pi s_w$, and a pseudopod radius of 0.1 µm, the value $F_s = 2.3$ µdyne is obtained. This suggests that only a small number of filaments is required to support cell extension, which agrees with an experimental assessment of pseudopodial filament numbers (Zhelev and Hochmuth, 1995). Thus, it is possible that forces of the order of microdynes may be sufficient to mediate neutrophil shape changes.

In the absence of an electric field, we propose that F_p is driven by ATP-actin. The *free* concentration of ATP ($\sim 10^7$ molecules/cell) is less than the total actin content ($\sim 10^8$/cell) and is near the K_d for ATP-actin interactions with microfilament. We hypothesize, therefore, that ATP concentration oscillations may entrain actin assembly oscillations, which may account for the binary switch model of actin assembly (Stossel, 1993) and oscillatory actin assembly (Wymann et al., 1990; Omann et al., 1995). This delicate biochemical balance in migratory cells, however, can be affected if an additional coherent force, phase matched to the ATP-actin assembly point of the cycle, is applied (Figure 3). One type of coherent external force is an electric field, F_e. When $F_p + F_e > F_s$, cell extension and net cytoskeletal assembly take place. Since F_p and F_s are both of the order of microdynes we speculate that a cell could detect a force of more than ~0.1–1% of the two counterbalancing forces (depending upon parameters chosen) to yield the biochemical and structural changes observed at the lowest field strengths. This idea could help to explain the threshold effect noted above; either $F_p + F_e > F_s$ or it is not, thus constituting a binary switch. When the barrier for neutrophil extension (F_s) is brought down by F_e, ATP-actin assembly into microfilaments is exaggerated. This, of course, may lead to exaggerated microfilament assembly and cell shape changes that were observed (see Figure 6). Since microfilament assembly is predominantly unidirectional (Petty, 1993), additional actin molecules are added to increase filament length during each field pulse. Consequently, exaggerated microfilament assembly draws large amounts of ATP-actin through the system, which represent a significant fraction of a leukocyte's total free ATP pool. We hypothesize that metabolic feedback circuitry compensates by driving metabolic amplitudes higher, thus causing metabolic resonance. In this system, initiation of microfilament extension and metabolic resonance are observed immediately; there is no 8- to 10-s lag as observed in high-field (800 V/m) studies (Franke and Gruler, 1994). This may be due to the fact that in the experiments presented here the weak electric field stimulus was phase matched with metabolic oscillations, whereas in the latter studies it was not.

7.7 Clinical abnormalities in neutrophil oscillators

If metabolic oscillations play a central role in cell locomotion as proposed above, then it should be possible to identify patients with neutrophil-related disorders who display aberrant metabolic oscillations. Pyoderma gangrenosum is one such disorder. It is an uncommon, destructive and poorly understood inflammatory disease, which has been reported to be heritable in pediatric cases. Pyoderma gangrenosum patient KN has had numerous skin lesions accompanied by ulceration and drainage in the absence of infection since birth. Her neutrophils are unable to acquire normal shape *in vitro*. KN's disorder was initially identified on the basis of integrin clusters and aberrant integrin-GPI-linked receptor interactions (Adachi *et al.*, 1998). No oscillatory CR4–uPAR RET signal was obtained. Phosphotyrosine Western blots showed heightened phosphorylation of the patient's cells. When NAD(P)H oscillations of KN's neutrophils were studied, incoherent changes in autofluorescence were observed (see Figure 2e). We therefore speculate that incoherent metabolic fluctuations did not permit normal cell shape change and migration.

Several physical perturbations affected NAD(P)H oscillations and the behavior of KN's neutrophils. We discovered that KN's phenotype was temperature dependent. As the temperature was reduced to $\cong 30\,^{\circ}$C, the metabolic oscillations returned to a sinusoidal waveform of normal frequency. Concomitantly, the integrins unclustered and the cells polarized normally. Although KN has had about three occasions of severe open lesions per year in the past, she has not had these problems on her extremities during the past 18 months as local hypothermia was employed. This suggests that knowledge gained concerning biological oscillators can be applied to clinical disease.

One conjecture, alluded to above, was that coherent metabolite (e.g., ATP) flux through the system can constitute a signal; whether ATP flux was promoted by 'signaling' molecules such as kinases or by actin was not relevant. For KN's neutrophils, we suggest that integrin clusters continually generate activation signals such as phosphorylation without dephosphorylation and cycling. When a pulsed DC electric field was applied at NAD(P)H oscillation minima, metabolic oscillations reverted to a sinusoidal waveform. Similarly, the exposed cells adopted normal shapes and migratory capacity. The effect was transient because removal of the external field led to incoherent metabolic oscillations within 3 min. We suggest that an external pulsed electric field caused metabolites to be drawn through the system via actin assembly, thus providing a surrogate signal.

A second pediatric-onset pyoderma gangrenosum patient (SC) with less severe disease was also evaluated (Shaya *et al.*, 1998). Although cell motility was defective, SC's neutrophils did not exhibit integrin clusters. Moreover, NAD(P)H oscillations were sinusoidal in SC's cells, although their frequency matched that of activated cells. Importantly, the CR4–uPAR RET oscillations exhibited a flyback sawtooth waveform (Figure 2f). That is to say, SC's neutrophils behaved like staurosporine-treated normal cells. This raised the interesting possibility that SC's neutrophils contained a less efficient kinase, thus leading to the observed waveform (e.g., compare also with individuals with diabetes in which the kinetic rate constant of the insulin receptor's kinase was diminished; Hubbard *et al.*, 1994). We next tested the hypothesis that phosphatase inhibitors would slow phosphatase kinetics, thus balancing the kinase/phosphatase pathways. Pervanadate returned SC's cells to a sinusoidal waveform whose period was doubled relative to that of untreated cells. This is consistent with cyclic phosphorylation/dephosphorylation mediated by the metabolic apparatus. Thus, topical application of phosphatase inhibitors may be a useful means of rational drug therapy in this case.

Another disorder exhibiting aberrant neutrophil chemotaxis is *Wiskott–Aldrich* syndrome (WAS; Badolato *et al.*, 1996; Remold-O'Donnell *et al.*, 1996). WAS genetic lesions have been traced to a single intracellular 60-kDa protein (WASP) associated with microfilaments. These molecular lesions lead to cytoskeletal defects including an inability to assemble microvilli. WASP participates in actin polymerization by anchoring microfilament assembly and lies at a crossroads between signaling and cytoskeletal machineries (Symons *et al.*, 1996). We hypothesized that WAS cells may display aberrant metabolic oscillations. Figure 2g shows NAD(P)H oscillations of WAS neutrophils. A shoulder was observed on the rising side of each oscillation. Similar results were obtained using cells from two related boys who express a Pro → Leu mutation at codon 39. These observations provided additional evidence for a link between intracellular oscillators, cell behavior and disease. We next tested the ability of WAS neutrophils to detect electric fields. Interestingly, metabolic resonance and exaggerated cell extension were not observed. As WAS cells express reduced amounts of CD43, a charged membrane sialoprotein, we employed higher field strengths (~ 1 V/m) without detecting metabolic resonance. Since WAS cells are deficient in pseudopod extension, we suggest that they are not subject to the force-balancing condition described above and cannot sense these electric fields. Thus, an intact cytoskeletal machinery is needed to support metabolic resonance and cytoskeletal assembly precedes metabolic resonance.

Depressed clinical inflammatory responses have been associated with an inability to acquire the 10-s NAD(P)H oscillation. Infection of fetuses and newborns is a major contributor to infant morbidity and mortality. One well-studied contributor to the depressed phlogistic potential of neonates is a broad spectrum of depressed neutrophil functions. These blunted functions include chemotaxis, extravasation, actin polymerization, ROM production and phagocytosis (e.g., Dos Santos and Davidson, 1993). To test the role of metabolic oscillations in the diminished function of neonatal neutrophils, we obtained fetal blood from umbilical cords: 20-s sinusoidal NAD(P)H oscillations were found for migrating neonatal neutrophils. However, attempted stimulation with 10^{-6} to 10^{-8} M FMLP did not affect the metabolic oscillation frequency. Thus, under nominally activating conditions, neonatal neutrophils were unable to respond normally. We speculate that this developmental delay in metabolic oscillations may account for neonatal disease susceptibility.

Another example of depressed innate immunity is the rapid progression of Ebola virus infections. The secretory glycoprotein (sGP) of Ebola virus binds to FcγRIIIB of neutrophils (Yang et al., 1998). Preliminary studies have indicated that sGP blocks acquisition of the 10-s metabolic oscillation. Cells stimulated with 10^{-6} to 10^{-8} M FMLP retained the 20-s NAD(P)H oscillation. Thus endogenous substances and products of infectious agents can affect neutrophil metabolic oscillations.

7.8 Discussion and conclusion

There is a growing awareness of the important role of biochemical oscillators in transmembrane signaling (O'Rourke et al., 1994; Goldbeter, 1996; Berridge, 1997). This is especially true for the Ca^{2+} oscillator. Ca^{2+} signals are thought to encode information in their frequency, amplitude, duration and number (e.g., Gu and Spitzer, 1995; Berridge, 1997; Dolmetsch et al., 1997). Recently, CaM kinase II was shown to decode Ca^{2+} oscillations in a frequency-dependent fashion (De Koninck and Schulman, 1998). Numerous physiological endpoints (gene expression, hormone secretion, etc.) have been correlated with the Ca^{2+} oscillator (Gilon et al., 1993; Goldbeter, 1996; Berridge, 1997). Although the present review did not focus on Ca^{2+} oscillations, these oscillations parallel NAD(P)H oscillations with constant phase angles (see Figures 2d and 3). However, some evidence suggests that Ca^{2+} oscillations may possibly be secondary to metabolic oscillations (Corkey et al., 1988; Pralong et al., 1990; O'Rourke et al., 1994).

In conclusion, we hypothesize that metabolic oscillations represent a physical as well as biochemical signal that regulates cell migration, spreading and phagocytosis. Furthermore, we propose that the *frequency, amplitude* and *phase information* contained in these oscillations is physiologically meaningful. Several independent lines of physical, chemical and genetic evidence support this concept. Although these lines of evidence strongly support the role of cellular oscillators in cell function, largely through correlative changes, they do not prove causality. There is, therefore, a small possibility that cellular oscillators may represent epiphenomena; they may reflect the mechanistic pathway without direct participation. However, we believe that this distinction is unimportant in the present context. In either case, the long-term goal of improved management of inflammatory disease could be achieved. These oscillators have already been used to characterize disease states and to search for novel anti-inflammatory substances. Further study of neutrophil oscillators may lead to both a mechanistic understanding of neutrophil activation and improved patient care.

Acknowledgments

This research has been supported by NIH grants AI27409 and CA74120, the J. P. McCarthy Foundation, and the American Heart Association of Michigan.

References

Adachi, Y., Kindzelskii, A. L., Ohno, N. and Petty, H. R. (1999) Amplitude and frequency modulation of metabolic signals in leukocytes: synergistic role of IFN-gamma in IL-6 and IL-2 mediated cell activation. *J. Immunol.* **163**: 4367–4374.

Adachi, Y., Kindzelskii, A. L., Shaya, S., Cookingham, G., Moore, E. C., Todd, R. F. and Petty, H. R. (1998) Aberrant neutrophil trafficking and metabolic oscillations in severe pyoderma gangrenosum. *J. Invest. Dermatol.* **111**: 259–268.

Albrecht, E. and Petty, H. R. (1998) Cellular memory: neutrophil orientation reverses during temporally decreasing chemoattractant concentrations. *Proc. Natl. Acad. Sci. USA* **95**: 5039–5044.

Albrecht-Buehler, G. (1990) In defense of 'nonmolecular' cell biology. *Int. Rev. Cytol.* **120**: 191–241.

Badolato, R., Malacarne, F., Bresciani, S., Cattaneo, R., Ugazio, A. G. and Notarangelo, L. D. (1996) Decreased chemotaxis but normal respiratory burst in response to MCP-1 in monocytes derived from Wiskott–Aldrich patients. *J. Leuk. Biol.* Supplement **60**: 17.

Berridge, M. J. (1997) The AM and FM of calcium signalling. *Nature* **386**: 759–760.

Chance, B., Pye, E. K., Ghosh, A. K. and Hess, B. (1973) *Biological and Biochemical Oscillators.* New York: Academic Press.

Corkey, B. E., Tornheim, K., Deeney, J. T., Glennon, M. C., Parker, J. C., Matschinsky, F. M., Ruderman, N. B. and Prentki, M. (1988) Linked oscillations of free Ca^{2+} and the ATP/ADP ratio in permeabilized RINm5F insulinoma cells supplemented with a glycolyzing cell-free muscle extract. *J. Biol. Chem.* **263**: 4254–4258.

De Koninck, P. and Schulman, H. (1998) Sensitivity of CaM kinase II to the frequency of Ca^{2+} oscillations. *Science* **279**: 227–230.

Dolmetsch, R. E., Lewis, R. S., Goodnow, C. C. and Healy, J. I. (1997) Differential activation of transcription factors induced by Ca^{2+} response amplitude and duration. *Nature* **386**: 855–858.

Dos Santos, C. and Davidson, D. (1993) Neutrophil chemotaxis to leukotriene B4 *in vitro* is decreased for the human neonate. *Pediatr. Res.* **33**: 242–246.

Ehrengruber, M. U., Coats, T. D. and Deranleau, D. A. (1995) Shape oscillations: a fundamental response of human neutrophils stimulated by chemotactic peptides? *FEBS Lett.* **359**: 229–232.

Evans, E. A. and Yeung, A. (1989) Apparent viscosity and cortical tension of blood granulocytes determined by micropipet aspiration. *Biophys. J.* **56**: 151–160.

Franke, K. and Gruler, H. (1994) Directed cell movement in pulsed electric fields. *Z. Naturforsch.* **49c**: 241–249.

Gerish, G. and Keller, H. H. (1981) Chemical reorientation of granulocytes stimulated with micropipettes containing f-Met-Leu-Phe. *J. Cell Sci.* **52**: 1–10.

Gilon, P., Shepherd, R. M. and Henquin, J.-C. (1993) Oscillations of secretion driven by Ca^{2+} as evidenced in single pancreatic islets. *J. Biol. Chem.* **268**: 22265–22268.

Goldbeter, A. (1996) *Biochemical Oscillations and Cellular Rhythms.* Cambridge: Cambridge University Press.

Gu, X. and Spitzer, N. C. (1995) Distinct aspects of neuronal differentiation encoded by frequency of spontaneous Ca^{2+} transients. *Nature* **375**: 784–787.

Hartman, R. S., Lau, K., Chou, W. and Coats, T. D. (1994) The fundamental motor of the human neutrophil is not random: evidence for local non-Markov movement in neutrophils. *Biophys. J.* **67**: 2535–2545.

Hess, B. and Boiteux, A. (1971) Oscillatory phenomena in biochemistry. *Annu. Rev. Biochem.* **40**: 237–258.

Hubbard, S. R., Wei, L., Ellis, L. and Hendrickson, W. A. (1994) Crystal structure of the tyrosine kinase domain of the human insulin receptor. *Nature* **372**: 746–754.

Hunter, T. (1987) A thousand and one protein kinases. *Cell* **50**: 823–829.

Jager, U., Gruler, H. and Bultmann, B. (1988) Morphological changes and membrane potential of human granulocytes under influence of chemotactic peptide and/or Echo-virus, type 8. *Klin. Wochenschr.* **66**: 434–436.

Kindzelskii, A. L., Eszes, M. M., Todd III, R. F. and Petty, H. R. (1997) Proximity oscillations of complement receptor type 4 and urokinase receptors on migrating neutrophils. *Biophys. J.* **73**: 1777–1784.

Kindzelskii, A. L., Laska, Z. O., Todd III, R. F. and Petty, H. R. (1996) Urokinase-type plasminogen activator receptor reversibly dissociates from complement receptor type 3 $(\alpha_M\beta_2)$ during neutrophil polarization. *J. Immunol.* **156**: 297–309.

Kindzelskii, A. L., Zhou, M. J., Haugland, R. P., Boxer, L. A. and Petty, H. R. (1998) Oscillatory deposition of reactive oxygen metabolites and pericellular proteolysis during neutrophil migration through model extracellular matrices. *Biophys. J.* **74**: 90–97.

Kruskal, B. A. and Maxfield, F. R. (1987) Cytosolic free calcium increases before and oscillates during frustrated phagocytosis in macrophages. *J. Cell Biol.* **105**: 2685–2693.

Liang, B. and Petty, H. R. (1992) Imaging neutrophil activation: analysis of the translocation and utilization of NAD(P)H-associated autofluorescence during antibody-dependent target oxidation. *J. Cell. Physiol.* **152**: 145–156.

Mogilner, A. and Oster, G. (1996) Cell motility driven by actin polymerization. *Biophys. J.* **71**: 3030–3045.

Omann, G. M., Rengan, R., Hoffman, J. F. and Linderman, J. J. (1995) Rapid oscillations of actin polymerization/depolymerization in polymorphonuclear leukocytes stimuated with leukotriene B4 and platelet-activating factor. *J. Immunol.* **155**: 5375–5381.

O'Rourke, B., Ramza, B. M. and Marban, E. (1994) Oscillations of membrane current and excitability driven by metabolic oscillations in heart cells. *Science* **265**: 962–966.

Petty, H. R. (1993) *Molecular Biology of Membranes.* New York: Plenum Press.

Petty, H. R. and Kindzelskii, A. L. (1997) Extremely low frequency electric fields promote metabolic resonance and cell extension during neutrophil migration. *J. Allergy Clin. Immunol.* **99**: S317.

Petty, H. R., Kindzelskii, A. L., Amit, A. and Jarvis, J. (1996) Effect of complement opsonization on immune complex-induced metabolic oscillations in migrating neutrophils. *FASEB J.* **10**: A1328.

Petty, H. R., Smith, B. A., Ware, B. R. and Rocklin, R. E. (1980) Alterations of the surface charge density of polymorphonuclear leukocytes by stimulated lymphocyte supernatants. *Cell. Immunol.* **54**: 435–444.

Petty, H. R. and Todd III, R. F. (1996) Integrins as promiscuous signal transduction devices. *Immunol. Today* **17**: 209–212.

Pralong, W.-F., Bartley C. and Wollheim, C. B. (1990) Single islet β-cell stimulation by nutrients: relationship between pyridine nucleotides, cytosolic Ca^{2+} and secretion. *EMBO J.* **9**: 53–60.

Rapp, P. E. (1979) An atlas of cellular oscillators. *J. Exp. Biol.* **81**: 281–306.

Remold-O'Donnell, E., Rosen, F. S. and Kenney, D. M. (1996) Defects in *Wiskott–Aldrich* syndrome blood cells. *Blood* **87**: 2621–2631.

Schienbein, M. and Gruler, H. (1995) Chemical amplifier, self-ignition mechanism, and amoeboid cell migration. *Phys. Rev. E* **52**: 4183–4197.

Shaya, S., Kindzelskii, A. L., Minor, J., Moore, E. C., Todd III, R. F. and Petty, H. R. (1998) Aberrant integrin (CR4, $\alpha_x\beta_2$; CD11c/CD18) oscillations in neutrophils in a mild form of pyoderma gangrenosum. *J. Invest. Dermatol.* **111**: 101–105.

Simon, D. I., Rao, N. K., Xui, H., Wei, Y., Majdic, O., Ronne, E., Kobzik, L. and Chapman, H. A. (1996) Mac-1 (CD11b/CD18) and the urokinase receptor (CD87) form a functional unit on monocytic cells. *Blood* **88**: 3185–3194.

Sitrin, R. G., Todd III, R. F., Petty, H. R., Brock, T. G., Shollenberger, S. B., Albrecht, E. and Gyetko, M. R. (1996) The urokinase receptor (CD87) facilitates CD11b/CD18-mediated adhesion of human monocytes. *J. Clin. Invest.* **97**: 1942–1951.

Smolen, J. E. and Weissmann, G. (1980) Effects of indomethacin 5,8,11,14-eicosatetraenoic acid, and *p*-bromophenacyl bromide on lysosomal enzyme release and superoxide anion generation by human polymorphonuclear leukocytes. *Biochem. Pharm.* **29**: 533–538.

Springer, T. A. (1990) The sensation and regulation of interactions with the extracellular environment: the cell biology of lymphocyte adhesion receptors. *Annu. Rev. Cell Biol.* **6**: 359–402.

Stossel, T. P. (1993) On the crawling of animal cells. *Science* **260**: 1086–1094.

Symons, M., Derry, J. M. J., Karlak, B., Jiang, S., Lemahieu, V., McCormick, F.,

H. R. Petty

Francke, U. and Abo, A. (1996) *Wiskott–Aldrich* syndrome protein, a novel effector for the GTPase CDC42Hs, is implicated in actin polymerization. *Cell* **84**: 723–734.

Vicker, M. G. (1994) The regulation of chemotaxis and chemokinesis in *Dictyostelium* amoebae by temporal signals and spatial gradient of cyclic AMP. *J. Cell Sci.* **107**: 659–667.

Weiss, S. J. (1989) Tissue destruction by neutrophils. *New Engl. J. Med.* **320**: 365–376.

Wymann, M. P., Kernen, P., Bengtsson, T., Andersson, T., Baggiolini, M. and Deranleau, D. (1990) Corresponding oscillations in neutrophil shape and filamentous actin content. *J. Biol. Chem.* **265**: 619–622.

Wymann, M. P., Kernen, P., Deranleau, D. A. and Baggiolini, M. (1989) Respiratory burst oscillations in human neutrophils and their correlation with fluctuations in apparent cell shape. *J. Biol. Chem.* **264**: 15829–15834.

Yang, Z.-Y., Delgado, R., Xu, L., Todd III, R. F., Nabel, E. G., Sanchez, A. and Nabel, G. J. (1998) Distinct cellular interactions of secreted and transmembrane Ebola virus glycoproteins. *Science* **279**: 1034–1037.

Zarewych, D. M., Kindzelskii, A. L., Todd III, R. F. and Petty, H. R. (1996) Lipopolysaccharide induces CD14 association with complement receptor type 3 which is reversed by neutrophil adhesion. *J. Immunol.* **156**: 430–433.

Zhelev, D. V. and Hochmuth, R. M. (1995) Mechanically stimulated cytoskeletal rearrangement and cortical contraction in human neutrophils. *Biophys. J.* **68**: 2004–2014.

8

Enzyme kinetics and nonlinear biochemical amplification in response to static and oscillating magnetic fields

JAN WALLECZEK AND CLEMENS F. EICHWALD

8.1 Introduction

The use of magnetic fields as a tool for influencing biological processes, which historically began with attempts to treat human disease, has had a long but checkered record. For many centuries following the discovery of the naturally magnetic material, magnetite (Fe_2O_3), the purported effects associated with this material were surrounded by superstition. In the first century AD, Pliny the Elder wrote about the apparently magical powers of 'lodestone', as magnetite was called then, such as the ability to heal the sick. The credible, scientific study of the biological effects of magnetism, however, has begun only in this century, and only in the 1960s were the first surveys of the laboratory evidence published (Barnothy, 1964, 1969).

Beginning in the 1970s, it was established that several animal species such as pigeon, salmon and honey bee were sensitive to even weak magnetic fields such as that of the Earth (for an overview, see Kobayshi and Kirschvink, 1995). This represents a remarkable sensitivity, since the magnetic flux density (B) of the Earth's magnetic field measured in units of tesla (T) is only about 50 microtesla (μT). For comparison, the magnetic field associated with a small, 1-cm toy magnet would be 1000-fold greater, for example, $B \approx 50$ millitesla (mT). In elegantly designed studies, scientists revealed that pigeons, salmon and bees were capable of sensing geomagnetic field lines as a way to orient themselves in their environment. The discovery of small amounts of magnetite in the biological tissue of these animals pointed to the role of magnetite as a potential element of a biological 'compass' for magneto-orientation. An understanding of the biomolecular mechanisms by which field interactions with magnetite could be translated into a biological change is, however, still lacking.

Until the late 1980s, with the exception of the behavioral effects in magneto-orientation, studies suggesting an influence of magnetic fields on other

193

biological endpoints were poorly accepted. This was despite the fact that many hundreds of studies had been published, including reports of the stimulation or inhibition by magnetic fields of enzyme activity and of biological signaling events, cell growth and metabolism, and tissue repair (for overviews, see Adey, 1981; Adey and Lawrence, 1984; Wilson *et al.*, 1990; Blank, 1995; Lacy-Hulbert *et al.*, 1998). One reason for this skepticism was the lack of independent replications of basic findings. In addition, studies were criticized for lacking sufficient methodological rigor, a criticism that appeared to be justified for a significant share of the earlier work. In the 1990s, however, reports appeared in the literature, suggesting the reproducibility between independent laboratories of several biological magnetic field effects such as on cellular signal transduction (Conti *et al.*, 1985; Walleczek and Budinger, 1992), isolated enzyme activity (Harkins and Grissom, 1994; Taoka *et al.*, 1997), and in clinical therapy (Ieran *et al.*, 1990; Stiller *et al.*, 1992). One specific insight that led to reproducible findings was the discovery that magnetic field effects, for example in lymphocytes, depended strongly on the *biological state* of the cells at the time of exposure (Walleczek, 1992, 1994; Eichwald and Walleczek, 1996a). The recent progress was echoed in the executive summary of a review by the US National Research Council (1997) which concluded, for example, that 'reproducible changes have been observed in the expression of specific features in the cellular signal-transduction pathways for magnetic-field exposures on the order of 100 µT and higher'. Although progress has been made, it is also clear that more work is needed to better establish the biological conditions under which robust effects can be routinely observed with stringent experimental protocols (e.g., Walleczek *et al.*, 1999).

8.1.1 The quest to understand magnetic field effects at the biomolecular level

Theoretical objections were another significant reason for the early skepticism concerning the experimental results. Physicists argued that most biological preparations do not contain any material that is magnetically susceptible with the rare, above-mentioned exception of magnetite potentially associated with certain neurosensory tissues. Therefore, the existence of an effective mechanism for direct magnetic field interaction with biological activity not related to sensory function seemed highly improbable. In 1992, we (Grundler *et al.*, 1992; Walleczek and Budinger, 1992) and McLauchlan (1992) hypothesized that the origins of some of the reported (nonsensory) biological magnetic field effects could be due to a mechanism already well-known in magnetochemistry for more than 20 years. This mechanism is known as the radical pair mechanism

(RPM). Similar proposals, although they went largely unnoticed then, appeared, for example, in the Russian scientific literature as early as 1976 (for a review, see Walleczek, 1995). A first critical step in the test of this hypothesis was to resolve the principal question of whether any biologically relevant activity could be affected as a consequence of this mechanism operating within a biological context. The answer came soon when a landmark study reported by Harkins and Grissom (1994) identified a magnetic field-dependent enzyme system for which its sensitivity to an applied magnetic field could conclusively be shown to be a direct consequence of the RPM. The studied enzyme was a vitamin B_{12}-dependent ammonia lyase, and, importantly, the paramagnetic radical pair species that served as the primary field coupling site could be identified as well (Harkins and Grissom, 1995). On the basis of these findings, in combination with knowledge about the details of the enzyme's reaction cycle, we constructed a biophysical model that could qualitatively reproduce the experimental results (Eichwald and Walleczek, 1996b). These developments represented a major advance in the long quest to understand how magnetic fields may interact with living systems: on the basis of established biophysical principles, the experimental findings, in combination with the modeling results, were capable of explaining a direct magnetic field effect on general biological (enzyme) activity at the molecular level.

8.1.2 A two-stage model for magnetic field interactions in biological systems

One major goal of our work is to help to establish the fundamental mechanisms of direct interactions between magnetic fields and living systems. A related goal is to develop magnetic fields as a novel tool for probing as well as for controlling biological activity for practical purposes, for example in biotechnology and biomedicine. At present, for both goal areas, the RPM provides the primary physical mechanism responsible for initial magnetic field detection, whereas enzyme-regulated and oscillatory biochemical reactions represent minimum models of secondary biological mechanisms. In our definition (see Walleczek, 1995; Walleczek *et al.*, 1999), the function of the *primary physical mechanism* is to enable the efficient coupling of the magnetic field signal to a molecular target in the presence of thermodynamic and other noise sources. The function of the *secondary biological mechanism* is to translate potentially small field-induced changes at the microscopic level into a biological change that is observable at the macroscopic level (Figure 1). In this chapter we discuss (1) the RPM as an example of a physical mechanism that is initially not limited by thermodynamic noise, and (2) possible features of

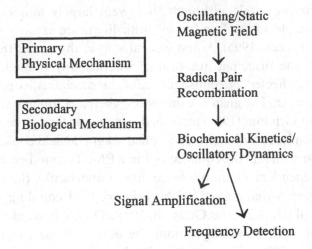

Figure 1. Proposed relationship between primary physical and secondary biological mechanisms during magnetic field interactions with biological systems. In this example, the primary physical mechanism describes magnetic field interactions with radical pair recombination events at the microscopic level, whereas the secondary biological mechanism represents a subsequent pathway leading to the induction of a macroscopic biological response.

enzyme-based biological mechanisms including signal frequency detection and nonlinear signal amplification (see Figure 1). Our findings will be discussed against the background of the role of self-organized biodynamical states in biological information processing in response to external stimulation and in regards to potential practical applications.

8.2 The radical pair mechanism in biological systems

The RPM is a well-established physical mechanism for describing how magnetic flux densities, e.g. of the order of 0.1 to 100 mT, can *nonthermally* affect chemical and biochemical reactions that involve transient free-radical states (e.g., Turro, 1983; Steiner and Ulrich, 1989; Grissom, 1995; Walleczek, 1995; Brocklehurst and McLauchlan, 1996). In contrast to electrical interactions (see Gailey, Chapter 7, this volume), magnetic field interactions through the RPM are not initially limited by thermal noise, because they result from direct field coupling to a coherent quantum-mechanical process that is not affected by thermal energy perturbations. During the past 20 years, research in magnetochemistry has identified several organic reaction systems that are sensitive to magnetic interactions through the RPM at millitesla intensities. For example, the application of magnetic fields of $B \approx 0.1$ to $1\,mT$ was found to change the reactivity of the dicyanobenzene/pyrene system (Weller *et al.*, 1983;

Batchelor *et al.*, 1993). Remarkably, in the case of this simple chemical reaction the lowest effective field intensity was close to the weak strength of the Earth's magnetic field. The systematic search for RPM-mediated magnetic field effects on biologically relevant systems, with the exception of photosynthetic reaction centers, has begun only recently (for reviews, see Grissom, 1995; Walleczek, 1995). One magnetically sensitive enzyme, a subject of research in our laboratory, is described further below. First, we need to briefly highlight basic physico-chemical principles that underlie the RPM. Since the subsequent description is restricted to a few key concepts only, for a detailed explanation we recommend the reviews by Steiner and Ulrich (1989) and McLauchlan and Steiner (1991).

8.2.1 Principles underlying the radical pair mechanism

Radicals are formed as reaction intermediates in many chemical and biochemical transformations. They are atoms or molecules with one or more free electrons, and they have, because of the magnetic spin moment of the free electron, *paramagnetic* properties (i.e., radical species are slightly attracted to a magnetic field gradient). Generally, radicals are created by cleavage of a covalent bond of a precursor molecule, M_p, as shown in Figure 2. Immediately after bond cleavage the two radical fragments, i.e., the *radical pair*, physically separate, for example by diffusion. They retain, however, their original electron spin orientation according to the Wigner spin conservation rule that describes the conservation of spin angular momentum. As a consequence of this quantum-mechanical process ('spin coherence'), despite physical separation, the spins remain correlated with each other for a short time period (e.g., 0.1–100 ns). To describe this special kind of correlated reaction intermediate the term *spin-correlated radical pair* was introduced (e.g., McLauchlan and Steiner, 1991). Effective magnetic field interactions may occur only during the brief time frame during which the radical pair spins are in their correlated state. This time frame is known as the *radical pair lifetime* (compare Figure 2).

The fate of the spin-correlated radical pair in the absence compared to the presence of an external magnetic field is illustrated in Figure 2. As shown in the top row in the Figure, the two members of the spin-correlated radical pair (R and R'), created by homolytic cleavage of the precursor molecule, M_p, may reencounter each other and – in the absence of an external magnetic field – recombine to form the original molecule, M_p. This process is called *radical pair recombination*, and the resulting species is known as the geminate or cage product. Recombination is, however, possible only if the radicals are *singlet* correlated; that is, if the two unpaired electron spins are oriented in an

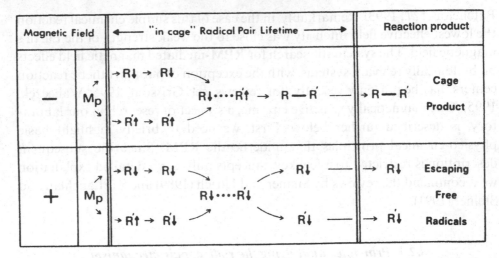

Figure 2. Illustration of the magnetic field effect on radical pair recombination. For details, see text in Section 8.2.1.

antiparallel fashion (R↓ and R′↑) as in the above example. The effect of the applied magnetic field on radical pair recombination is shown in the bottom row in Figure 2. The field changes the probability of conversion from the singlet-correlated spin state into an overall triplet configuration; that is, with the spins now in a parallel orientation (R↓ and R′↓). Unlike for singlet-correlated radical pairs, recombination of the *triplet*-correlated radical pairs is spin forbidden because of the Pauli exclusion principle; hence, as is illustrated in the figure, the two members of the radical pair *cannot* recombine, but diffuse away as free radicals. It is apparent from the above description that the magnetic field does not affect the chemical nature of the involved species; however, the external field does alter the ratio between recombined cage product and escape product (free radical) yields. Importantly, it has been shown that a related situation may occur in enzyme systems whose activity depends on the formation of a radical pair intermediate state during the enzyme reaction cycle (Grissom, 1995). Thus, an external magnetic field may be used as a tool to alter the ratio between cage and escape product formation and, consequently, to change the net forward flux within an enzyme reaction cycle (see Section 8.2.2).

As illustrated in Figure 2, the physical basis for the sensitivity of radical recombination to applied magnetic fields is the field's capacity to affect the interconversion between singlet and triplet electron spin states. Briefly, two mechanisms can be distinguished by which this may occur: the hyperfine coupling (HFC) and the Δg mechanism. The HFC mechanism describes the field's influence on interactions between the magnetic moments of the nuclear and electron spins, and it operates typically at low-to-moderate flux densities, e.g., $B \approx 0.1$ to $50\,mT$. For greater flux densities ($B \approx 0.1$ to $1\,T$) magnetic field effects can usually be accounted for by the Δg mechanism. The latter mechanism describes the interference of an external magnetic field with intrinsic magnetic interactions between the spin moment of the electron and its orbital motion around the nucleus, termed spin–orbit coupling. It is important to note that the effects due to the two mechanisms are opposite in sign: the HFC mechanism *decreases* the rate of singlet–triplet interconversions, whereas this rate is always *increased* as a result of the Δg mechanism. Consequently, *biphasic* effects may be observed for reaction systems in which both mechanisms operate; that is, field effects in opposite directions may result between exposures to lower compared to higher magnetic flux densities. An experimental example of this phenomenon for enzyme activity involving radical pair recombination is the subject of the next section.

8.2.2 Static magnetic field effects in enzyme kinetics

Recent *in vitro* experiments have demonstrated a substantial magnetic field effect on the B_{12}-dependent enzyme ammonia lyase, as was mentioned in Section 8.1.1 (Harkins and Grissom, 1994). Harkins and Grissom observed that the effect of the external field on the enzyme was *biphasic* in nature; that is, an inhibitory effect of the order of 25% was observed at lower flux densities and a return to zero-field conditions or even a slight increase at higher flux densities (compare Figure 3). Importantly, the original observation could be subsequently confirmed by another laboratory (Taoka *et al.*, 1997). This robust field effect, together with detailed knowledge of the enzyme's reaction cycle, presented us with a unique opportunity (1) to model the molecular mechanisms that may underlie this field effect and (2) to gain insight into the role of enzyme kinetic features in mediating this interaction. For this purpose we developed a prototypical model of a magnetic field-sensitive enzyme system. The model appropriately integrated the slower times scales of Michaelis–Menten-type enzyme kinetics with the much faster time scales of magnetic field-dependent electron spin kinetics associated with the paramagnetic enzyme radical pair species (for details, see Eichwald and Walleczek, 1996b). As illustrated in Figure 3, which compares the experimental results with those obtained from the simulations, our model was able to qualitatively reproduce the *biphasic* magnetic field effect on B_{12}-dependent ammonia lyase activity. In particular, the model simulations revealed that the size of the magnetic field effect was a function of the specific relations between the

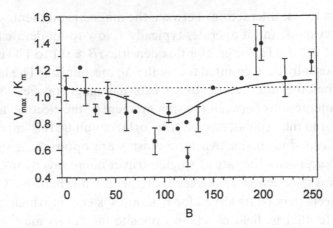

Figure 3. Comparison between experimental data and the model simulation of the biphasic magnetic field effect on V_{max}/K_m of coenzyme B_{12}-dependent ethanolamine ammonia lyase *in vitro* activity. The data points represent the experimental results obtained at different flux densities. The solid line indicates the result from the computer simulation. The magnetic flux density, B, is in mT. (Reproduced with permission from the Biophysical Society (Eichwald and Walleczek, 1996b).)

different rate constants such as (1) the ratio between radical pair lifetime and the rate of magnetic field-sensitive spin conversion induced by the HFC and the Δg mechanisms and (2) the chemical rate constants of the enzyme reaction cycle (Eichwald and Walleczek, 1996b).

It is a common assumption that, even if there exist magnetic field-sensitive biochemical pathways in cells as a consequence of the RPM, biologically significant effects are unlikely to occur on the basis of this mechanism. The reason is the usually small change in radical pair recombination probability due to a weak magnetic field that typically translates into a chemical rate change of only $\sim0.1\%$ to 5%. Biochemical rate changes of this size are not normally expected to play any significant role in shaping biological function. Our simulations suggested, however, another possibility: the field-sensitive enzyme itself may act as a *biochemical amplifier*. Specifically, we found that the specific relationships of the kinetic coefficients that govern the enzyme reaction cycle provide a means for amplifying small initial changes in radical pair recombination into disproportionally large changes in enzyme activity. We derived an amplification factor, η, from the specific relations between the different rate constants. This factor quantitatively characterizes the amplification properties of the single enzyme molecule, and accounts for the fact that – although the magnetic field-induced change in radical pair recombination probability is small – the effect on the enzyme reaction rate can be considerably larger, for example by a factor of 10 (see Figure 4). Thus, we suggest that

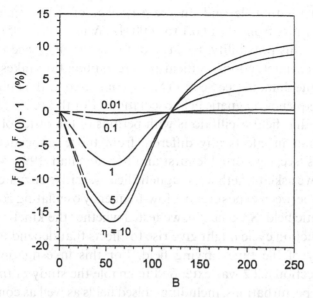

Figure 4. Dependence on the amplification factor, η, of the magnitude of the magnetic field effect on enzyme reaction rate, v^F. The magnetic flux density, B, is in mT. The dashed line extrapolates the behavior for $B \to 0$. (Reproduced with permission from the Biophysical Society (Eichwald and Walleczek, 1996b).)

enzyme-based kinetic amplification may have played a significant role in the induction of the reported 25% field effect on ammonia lyase activity (Harkins and Grissom, 1994). Our computational approach revealed that the combined analysis of physical and biochemical mechanisms (magnetic field modulation of radical pair recombination probability and amplifying kinetic processes within the enzyme reaction cycle, respectively) is a prerequisite for a qualitative as well as quantitative interpretation of magnetic field effects on enzyme activity (Eichwald and Walleczek, 1996b). In summary, despite the fact that magnetic modulation of RPM-dependent reaction rates may appear to be negligibly small, we propose that, when such interactions occur within an appropriate biological context, they may still lead to biochemical changes substantial enough to significantly affect biological function. Another mechanism for biochemical amplification is discussed in Section 8.3.2.

8.2.3 Oscillating magnetic field effects in enzyme kinetics

After successfully modeling the effect of a time-invariant, static magnetic field on enzyme kinetics (Figure 3), we set out to study the effect of a magnetic field whose intensity varies with time. The motivation for this work was our previous hypothesis that RPM-mediated magnetic field effects could occur in biological systems that depend on the *frequency* of field oscillations in the low-frequency range from about 0.1 to 1000 Hz (Walleczek, 1995). Others have argued against this possibility, because of the large difference in time scales involved in such interactions; radical pair recombination takes place in the nanosecond time domain (see Section 8.2.1), compared to the millisecond time scale of low-frequency magnetic field oscillations. For this reason it had always been assumed that field oscillations with periods of the order of milliseconds would not result in effects any different from those associated with static magnetic fields (see, e.g., Brocklehurst and McLauchlan, 1996; Valberg *et al.*, 1997). Hence, we asked whether a magnetic field-sensitive enzyme could detect any difference between exposure to a low-frequency oscillating compared with a static magnetic field. Specifically, we tested whether the kinetic properties of the enzyme reaction cycle might give rise to effects that depend on the oscillation frequency of the time-varying field. For this investigation the model discussed in Section 8.2.2 was extended to enable the study of *time-dependent* magnetic field perturbations, including pulsed fields as well as combinations of static and sinusoidally oscillating fields (for details, see Eichwald and Walleczek, 1997). One representative result from the computer simulations with *pulsed* magnetic fields is displayed in Figure 5. The figure illustrates that the magnetic field-exposed enzyme can indeed act as a *frequency sensor* that is responsive at lower field frequencies but less responsive at frequencies that are

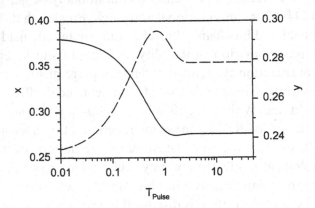

Figure 5. Relaxation behavior of the magnetic field-sensitive enzyme under the influence of pulsed magnetic fields. Time evolution of the variables *x* and *y* as a function of the duration of the magnetic field pulse. The variable *x* (solid line) is the fraction of the enzyme formed as enzyme–substrate complex prior to radical pair generation, and *y* (dashed line) representing the enzyme–substrate complex prior to product release (Eichwald and Walleczek, 1997). (Reprinted with permission from Eichwald, C. F. and Walleczek, J., Low-frequency-dependent effects of oscillating magnetic fields on radical pair recombination in enzyme kinetics. *J. Chem. Phys.* **107**: 4943–4950. Copyright 1997 American Institute of Physics.)

faster than the time scales inherent in the kinetic properties of the enzyme reaction cycle. Furthermore, a transition region in the frequency domain could be characterized that reflected the enzyme's relaxation behavior to *sinusoidally oscillating* magnetic fields, $B(t) = B_{AC} \cos(\omega_{AC} t)$: Figure 6a reveals a characteristic transition region near $\omega \approx 1$, where the oscillation amplitudes of two enzyme intermediate states corresponding to variables x and y decrease drastically, and Figure 6b illustrates the phase shift between the temporal variations of the applied field and the variations in the variables x and y.

The model simulations also suggested that specific combinations of static and sinusoidally oscillating magnetic fields could critically determine the temporal variations in the enzyme–substrate states as well as the overall enzyme reaction rate. Importantly, we determined that – at higher field frequencies – the application of oscillating magnetic fields caused changes that could not be predicted by knowledge of the effect of the static magnetic field component only. The simulations also provided evidence for the effectiveness of the oscillating magnetic field, even in the presence of a relatively stronger static magnetic field. It is important to note again that these effects reflect the kinetic properties of the underlying biological system; that is, the characteristic response behavior of the magnetic field-exposed enzyme, and not any low-frequency sensitivity of the RPM itself (Walleczek, 1995, 1999; Eichwald and Walleczek, 1997). In summary, our computer simulations confirmed the

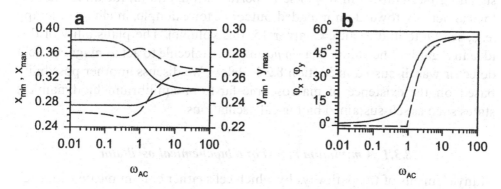

Figure 6. Frequency dependence of the response behavior of the enzyme to sinusoidally oscillating magnetic fields, $B(t) = B_{AC} \cos(\omega_{AC} t)$. The variables x (solid lines) and y (dashed lines) are defined as in Figure 5. (a) Oscillation amplitudes of x and y. (b) Phase shift ϕ_x and ϕ_y between oscillations in the variables x and y and the applied magnetic field. The phase shift ϕ_x (ϕ_y) is defined as the phase difference between the maximal amplitude of the oscillating magnetic field at $t = 0$ and the minimum (maximum) of the variable x (y). (Reprinted with permission from Eichwald, C. F. and Walleczek, J., Low-frequency-dependent effects of oscillating magnetic fields on radical pair recombination in enzyme kinetics. *J. Chem. Phys.* **107**: 4943–4950. Copyright 1997 American Institute of Physics.)

feasibility of our proposal that a magnetic field-dependent enzyme may sense
the frequency information contained in a low-frequency signal (Walleczek,
1995). Additionally, this work also indicated the possibility that an enzyme
may act as a *frequency-specific amplifier* of initially small RPM-mediated
effects induced by an oscillating magnetic field (Eichwald and Walleczek,
1997). An experimental test of this prediction is the subject of a future investigation.

8.3 Magnetic field stimuli as a tool for controlling self-organized biological dynamics

The discovery that certain enzyme-regulated biochemical reactions may provide a molecular coupling target for magnetic fields in cells raised an interesting possibility: the application of oscillating and static magnetic fields could be
developed into a minimally invasive tool for controlling biochemical and
biological activity. In particular, given the often cyclic or oscillatory features of
biological activity, we proposed the concept of influencing biological activity
with specific dependence on the *frequency information* contained in an oscillatory magnetic field (Walleczek, 1995; Eichwald and Walleczek, 1998). As was
shown before for other time-varying physical stimuli such as oscillating electric
stimuli, our approach could lead to the development of magnetic field perturbations as a tool (1) for studying the dynamical properties of biological
signaling mechanisms in response to perturbation, and (2) for directing biological activity toward an intended outcome, for example, in clinical therapy
(e.g., see Ditto and Spano, Chapter 15, this volume). The plausibility of this
idea in regard to the ability of a *single* enzyme molecule to act as magnetic field
detector was discussed in Section 8.2.3. Below, we discuss another possibility
based on the existence of macroscopic far-from-equilibrium biodynamical
states such as self-sustained biological oscillations.

8.3.1 A minimum model of a biochemical oscillator

Many elements of the pathways by which cells either communicate with each
other or receive and transduce signals from the environment are controlled
dynamically. This is reflected, for example, by the observation that the cellular
concentration of chemical messengers such Ca^{2+}, or of metabolites such as
nicotinamide adenine dinucleotide phosphate (NADPH), may oscillate spontaneously or in response to external stimulation. After years of speculation,
experimental evidence has finally emerged in confirmation of the idea that the
frequency of a biological oscillator may encode dynamical information that
controls basic cellular activity (e.g., see Petty, Chapter 7, this volume). In

particular, work with human lymphocytes has now firmly established that the degree of gene activation can be critically determined by the *frequency* of cytoplasmic Ca^{2+} oscillations in these cells (Dolmetsch *et al.*, 1998; Li *et al.*, 1998). Within this context we refer the reader to our earlier work, not discussed here, which specifically addresses the possibility of electromagnetic field perturbation of cytoplasmic Ca^{2+} oscillations (Eichwald and Kaiser, 1993, 1995).

Here we describe a model that we have developed for testing the feasibility of nonlinear frequency and amplitude control of biodynamical states with magnetic fields through the RPM (for details, see Eichwald and Walleczek, 1998). As a minimum model of a biodynamical state we chose a biochemical oscillator consisting only of coupled two-enzyme reactions (Figure 7). The system represents a cyclic reaction wherein substrate (S) is converted into product (P) catalyzed by reaction E_1, and product is subsequently converted back into substrate catalyzed by reaction E_2. It is assumed that one of the reactions includes an energy conversion step to drive the reaction against a concentration gradient. The designation substrate and product are of course interchangeable because of the cyclic nature of the system. The following specific case was investigated: (1) the activity of one of the enzymes (E_1) is controlled by substrate inhibition kinetics, and (2) the activity of the other enzyme (E_2) exhibits magnetic field sensitivity through the RPM in the manner described in Section 8.2. The first condition enables the inclusion of a regulatory feedback mechanism. For certain boundary conditions, this results in a system that is capable of sustained oscillatory behavior, since substrate inhibition kinetics offers a pathway for inducing enzyme-regulated biochemical oscillations (Shen and Larter, 1994). The second condition allows the integration of the magnetic

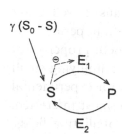

Figure 7. Scheme of the model of two coupled-enzyme controlled reactions. Substrate, S, is converted into product, P, catalyzed by enzyme reaction E_1 and, subsequently, P is converted back into S by enzyme reaction E_2. It is implicitly assumed that one of the reactions includes an energy conversion step, e.g., by hydrolysis of adenosine triphosphate (ATP), to drive the reaction against a concentration gradient. S is supplied at a net rate $\gamma (S_0 - S)$. The activity of enzyme E_1 is controlled by substrate inhibition kinetics, as indicated by the minus sign in the circle in the figure. Enzyme reaction E_2 exhibits magnetic field sensitivity through the radical pair mechanism. (Reprinted from *Biophys. Chem.* **74**, Eichwald, C. F. and Walleczek, J., Magnetic field perturbations as a tool for controlling enzyme-regulated and oscillatory biochemical reactions, pp. 209–224, Copyright 1998, with permission from Elsevier Science.)

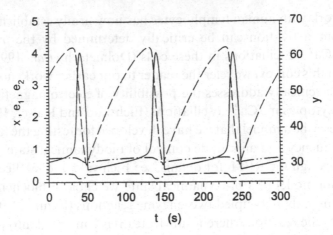

Figure 8. Oscillation diagram of the substrate and product concentrations and of two intermediate enzyme–substrate complexes. The lines represent the temporal evolution of substrate concentration (solid line, x), product concentration (dashed line, y), concentration of intermediate enzyme–substrate complex corresponding to enzyme E_1 (dashed-dotted line, e_1), and concentration of intermediate enzyme–substrate complex corresponding to enzyme E_2 (dashed-dot-dotted line, e_2). (Reprinted from *Biophys. Chem.* **74**, Eichwald, C. F. and Walleczek, J., Magnetic field perturbations as a tool for controlling enzyme-regulated and oscillatory biochemical reactions, pp. 209–224, Copyright 1998, with permission from Elsevier Science.)

field coupling step into the oscillatory biochemical reaction based on our previous model (Eichwald and Walleczek, 1996b, 1997). Figure 8 shows an example of the natural oscillation pattern of the coupled two-enzyme reaction system in the *absence* of magnetic field perturbation.

8.3.2 *Oscillating and static magnetic field control of the biochemical oscillator*

First, to characterize the amplification properties of the enzyme oscillator, the effects of static magnetic fields were studied. In Figure 9 the size of the primary field effect on radical recombination probability (right axis) is shown in comparison to the corresponding effect size on the oscillator period (left axis). For example, at $B = 25\,\text{mT}$ one can observe that, while the field caused a reduction in radical recombination probability by 6%, the oscillation period was reduced by about 60%. This finding demonstrates that the kinetic properties of the coupled enzyme reactions provide a source for greatly amplifying small initial changes, i.e., by a factor of 10 (for a comparison with experimental results, see Section 8.4.1). In this theoretical study we were most interested, however, in the response of the oscillator to periodically oscillating fields,

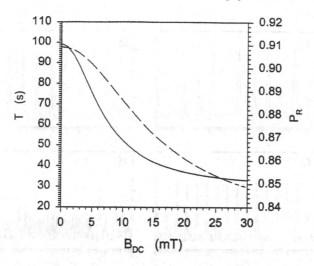

Figure 9. Response behavior of the enzyme oscillator toward static magnetic field perturbations. The solid line refers to the oscillation period (T) in seconds. The dashed line represents the probability for radical pair recombination (P_R). Note the different axis scaling. For an interpretation in regards to biochemical amplification see Section 8.3.2. (Reprinted from *Biophys. Chem.* **74**, Eichwald, C. F. and Walleczek, J., Magnetic field perturbations as a tool for controlling enzyme-regulated and oscillatory biochemical reactions, pp. 209–224, Copyright 1998, with permission from Elsevier Science.)

because in this case the oscillator's frequency-sensitive features may be directly affected. A range of different field frequencies was employed to study the oscillator's response towards stimulation (for details, see Eichwald and Walleczek, 1998). Some representative oscillation patterns are shown in Figure 10, where the flux density of the oscillating magnetic field was held fixed at 12 mT and only the field frequency was varied. For easy comparison, note that the natural frequency of the *unperturbed* oscillator ($\omega_{AC} = 0.063\,\text{s}^{-1}$) is close to the one shown in panel (b) in Figure 10. The panels (c)–(f) in the figure illustrate changes in the oscillatory patterns in response to field perturbation at increasing frequencies ($\omega_{AC} = 0.25, 0.47, 0.6$ and $1.0\,\text{s}^{-1}$). Panel (a), on the other hand, provides an example of the oscillator response to a field frequency ($\omega_{AC} = 0.03\,\text{s}^{-1}$) smaller than the natural, unperturbed oscillator frequency. It is evident from these results that both the amplitude as well as frequency of the biochemical oscillator can be drastically altered in strict dependence on magnetic field frequency. Remarkably, field frequencies much greater than the natural oscillator frequency, for example by a factor of 100, still induced nonlinear resonant responses (see Eichwald and Walleczek, 1998). The simulations, thus, support the notion of nonlinear frequency control of cellular activity by an oscillating magnetic field assuming that (1) the oscillator

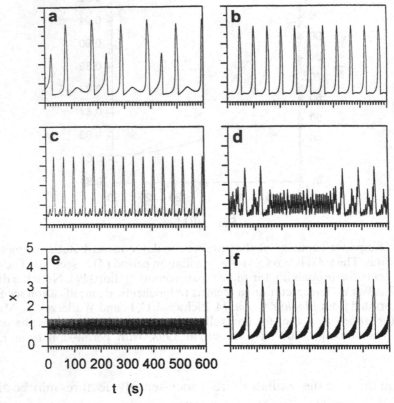

Figure 10. Frequency dependence of the enzyme oscillator to oscillating magnetic field perturbations, $B(t) = B_{AC} \cos(\omega_{AC} t)$. (a) $\omega_{AC} = 0.03 \, s^{-1}$, (b) $\omega_{AC} = 0.063 \, s^{-1}$, (c) $\omega_{AC} = 0.25 \, s^{-1}$, (d) $\omega_{AC} = 0.47 \, s^{-1}$, (e) $\omega_{AC} = 0.6 \, s^{-1}$, (f) $\omega_{AC} = 1.0 \, s^{-1}$. For details, see text in Section 8.3.2. (Reprinted from *Biophys. Chem.* **74**, Eichwald, C. F. and Walleczek, J., Magnetic field perturbations as a tool for controlling enzyme-regulated and oscillatory biochemical reactions, pp. 209–224, Copyright 1998, with permission from Elsevier Science.)

dynamics is dependent on a field-sensitive reaction step and (2) the oscillator frequency controls cellular activity as was observed, for example for cytoplasmic Ca^{2+} oscillations in lymphocytes (Dolmetsch *et al.*, 1998; Li *et al.*, 1998).

8.4 Nonlinear biochemical amplification in response to weak perturbation

In order to experimentally test our model predictions we implemented the oscillatory peroxidase–oxidase enzyme reaction as an experimental paradigm of a small biochemical reaction network (BRN). Because this enzyme reaction represents a cell-free system of a simple BRN consisting of only two coupled feedback loops, it yields basic information in regard to nonlinear signal amplification processes without interference from the complexity of biochemical interactions in whole cells. (For a detailed description of this oscillator, which is controlled by the enzyme horseradish peroxidase, see Larter *et al.*, Chapter 2, this volume). The experimental results shown here are chosen to demonstrate the *amplification properties* of an enzyme-regulated biochemical oscillator. The degree of amplification is expressed in terms of an *amplification factor*, A_S. A_S is defined as the ratio of an observed change in output response to the change in the input stimulus,

$$A_S = \frac{(\phi_f - \phi_i)/(\phi_i)}{(S_f - S_i)/(S_i)},$$

where ϕ represents the output response, S the input stimulus, and the subscripts i and f refer to the initial and final values, respectively (Goldbeter and Koshland, 1982). Our approach allowed the direct, real-time observation of the nonlinear amplification properties of a BRN in a continuously stirred, open-flow tank reactor (Figure 11). A change in the concentration of one of the enzyme substrates, nicotinamide adenine dinucleotide (NADH), served as the perturbing input signal. As illustrated in Figure 11, a small ($\sim 8\%$) change in the input signal at $t = 4000$ s was greatly and rapidly amplified by the BRN to yield a large change in the output response. This is apparent from the increase in the oscillation amplitude ($\sim 200\%$) as well as the oscillation period

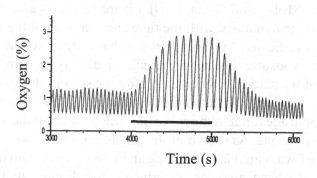

Time (s)

Figure 11. Experimental demonstration of the amplification properties of a nonlinear enzyme oscillator in response to weak substrate perturbation. The bar in the figure indicates the time during which the perturbation was applied. The perturbation consisted of a small change in the inflow concentration of substrate (NADH) from 4.0 nmol/s to 3.67 nmol/s to the reaction mix in a continuously stirred open-flow tank reactor (CSTR). The perturbation response of the peroxidase–oxidase enzyme oscillator was monitored with an oxygen electrode placed inside the CSTR (J. J. L. Carson and J. Walleczek, unpublished results). For details, see text in Section 8.4.

($\sim 100\%$); upon cessation of the perturbation at $t = 5000\,\text{s}$, the oscillator returned to its original dynamics within a few minutes. From the time series in Figure 11 we calculated that $A_S \approx 23$ and $A_S \approx 10$ for the effects on the enzyme oscillator amplitude and frequency, respectively. This finding clearly demonstrated the great capacity of an oscillatory BRN to amplify the effects from weak perturbations. The degree of amplification was similar to the magnitude ($A_S = 10$ to 20) suggested by our minimum model of a biochemical oscillator (see Figure 9). Finally, experiments with magnetic field perturbations, instead of chemical perturbations, of the peroxidase–oxidase oscillatory BRN are briefly discussed.

8.4.1 *Magnetic field control of an enzyme-regulated biochemical oscillator*

Stopped-flow kinetic measurements have recently revealed that there exist intermediate reaction steps in the redox cycle of the peroxidase enzyme that are sensitive to magnetic fields of the order of 1 to 100 mT in accord with the RPM (Taraban *et al.*, 1997). This finding compelled us to construct the minimum model described in Section 8.3.1 that integrates magnetic field-dependent chemical kinetics with oscillatory enzyme dynamics. The resulting computer simulations predicted that exposures to magnetic fields of $B \approx 1$ to 100 mT should be able (1) to affect the amplitude of the peroxidase–oxidase oscillatory system and (2) to cause transitions of the oscillatory dynamics from complex oscillations to periodic ones or vice versa (Eichwald and Walleczek, 1998). Recent experiments have confirmed both predictions (Carson *et al.*, 1999; Christine-Møller and Olsen, 1999). Figure 12 shows a time series of the oscillatory enzyme dynamics, with the time and duration of the applied static magnetic fields indicated by the horizontal bars. Prior to the onset of the magnetic field exposure, the enzyme oscillator displays a complex, possibly chaotic, oscillatory pattern. Immediately after the start of the 50-mT magnetic field exposure, the complex state switches to a periodic one of much smaller oscillation amplitude; after removal of the field, the periodic state switches back to a complex one. As shown in the graph, even the lower magnetic flux density of 25 mT was capable of inducing an inhibitory effect on the oscillation amplitude as had been predicted in principle (see Figure 9). This real-time evidence for magnetic field effects on a small BRN demonstrates the feasibility of the use of magnetic fields as a minimally invasive tool to influence and control biological activity. Importantly, these experimental findings, in combination with our modeling results (Eichwald and Walleczek, 1998), confirm that magnetic field effects on nonlinear enzyme dynamics can be understood on a theoretical basis.

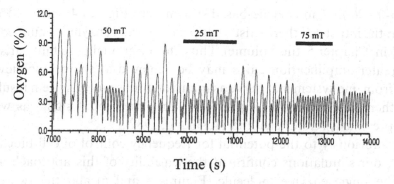

Figure 12. Real-time evidence for the magnetic field control of an enzyme-regulated, oscillatory state. The bars in the figure indicate the time during which the magnetic field was applied. The perturbation response of the peroxidase–oxidase enzyme oscillator was monitored with an oxygen electrode placed inside the continuously stirred open-flow tank reactor (Carson *et al.*, 1999). For details, see text in Section 8.4.1.

8.5 Conclusions and outlook

In recent years the science of bioelectromagnetics, i.e., the study of electromagnetic field interactions with biological systems, has offered significant new insights. For instance, this chapter has explained how direct magnetic field effects on enzyme activity can now be understood as a consequence of well-established biophysical principles. The century-old claims mentioned in Section 8.1 about possible beneficial health effects of strong, static magnetic fields thus might not be without mechanistic basis. In any case, the recent insights have opened the door for developing oscillating and static magnetic fields as a tool for studying, as well as for controlling, biological activity. Our findings in regard to *biochemical amplification* indicate that biological systems could be more sensitive than previously thought to weak external perturbations such as electromagnetic stimuli. While the sensitivity of a living cell to natural and artificial stimuli depends, of course, on the existence of a molecular target that is sensitive to stimulation at the microscopic level, it is the amplification properties of the underlying cell biochemistry that ultimately determine the *magnitude* and *characteristics* of the observable biological response (compare Figure 1). In this contribution, we have discussed two possible pathways for amplifying initially small effects: (1) the kinetic properties of single enzyme molecules (Eichwald and Walleczek, 1996b, 1997), and (2) enzyme-regulated, nonlinear biochemical oscillators operating under far-from-equilibrium conditions (Eichwald and Walleczek, 1998). Both the results from the computer simulations and those from the experiment with the peroxidase–oxidase oscillator demonstrated that small perturbations could be rapidly amplified, at least *10- to 20-fold*, in enzyme-based systems (see Figures 4 and 11). Furthermore, in the intact cell there exist multiply connected enzyme-regulated BRNs (see Arkin, Chapter 5, this volume). Thus, the additional possibility arises that even greater amplification gains may be achieved at the cellular level: the output from one system may serve as input for a subsequent system resulting in the further amplification of a previously amplified input signal as was discussed previously (Goldbeter and Koshland, 1982).

With previously to the potential for frequency control of cell biochemical activity, our simulations confirmed the feasibility of this approach as well. Both, the single enzyme molecule (Figures 5 and 6) and the two-enzyme coupled biochemical oscillator (Figure 10) were found to be capable of discriminating between different frequencies of magnetic field oscillations. These results may thus also provide a starting point for explaining previously reported field effects on biological systems that depended on the frequency of the applied magnetic field. For example, Bawin *et al.* (1996) reported that oscillatory electrophysiological activity in hippocampal brain tissue was capable of discriminating between 1-Hz and 60-Hz magnetic field exposures (for details, see Engström *et al.*, Chapter 9, this volume).

In summary, we have presented a theoretical framework based on the integration of concepts from bioelectromagnetics and nonlinear dynamics. The results from this cross-fertilization are encouraging because they have already revealed novel pathways for influencing essential biological processes. Our work may enable, for example, the development of an advanced magnetic field technology for the minimally invasive control of self-organized biodynamical processes at the cell biochemical level. At present, the kinds of static or oscillating magnetic field employed in the treatment of human disease such as in tissue repair are almost exclusively derived from empirical work without any firm theoretical foundations. In the future, research such as ours may enable the *rational design* of optimal magnetic field exposure protocols as a result of new insights into the sensitivity and response patterns of the underlying biodynamical mechanisms. Therefore we are optimistic that new approaches, based on the concepts proposed above, may soon assist the development of more effective electromagnetic technologies for clinical diagnosis and therapy.

Acknowledgment

Work at the Bioelectromagnetics Laboratory is supported by the US Department of Energy and the Fetzer Institute. Clemens F. Eichwald was a recipient of a Fetzer Institute Post-doctoral Fellowship.

References

Adey, W. R. (1981) Tissue interactions with non-ionizing electromagnetic fields. *Physiol. Rev.* **61**: 435–514.

Adey, W. R. and Lawrence, A. F. (eds.) (1984) *Nonlinear Electrodynamics in Biological Systems*. New York: Plenum Press.

Barnothy, M. F. (ed.) (1964) *Biological Effects of Magnetic Fields*. Vol. I. New York: Plenum Press.

Barnothy, M. F. (ed.) (1969) *Biological Effects of Magnetic Fields*. Vol. II. New York: Plenum Press.

Batchelor, S. N., Kay, C. W. M., McLauchlan, K. A. and Shkrob, I. A. (1993) Time-resolved and modulation methods in the study of the effects of magnetic fields on the yields of free-radical reactions. *J. Phys. Chem.* **97**: 13250–13258.

Bawin, S., Satmary, W., Jones, R., Adey, W. R. and Zimmermann, G. (1996) Extremely-low-frequency magnetic fields disrupt rhythmic slow activity in rat hippocampal slices. *Bioelectromagnetics* **17**: 388–395.

Blank, M. (ed.) (1995) *Electromagnetic Fields: Biological Interactions and Mechanisms*. Advances in Chemistry, No. 250, Washington, DC: American Chemical Society.

Brocklehurst, B. and McLauchlan, K. A. (1996) Free radical mechanism for the effects of environmental magnetic fields on biological systems. *Int. J. Radiat. Biol.* **69**: 3–24.

Carson, J. J. L., Maxim, P. G. and Walleczek, J. (1999) Real-time evidence for magnetic field-induced modulation and state transition in the oscillatory peroxidase–oxidase system. In *Proceedings of the 21st Annual Meeting of the Bioelectromagnetics Society*, 20–24 June, Long Beach, California, pp. 72–73.

Christine-Møller, A. and Olsen, L. F. (1999) Effect of magnetic fields on an oscillating enzyme reaction. *J. Am. Chem. Soc.* **121**: 6351–6354.

Conti, P., Gigante, G. E., Cifone, M. G., Alesse, E., Fieschi, C. and Angeletti, P. U. (1985) A role for calcium in the effect of very low frequency electromagnetic field on the blastogenesis of human lymphocytes. *FEBS Lett.* **181**: 28–32.

Dolmetsch, R. E., Xu, K. and Lewis, R. S. (1998) Calcium oscillations increase the efficiency and specificity of gene expression. *Nature* **392**: 933–936.

Eichwald, C. F. and Kaiser, F. (1993) Model for receptor-controlled cytosolic calcium oscillations and for external influences on the signal pathway. *Biophys. J.* **65**: 2047–2058.

Eichwald, C. F. and Kaiser, F. (1995) Model for external influences on cellular signal transduction pathways including cytosolic calcium oscillations. *Bioelectromagnetics* **16**: 75–85.

Eichwald, C. F. and Walleczek, J. (1996a) Activation-dependent and biphasic electromagnetic field effects: model based on cooperative enzyme kinetics in cellular signaling. *Bioelectromagnetics* **17**: 427–435.

Eichwald, C. F. and Walleczek, J. (1996b) Model for magnetic field effects on radical pair recombination in enzyme kinetics. *Biophys. J.* **71**: 623–631.

Eichwald, C. F. and Walleczek, J. (1997) Low-frequency-dependent effects of oscillating magnetic fields on radical pair recombination in enzyme kinetics. *J. Chem. Phys.* **107**: 4943–4950.

Eichwald, C. F. and Walleczek, J. (1998) Magnetic field perturbations as a tool for controlling enzyme-regulated and oscillatory biochemical reactions. *Biophys. Chem.* **74**: 209–224.

Goldbeter, A. and Koshland, D. E. (1982) Sensitivity amplification in biochemical systems. *Quart. Rev. Biophys.* **15**: 555–591.

Grissom, C. B. (1995) Magnetic field effects in biology: a survey of possible mechanisms with emphasis on radical-pair recombination. *Chem. Rev.* **95**: 3–24.

Grundler, W., Keilmann, F., Kaiser, F. and Walleczek, J. (1992) Mechanisms of electromagnetic interaction with cellular systems. *Naturwissenschaften* **79**: 551–559.

Harkins, T. T. and Grissom, C. B. (1994) Magnetic field effects on B_{12} ethanolamine ammonia lyase: evidence for a radical mechanism. *Science* **263**: 958–960.

Harkins, T. T. and Grissom, C. B. (1995) The magnetic field dependent step in B_{12} ethanolamine ammonia lyase is radical-pair recombination. *J. Am. Chem. Soc.* **117**: 566–567.

Ieran, M., Zaffuto, S., Bagnacani, M., Annovi, M., Moratti, A. and Cadossi, R. (1990) Effect of low frequency pulsed electromagnetic fields on skin ulcers of venous origin in humans: a double-blind study. *J. Orthopaedic Res.* **8**: 276–282.

Kobayashi, A. and Kirschvink, J. L. (1995) Magnetoreception and electromagnetic field effects: sensory perception of the geomagnetic field in animals and humans. In *Electromagnetic Fields: Biological Interactions and Mechanisms* (ed. M. Blank), *Advances in Chemistry*, No. 250, pp. 367–394. Washington, DC: American Chemical Society.

Lacy-Hulbert, A., Metcalfe, J. C. and Hesketh, R. (1998) Biological responses of electromagnetic fields. *FASEB J.* **12**: 395–420.

Li, W. H., Llopis, J., Whitney, M., Zlokarnik, G. and Tsien, R. Y. (1998) Cell-permeant caged $InsP_3$ ester shows that Ca^{2+} spike frequency can optimize gene expression. *Nature* **392**: 936–941.

McLauchlan, K. A. (1992) Are environmental magnetic fields dangerous? *Physics World* **92**: 41–45.

McLauchlan, K. A. and Steiner, U. E. (1991) The spin-correlated radical pair as a reaction intermediate. *Mol. Phys.* **73**: 241–263.

National Research Council (1997) *Possible Health Effects of Exposure to Residential Electric and Magnetic Fields.* Washington, DC: National Academy Press.

Shen, P. and Larter, R. (1994) Role of substrate inhibition kinetics in enzymatic chemical oscillations. *Biophys. J.* **67**: 1414–1428.

Steiner, U. E. and Ulrich, T. (1989) Magnetic field effects in chemical kinetics and related phenomena. *Chem. Rev.* **89**: 51–147.

Stiller, M. J., Pak, G. H., Shupack, J. L., Thaler, S., Kenny, C. and Jondreau L. (1992) A portable pulsed electromagnetic field device to enhance healing of recalcitrant venous ulcers: a double-blind, placebo-controlled clinical trial. *Brit. J. Dermatology* **127**: 147–154.

Taoka, S., Padmakumar, R., Grissom, C. B. and Banerjee, R. (1997) Magnetic field effects on B_{12}-dependent enzymes: validation of ethanolamine ammonia lyase results and extension to methylmalonyl CoA mutase. *Bioelectromagnetics* **18**: 506–513.

Taraban, M. B., Leshina, T. V., Anderson, M. A. and Grissom, C. B. (1997) Magnetic field dependence of electron transfer and the role of electron spin in heme enzymes: horseradish peroxidase. *J. Am. Chem. Soc.* **119**: 5768–5769.

Turro, J. N. (1983) Influence of nuclear spin on chemical reactions: magnetic isotope and magnetic field effects. *Proc. Natl. Acad. Sci. USA* **80**: 609–621.

Valberg, P., Kavet, R. and Rafferty, C. N. (1997) Can low-level 50/60 Hz electric and magnetic fields cause biological effects? *Radiat. Res.* **148**: 2–21.

Walleczek, J. (1992) Electromagnetic field effects on cells of the immune system: the role of calcium signaling. *FASEB J.* **6**: 3177–3185.

Walleczek, J. (1994) Immune cell interactions with extremely-low-frequency magnetic fields: experimental verification and free radical mechanisms. In *On the Nature of Electromagnetic Field Interaction with Biological Systems* (ed. A. H. Frey), pp. 167–180. Austin, TX: R. G. Landes.

Walleczek, J. (1995) Magnetokinetic effects on radical pairs: a paradigm for magnetic field interactions with biological systems at lower than thermal energy. In *Electromagnetic Fields: Biological Interactions and Mechanisms* (ed. M. Blank), *Advances in Chemistry*, No. 250, pp. 395–420. Washington, DC: American Chemical Society.

Walleczek, J. (1999) Low-frequency-dependent magnetic field effects in biological systems and the radical pair mechanism. In *Electricity and Magnetism in Biology and Medicine* (ed. F. Bersani), pp. 363–366. New York: Plenum Press.

Walleczek, J. and Budinger, T. F. (1992) Pulsed magnetic field effects on calcium signaling in lymphocytes: dependence on cell status and field intensity. *FEBS Lett.* **314**: 351–355.

Walleczek, J., Shiu, E. C. and Hahn, G. M. (1999) Increase in radiation-induced HPRT gene mutation frequency after nonthermal exposure to nonionizing 60 Hz electromagnetic fields. *Radiat. Res.* **151**: 489–497.

Weller, A., Nolting, F. and Staerk, H. (1983) A quantitative interpretation of the magnetic field effect on hyperfine-coupling-induced triplet formation from radical ion pairs. *Chem. Phys. Lett.* **96**: 24–27.

Wilson, B. W., Stevens, R. G. and Anderson, L. E. (eds.) (1990) *Extremely Low Frequency Electromagnetic Fields: The Question of Cancer*. Columbus, OH: Batelle Press.

9

Magnetic field sensitivity in the hippocampus

STEFAN ENGSTRÖM, SUZANNE BAWIN
AND W. ROSS ADEY

9.1 Introduction

We think of the brain as our very own computer, a never-ending process of signaling, providing overall guidance for the body and ultimately defining who we are. A multitude of signaling means has evolved for this task, combining mechanical, chemical and electrical communication into a complex network. As we consider the diversity among the types of signal present in the brain, it is not surprising that dynamical processes are also utilized to deliver its messages. Signals of electrical origin are particularly suitable for modulation in time. We see direct evidence of this by examining an electroencephalogram (EEG) or its magnetic counterpart, the magnetoencephalogram (MEG). These techniques provide ways of visualizing the natural rhythms occurring during various brain activities.

The hippocampus is an important structure in the brain that is particularly interesting within the context of dynamical signaling. It is central to the seemingly disparate phenomena of memory function and epilepsy. These two expressions of hippocampal activity are associated with normal and pathological behavior, both of which turn out to be responsive to manipulation with drugs as well as magnetic fields. *In vitro* studies of the rat hippocampus led us to conclude that an externally applied magnetic field with a steady low-frequency oscillation is capable of mimicking the function of nitric oxide (NO) in this model system. Rhythmic slow activity (RSA, or theta rhythm), which is correlated with learning and attention, is modified when the NO pathway is manipulated, either with drugs or with a low-frequency magnetic flux density of about 50 to 500 microtesla (μT). A second type of hippocampal rhythm, epileptiform activity, is similarly affected by NO manipulations and by the magnetic field exposure. These stimuli are capable of reducing the rate of occurrence of ictal episodes (electrical seizures) in a hippocampal slice, or have them cease entirely.

216

The prospect of using magnetic fields as a clinical tool is the ultimate motive for the studies presented in this chapter. In the future it may be possible to develop a therapeutic magnetic field technology by analyzing the magnetic field action within the context of better-understood drug applications. If so, the obvious therapeutic advantage would be the noninvasive nature of a magnetic field treatment. Surgery or scalp electrodes would not be necessary to deliver the stimulus, and it might even be possible to use existing equipment for transcranial stimulation to obtain the suggested low-level exposures.

The remainder of this section provides a brief overview of hippocampal stimulation by electric and magnetic fields, followed by two sections that describe our own parallel pharmacological and magnetic field experiments. We also review a set of experiments that examines a dynamical hypothesis for understanding the magnetic field detection mechanism at work in this system. In the final section we summarize our conclusions and offer some speculative thoughts suggesting future investigative paths.

9.1.1 Electric and magnetic field stimulation of the brain

Brain function relies heavily on electrical signaling, and it is not surprising that exogenous electric fields of sufficient strength are able to influence its operation. Magnetic field stimulation at high magnetic flux densities is also well known to influence many types of brain function, usually with the mechanistic understanding that a strong oscillating magnetic field induces substantial electric currents in the tissue (Markwort *et al.*, 1997). At the high-field end of magnetic stimulation as used in transcranial magnetic stimulation (TMS) and magnetic resonance imaging (MRI), magnetic flux densities are of the order of a few tesla. These high-field levels correspond to some tens of thousands times the naturally occurring geomagnetic field, a flux density comparable to the exposures employed in our experiments.

It is clear, then, that the present study is concerned with relatively weak magnetic fields – so weak that the corresponding induced electric fields are not likely to be sufficiently large to influence the normal electrical processes in the brain. About this level of interaction, there is much less information in the literature (Bell *et al.*, 1992a,b; Lyskov *et al.*, 1993), and the mechanism by which the tissue detects the magnetic field is not known. Since only the time-varying part of an oscillating magnetic field is responsible for inducing electric fields in living tissue, observations of static field effects on the brain immediately establishes a magnetic field receptor. The observation of static magnetic field effects is not definitive for human subjects, but is well established in several animal orientation models (Able, 1994).

The normally operating hippocampus is crucial for successful learning and memory retrieval (Berry and Thompson, 1978; Winson, 1978). RSA is a key component in this function, and was chosen for the first part of the study described below. Magnetic field influences have been implicated in behavioral experiments (Stern, 1995; Baker-Price and Persinger, 1996; Kavaliers *et al.*, 1996; Persinger, 1997; Lai *et al.*, 1998; Sienkiewicz *et al.*, 1998), and a plausible mechanistic explanation of these results may well include magnetic field sensitivity of the hippocampus.

A hippocampal epileptic seizure is a pathological condition directly related to excessive synchronous neuronal bursting. This activity can be triggered by visual or auditory stimuli. Electric currents injected directly into the hippocampus can induce or reverse an epileptic seizure (Durand and Warman, 1994). The mechanism by which these substantial electric currents affect the process is well understood in terms of hyperpolarization of the participating neurons. Fairly weak electric fields are also capable of affecting this system, as we have reported before (Bawin *et al.*, 1984). Control of hippocampal bursting patterns at a very sophisticated level is possible if the chaotic, but locally deterministic, system is properly modeled (Schiff *et al.*, 1994; for details, see Ditto and Spano, Chapter 15, this volume).

The strong dynamical character of both RSA and epileptiform activity is a common key to a deeper understanding of these processes. It is sometimes possible to avert an epileptic seizure, if during the aural period (immediately preceding the seizure), some kind of sudden distraction is introduced. This distraction, or perturbation in dynamical terms, can be as simple as a sudden noise. This auditory stimulus is probably only weakly coupled to the hippocampal rhythm – but a small change in the dynamical state is sometimes all it takes to deflect the system's descent into the highly synchronous and potentially destructive state of an epileptic seizure. Simply acknowledging that epilepsy belongs to the class of *dynamical diseases* allows a better understanding of its process, thus opening the door to subtle dynamical ways of therapy (see Milton, Chapter 16, this volume).

9.2 Parallel pharmacological and magnetic field studies

The sequence of experiments below is outlined in approximately the order in which they were performed and reported at scientific meetings over the past five years. Subtitles provide the main conclusions and supporting details are given in the text. In order to render the text more accessible, details of the experimental methodology are presented in the captions to figures and tables.

A number of pharmacological agents were used, mainly to probe the nature

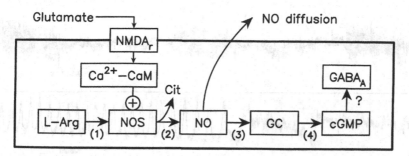

Figure 1. Overview of the nitric oxide/cyclic GMP/γ-aminobutyric acid (NO/cGMP/ GABA) pathway. (1), (2) L-Arginine is converted by the enzyme nitric oxide (NO) synthase (NOS) into NO and citrulline. This conversion may be blocked by agents that compete with L-arginine for binding sites on the enzyme, without being converted to NO and a residue. The available amount of NO can also be increased by adding NO donors, or decreased with NO chelating agents (see text for details). (3), (4) One of the main roles of NO in the brain is to activate guanylyl cyclase (GC) which is responsible for the formation of cyclic guanosine monophosphate (cGMP). Stable analogs of cyclic GMP can be added to the tissue to bypass blockade of NOS or GC. Cyclic GMP has been shown to modulate GABA receptor functions in cell cultures. We used agonists of $GABA_A$ to find out whether the magnetic field affects this portion of the pathway in our preparation. $NMDA_r$, *N*-methyldopamine receptor.

of a pathway initially involving NO (see Figure 1). Our story, as told below, is a parallel investigation of drug application and magnetic field exposure. These apparently diverse stimuli have similar effects on the two different aspects of hippocampal function that we chose to study. The most intriguing dynamical aspect of this investigation is the apparent frequency specificity of the response to magnetic field stimulation. Our studies show that a 1-Hz oscillating magnetic field, but not static or 60-Hz fields, are capable of interfering with the normal course of the hippocampal rhythms.

9.2.1 Establishing the magnetic field effect: frequency and amplitude response

Repetitive short epochs of RSA were induced in hippocampal slices by carbachol perfusion. RSA consists of episodes of rhythmic neuronal activity (8–15 Hz oscillations) of 7–15 s duration, separated by 20–40 s of relative quietude (see Figure 2). After a steady baseline of RSA intervals was established, slices were exposed to sham or field conditions for 10 min. During this interval, 20% of all slices destabilized spontaneously, if exposed only to ambient static magnetic fields. By 'steady baseline' we mean a regular behavior in terms of the seizure interval. Destabilization simply implies that these intervals become irregular, or cease altogether. When exposed to a 1-Hz, 56-μT_{rms} magnetic field (rms indicating root mean square), 67% of all slices

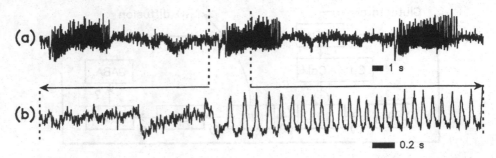

Figure 2. Rhythmic slow activity (RSA) in rat hippocampal slices. (a) Amplitude trace, as a function of time, of three episodes of RSA. (b) Magnification of the initial part of the second episode. RSA was induced by perfusion with carbachol, a stable analog of acetylcholine (Konopacki *et al.*, 1987). Carbachol is added to the perfusion solution and slices are perfused on average for an additional 40 min. RSA develops in about 80% of all slices treated in this way. If, at this point, a slice displays a steady baseline, it is chosen for inclusion in the study. We define a 'steady baseline' as a 10-min sequence of RSA episodes where no single interval between episodes deviates by more than two standard deviations (SDs) from the mean over the sequence. This baseline is later used to determine when the RSA becomes destabilized. When the RSA interval between two consecutive episodes deviates by more than six SDs from the baseline average, we record this time as the destabilization time. For carbachol-induced RSA experiments, Table 1 reports the median destabilization time and the number of slices that did destabilize during the first 10 min of field or sham exposure. Time scales are given by the black bars.

destabilized during exposure. The null hypothesis that these results could have been obtained purely by chance was rejected ($p < 0.02$). This experiment establishes that the magnetic field affected the process controlling RSA interval stability. An additional field condition using a 60-Hz, 56-μT_{rms} magnetic field destabilized 47% of the slices during the first 10 min of exposure. Although this result was not significantly different from the control experiment, we note that it elicited a response between that of the 1-Hz exposure and the control (Bawin *et al.*, 1996; see Table 1, nos. 4–6).

The effect of the 1-Hz, 56-μT_{rms} magnetic field is probably due to direct magnetic field detection, since the maximal induced electric field in the slices is estimated to be less than 0.4 $\mu V/m$. This level is well below reported effects of biological electric field stimulation, with the exception of the low-field end of some very specialized marine animals such as sharks (Kalmijn, 1997). Exploring a modest exposure parameter matrix, we tested the effects of varying the strength of the magnetic field. Apart from control experiments (zero AC field), fields at 5.6, 56 and 560 μT_{rms} were applied to the hippocampal slices after a stable baseline had been obtained (see Table 1, nos. 1–9). The lowest field level did not destabilize the RSA intervals to any degree different from the control

Table 1. *Summary of results from magnetic field exposures of rat hippocampal slices collected during the past four years*

No.	Prep.	Field exposure	Time (min.)	Fraction (%)	Statistics	Exp. no.
1	C	Control	19.2	20	NA	15
2	C	$5.6 \mu T_{rms}$, 1 Hz	15.0	47	NS	15
3	C	$5.6 \mu T_{rms}$, 60 Hz	10.0	47	NS	15
4	C	Control	17.9	20	NA	30
5	C	$56 \mu T_{rms}$, 1 Hz	7.3	67	$p < 0.02$	30
6	C	$56 \mu T_{rms}$, 60 Hz	10.4	47	NS	30
7	C	Control	18.9	13	NA	15
8	C	$560 \mu T_{rms}$, 1 Hz	5.0	80	$p < 0.05$	15
9	C	$560 \mu T_{rms}$, 60 Hz	12.3	33	NS	15
10	K	Control		10	NA	30
11	K	$560 \mu T_{rms}$, 1 Hz		37	$p < 0.05$	30
12	K	Control		5	NA	20
13	K	1.6 mT, static		12	NS	8
14	K	3.2 mT, static		17	NS	12
15	K + M	Control		25	NA	24
16	K + M	$560 \mu T_{rms}$, 1 Hz		62	$p < 0.02$	24
17	K + M	$1.6 mT_{p-p}$, PTF		25	NS	24

Different experimental protocols have been used throughout the time course of our study. The main differences lie in the type of hippocampal activity being studied, primarily influencing the perfusion preparation and the measures of activity. In all of the studies reported here, hippocampi were dissected from Sprague–Dawley rats (25–50 days old) and placed into cold (4 °C) artificial cerebrospinal fluid (ACSF; see Table 2). Slices are cut (500 µm thick) perpendicular to the long axis of the hippocampus. Six slices are then transferred to a recording chamber where they rest on a nylon mesh and are perfused from below with warmed (33–35 °C) and gassed (95% O_2, 5% CO_2) ACSF. A constant stream of a humidified O_2/CO_2 gas mixture flows over the slices (Bawin *et al.*, 1996). After equilibrating to these conditions, recording and stimulating electrodes are positioned in the cell layer of the slices (in the CA3 region). One or two slices are selected for study on the basis of their response to a test pulse. Once suitable slices are found, the stimulating electrodes are removed and the pharmacological preparation commences. All further pharmacological treatments are delivered by addition to the perfusion solution. All experiments are done in the ambient laboratory static field background of 45 µT (at an angle of 66° from the horizontal). Alternating vertical magnetic fields are generated with a single horizontal double-wound coil mounted around the perfusion chamber, and level with the slices. Prep., for slice preparation: C, carbachol; K, high [K^+]; M, muscimol. Field exposure: all controls are independent sham-field exposures (coils are powered with equal antiparallel currents). PTF, phase-tracking field (see Section 9.3 for details). Time: median time of destabilization onset. Fraction: fraction of slices which destabilized during a set interval. Statistics: statistical confidence for rejecting null hypothesis is given; NS, not significant; NA, not applicable. Exp. no., number of tested hippocampal slices.

samples. As mentioned above, the 56-μT_{rms} magnetic field destabilized RSA in 67% of the slices. When exposed to 560 μT_{rms}, 90% of all slices destabilized during the 10-min exposure. This experiment therefore establishes a lower threshold for the observed magnetic field effect, and also supports the notion of a 'normal' dose–response relationship with respect to field amplitude: a larger stimulus produces a larger response. These experiments were also carried out with 60-Hz instead of 1-Hz fields of the same flux densities. The higher frequency produced the same trends in terms of dose response, but they were not significantly different from the control experiments (see Table 1).

9.2.2 The presence of NO destabilizes RSA intervals

Previous work in our laboratory showed that an increasing amount of NO tends to destabilize the RSA intervals, and, conversely, decreasing its availability promotes stability (Bawin *et al.*, 1994). We found that hemoglobin, which chelates extracellular NO, stabilizes the system. The addition of nitro-L-arginine (NLA) or methyl-L-arginine (MLA), which decrease the intracellular production of NO by inhibiting NO synthase, similarly regularizes the RSA intervals. In contrast, sodium nitroprusside (SNP), a NO donor, destabilizes RSA intervals. For better overview, Table 2 summarizes the employed pharmacological interventions. Our experimental findings are consistent with the suspected action of NO in this system and are in general agreement with results reported in the literature (Pape and Mager, 1992; Vincent and Hope, 1992; Zorumski and Izumi, 1993). All experiments were qualitative in the sense that no attempt was made to find optimal drug concentrations, but rather they were used to infer the type of response elicited with any particular substance. Figure 1 shows pertinent aspects of the NO pathway in our system.

9.2.3 Hippocampal magnetic field response is dependent on NO

One hypothetical magnetic field coupling target is the radical-dependent enzyme NO synthase. NO synthase activity has been proposed as a potential candidate for a magnetically sensitive enzyme reaction, although no direct measurements have been made (Walleczek, 1995). As noted above, addition of NLA to the perfusion solution inhibits NO synthase and stabilizes the system. Furthermore, in the presence of NLA, exposure of the slices to the 1-Hz, 56-μT_{rms} magnetic field does not destabilize the RSA intervals. However, after removal of NLA from the slices, the 1-Hz field may again destabilize the intervals, suggesting that the field interaction is dependent on NO synthase activity (Bawin *et al.*, 1996).

Table 2. *Summary of types of chemical agent used during the study*

Abbreviations	Compound	Comments
ACSF	Artificial cerebrospinal fluid	Physiological perfusion solution
	Carbachol (20 μM)	Induces rhythmic slow activity (RSA)
K^+	Potassium ions (8.5 mM)	Induces ictal activity
NLA	N^G-nitro-L-arginine (100 μM)	NO synthase blocker
MLA	N^G-methyl-L-arginine (100 μM)	NO synthase blocker
SNP	Sodium nitroprusside (100 μM)	NO donor
CGMP	8-Bromo-cGMP (10 μM)	Cyclic GMP analogue
Hb	Hemoglobin (100 μM)	Chelates extracellular NO
	Muscimol (0.6 μM)	$GABA_A$ receptor agonist

All pharmacological treatments were supplied through the perfusion system. The ACSF (pH = 7.3) consisted of: 2.0 mM $CaCl_2$, 3.75 mM KCl, 1.2 mM $MgSO_4$, 1.25 mM NaH_2PO_4, 124 mM NaCl and 10 mM glucose. Drugs are generally added once the slice exhibits stable baseline behavior.

Finally, the addition of hemoglobin to chelate extracellular NO still allowed the 1-Hz, 560-$μT_{rms}$ field to destabilize the RSA intervals. This finding indicates that the field detection process is not dependent on freely diffusing extracellular NO, lending further support to the hypothesis that an intracellular process such as NO synthase activity is directly involved in the field detection process.

9.2.4 Epileptiform activity is also a NO-dependent process

NO is a general signaling substance, affecting not only RSA but many other neuronal processes as well (Vincent and Hope, 1992). Therefore, we explored the possibility that NO can modulate hippocampal rhythmic activities other than RSA. Our next experimental paradigm, *epileptiform activity*, again turns out to respond to both NO manipulations and magnetic field stimulation. It is also an experimental model with direct clinical relevance as a prototype for epilepsy. To provoke the induction of paroxysmal (epileptiform) activity, which is characterized by ictal episodes (seizures) separated by interictal bursting (see Figure 3), we perfused the hippocampal slices with a high (8.5 mM) concentration of potassium ions ([K^+]). For experimental details see the caption to Figure 3.

NO has been suggested as an anticonvulsant from *in vivo* experiments (Kirkby *et al.*, 1996), and the expected behavior to pharmacological treatment is indeed observed in our system: the interfit intervals between ictal episodes

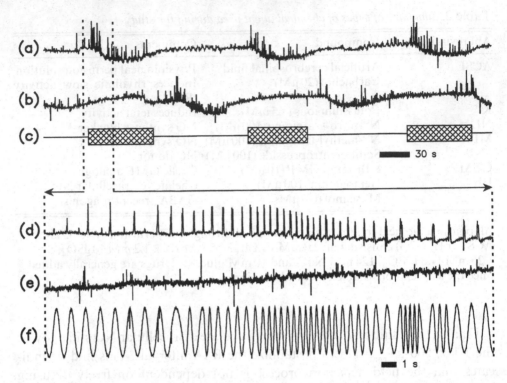

Figure 3. Epileptiform (ictal) activity in rat hippocampal slices. (a) Three ictal episodes in the 'target' slice are shown. (b) Simultaneous trace of the 'control' slice. (c) The phase tracking is active during the periods of ictal activity in the 'target' slice (shown as hatched boxes); at all other times (connecting lines), the magnetic field signal reverts to the constant frequency of 1 Hz. (d)–(f) show a magnification of traces (a)–(c) for the indicated regions. Epileptiform activity was induced by perfusion at a high $[K^+]$, which generates paroxysmal bursting by activation of the glutamate receptor (Korn *et al.*, 1987). The slices require perfusion at a high $[K^+]$ (increased to 8.5 mM from 3.5 mM) for 90 min to exhibit steady ictal episodes. A slice is accepted for experimentation if it displays stable baseline interictal intervals for 20 min. For the early studies on epileptiform activity, a measure similar to that used for quantifying the rhythmic slow activity intervals was used. Baseline behavior was established and the first interictal interval to deviate more than six standard deviations from the established average was taken as a measure of the onset of destabilization. In later experiments, muscimol was added to the perfusion solution to bias the ictal response. In these cases, the destabilizing agents tend to completely shut down the ictal activity, and the time of complete inhibition was used as a measure of efficacy of the drug or field treatment.

were lengthened by addition of the NO donor SNP, and shortened by NLA, an inhibitor of NO synthase (see Table 2).

9.2.5 Epileptiform activity is inhibited by magnetic field exposure and by GABA agonists

The interfit intervals are prolonged by the 1-Hz, 560-μT_{rms} magnetic field, a behavior analogous to our previous observations of the RSA intervals. The similarity in response to pharmacological treatment suggests that this system too may be manipulated via the NO pathway (see Table 1, nos. 10–11). Recent pharmacological studies in our laboratory suggest that interfit intervals are regulated by γ-aminobutyric acid (GABA) through the GABA$_A$ receptor. The receptor's activity has been shown to be modulated by NO through a cyclic GMP-dependent process (compare Figure 1; Bradshaw and Simmons, 1995; Wexler *et al.*, 1998). Thus, as expected, the addition of the GABA$_A$ agonist muscimol increased the interfit intervals, or inhibited them altogether if given in sufficiently high concentrations. Therefore, we employed muscimol as an agent to potentiate the slices for the magnetic field stimulation, a procedure that allowed us to bias the slices' response to other influences. This technique was used in the series of experiments described in Section 9.3 below.

9.2.6 Static or variable-frequency magnetic fields do not affect epileptic activity

Static magnetic fields of the same and twice the amplitude as the peak-to-peak amplitude of the 1-Hz sinusoidal magnetic field ($B = 1.6$ and $3.2\,mT$) failed to influence the interval between seizures in hippocampal slices perfused with a high $[K^+]$. These results are summarized in Table 1 (nos. 12–14) and suggest that the oscillating character of the 1-Hz signal is crucial for influencing this NO-dependent system. Thus, we now focus on the observation that the 1-Hz, but not the 60-Hz or the static magnetic field was effective. One can view this as general frequency dependence, probably stemming from a dynamical response at, or following, the magnetic field transduction step. A more specialized conjecture is that there exists an oscillatory dynamical system in hippocampal tissue, with an intrinsic frequency near 1 Hz, making it particularly sensitive to signals similar to its own oscillation period. This idea of resonant coupling could provide us with the needed leverage to understand how the relatively weak field could have an effect on the system.

Assuming that this intrinsic oscillator exists and that it manifests itself directly or indirectly in the electrophysiological recordings, we hypothesized that applying the magnetic field *in-phase* with the observed signal would affect the system to a greater degree. The details and logic of this experiment are somewhat involved and will be deferred until Section 9.3. Suffice to say here that this hypothesis appears to be rejected by our observations, and that a possible resolution is that the detailed timing and frequency of the ictal episodes is not a direct indicator of any system manipulated by the magnetic field.

With hindsight, another experimental observation from the previous studies appears to have some relevance within this context. In no case where the magnetic field affected the slice behavior was the signal amplitude, burst rate or fit duration affected. Only the intervals between episodes changed in response to the magnetic field. This may indicate that the interaction occurs on a higher level of regulatory control of these rhythmic activities.

9.3 Dynamical studies: phase-tracking experiments

Our most recent magnetic field experiments are especially pertinent to the scope of this book. As we outlined above, our experiments revealed that sinusoidal 1-Hz magnetic fields at 56 and 560 μT_{rms} are capable of inhibiting normal as well as paroxysmal activity in hippocampal slices. The frequency character of this field was found to be important: the same strength field at 60 Hz, or a static field of the same amplitude, failed to induce a significant change.

Our objective was to test whether the 1-Hz signal is somehow linked to the comparable frequencies we obtained from the time series of the electrophysiological recordings. As mentioned before, the natural frequency of RSA is 5–13 Hz, while the burst interval in an ictal episode is of the order of 0.5–3 Hz (see Figures 2 and 3). We initially assumed that the extracellular recording from the CA3 region reflects an intrinsic oscillator that is directly influenced by the element that mediates the magnetic field action. Given this assumption, we hypothesized that, by providing a stimulus that closely follows this observed bursting pattern in frequency and phase, we could increase the field's impact on the underlying neuronal processes. Three exposure conditions were tested: (1) 1-Hz, 1.6-mT_{p-p} (p–p indicating peak to peak) sinusoidal magnetic field, (2) 1.6-mT_{p-p} magnetic field, with the frequency varying in the range 0.5–3.0 Hz, attempting to match the phase of the target slice signal, and (3) zero AC field.

We monitored the extracellular electrophysiological signals in two hippocampal slices called 'target' and 'control' (see Figure 4). To achieve the phase-tracking experimental condition (2), the electrophysiological signal from the 'target' slice was processed in real time to predict the occurrence of the

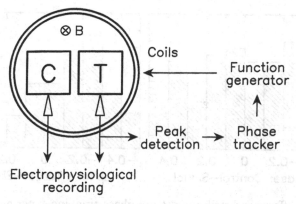

Figure 4. Sketch of the exposure strategy for the phase-tracking experiments. Extracellular signals were monitored in two hippocampal slices called 'target' (T) and 'control' (C). To achieve phase tracking, data from the 'target' slice are processed in real time and used to control a function generator. This, in turn, drives a set of coils to produce a magnetic field, which is close in frequency and phase to the experimental signal in the 'target' slice. The 'control' slice experiences an irregular field with no correlation to its own bursting activity. Slices were exposed to the sham or field stimulus for 10 min. A preparation was considered to respond to the field stimulation if the fit activity stopped within 30 min following the onset of the sham/field exposure.

next burst. Once an estimate of the next burst is available, the frequency of a function generator is modified in an attempt to synchronize the phase of the applied magnetic field oscillations with the occurrence of the next burst. During the quiescent periods between ictal episodes, the function generator reverts to the 1-Hz field used throughout the exposure condition (1).

The occurrence of each individual burst is a stochastic event, and the phase tracking is not perfect. We monitored a five-event window to achieve the tracking, and this number of bursts was enough to initiate adequate phase tracking. Figure 5 gives an idea of how well our phase-tracking device performed. While tracking the 'target' slice, the 'control' slice experiences an irregular field with no correlation to its own bursting activity, providing a simultaneous out-of-phase reference. This fact turns out to be relevant for interpreting the outcome of this series of experiments.

9.3.1 Interpretation of phase-tracking experiments

The 1-Hz sinusoidal signal in exposure condition (1) stopped the seizures within 30 min in 62% of the slices. On the other hand, exposure condition (2), using a frequency chosen to match the signal phase to that of the bursting pattern in the fits, led to results no different from the control condition (3): 25% of slices had ceased their seizure activity in both cases (see Table 1, nos. 15–17).

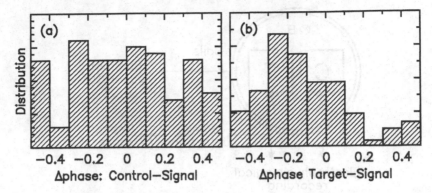

Figure 5. Phase difference distributions from phase-tracking experiments. (a) Relative phase between the 'control' and the phase-tracking signal generated by monitoring the burst activity in the 'target' slice. There is no correlation and a uniform distribution is expected. (b) Ideal phase tracking would be a single spike at Δphase = 0. We obtain a broadened peak due to the stochastic nature of the individual bursts, and a lag because we are predicting the next burst from data derived from immediately preceding events. The relative phase is defined as the phase difference divided by the current period of the signal.

In no instance was there a difference between the 'target' and 'control' slices. No strong conclusions are possible besides the obvious rejection of the hypothesis that seizure inhibition would improve by phase tracking the individual bursts. The simplest explanation is that the system where magnetic field transduction occurs is not directly associated with the signal represented by the electrophysiological measurements. The transduction step, however, may still be resonating with an intrinsic time scale of the order of 1 Hz, although affecting a higher level of control.

We can choose to view the phase-tracking field in the 'target' slice as a 1-Hz field interrupted by episodes of the same amplitude but irregular frequency. Formulated in this way, our results resemble those of Litovitz *et al.* (1997), where the constancy of the field over a certain period of time is the most important aspect of the field exposure. However, the required time over which the character of the field should not change differs from their studies. Our system appears to require constancy over the range 1–10 min, whereas Litovitz *et al.* (1997) observed 1–10 s as the required time interval during which exposure parameters should not change.

9.4 Conclusions, discussion and a speculative outlook

The experiments reviewed here led us to conclude that relatively weak 1-Hz oscillating magnetic fields are capable of modulating normal and paroxysmal rhythmic actions of the rat hippocampus. We observed an apparent effect threshold somewhere in the magnetic flux density range from 5.6 to $56\,\mu T_{rms}$. In contrast, static fields or fields oscillating at 60 Hz did not induce a significant effect, although the 60-Hz field exposures exhibited the same trends as a function of field amplitude as did the 1-Hz exposures. We also found that pharmacological manipulation of the NO/cyclic $GMP/GABA_A$-pathway affected the studied hippocampal rhythms. In the instances where drug and magnetic field studies interlock, our results support the hypothesis that these two types of stimulus are acting upon the same system. The transduction step for the magnetic field sensitivity is still largely undefined in this system. What we do know is that the detector is probably sensing the magnetic field directly, as opposed to detecting secondary electric fields or currents that are induced by the time-varying component of the field. We also know that the frequency information is critical to the detection process, since static and 60-Hz fields did not elicit significant responses. At this point in the discussion, however, we are forced to move into the realm of conjecture.

As mentioned above, there is minimal information pointing to the type of the primary field interaction step. One possibility concerns the prediction that a mechanism based on magnetite could provide a reduced effect magnitude at the higher frequency as observed in our system (Kirschvink *et al.*, 1992a,b). Magnetite has been reported to exist in the human hippocampus, although its functional and structural organization has not been demonstrated in this context (Dunn *et al.*, 1995). On the other hand, static magnetic fields might be expected to affect a magnetite-controlled system as well, but dynamical events at, or after, the initial transduction step could, in theory, circumvent such objections.

Another transductive mechanism concerns the possibility of affecting the recombination of geminate or spin-correlated radical pairs in accord with the radical pair mechanism (RPM), especially within the context of enzyme kinetics (Walleczek, 1995; Eichwald and Walleczek, 1996). One of the objections to the RPM is the expected small change in enzyme activity to magnetic fields at the field levels with which we are concerned here. The proposed enzyme model suggests solutions to some of these objections. For example, it has been proposed that an enzyme may act as an effective amplifier of initially small RPM-dependent changes (Eichwald and Walleczek, 1996). Consistent with our observation of a reduced effect at the higher frequency, the enzyme model predicts that only magnetic fields oscillating at sufficiently low frequencies would be able to affect enzyme activity (Eichwald and Walleczek, 1997). These models have recently been extended to account for field effects on nonlinear dynamical enzyme activity, which provides an additional foundation for signal frequency specificity and amplification (Eichwald and Walleczek, 1998; see

Walleczek and Eichwald, Chapter 8, this volume). Signal cascades may achieve amplification and it is possible that the signaling pathway in our system is capable of providing this function (see Figure 1).

As was mentioned in Section 9.1, a magnetoencephalogram (MEG) reveals global spatiotemporal dynamics during brain function (Bamidis *et al.*, 1995; Kawamichi *et al.*, 1998). The low-frequency magnetic fields detected by the MEG are a consequence of the underlying neuroelectrical signaling processes. It is unlikely that these extremely weak biogenic magnetic fields may themselves play a role in the signaling scheme of the brain, although this possibility cannot be completely ruled out. For example, it might still be possible that these fields are sufficiently large *locally* to be able to modulate some magnetically sensitive neurochemical process. At present, however, no molecular targets or mechanisms are known that would be sensitive to the average magnetic flux densities associated with neuroelectrical dynamics.

As to the dynamical component of the magnetic field transduction, our only solid observation is that the phase-tracking strategy we adopted was not successful. Here, we provide the three seemingly most plausible explanations for this result:

(1) The electrophysiological recording is not directly reflecting any property of the system that is responsible for detecting the 1-Hz sinusoidal signal.
(2) The detection system requires constancy over a certain period of time in order to function.
(3) The phase-tracking device was using an incorrect phase-angle difference and it might have succeeded if this parameter were appropriate for the system.

The observation that drug or field treatment does not affect the structure of individual bursts or the length of the ictal episode lends some support to explanation (1) above (see Sections 9.2.6 and 9.3.1).

Another important topic relevant to this discussion is the role of neuronal communication pathways in determining the dynamics of these processes. Chemical synapses are adequate to model the ictal events we are discussing, at least for understanding synchronization down to the level of several tens of milliseconds (Traub and Richard, 1991). Epileptic discharges are synchronized, however, on much faster time scales, and thus there must be more to the interaction network than synapses alone. An independent indication is the fact that synchronous bursting is possible in the hippocampus even while synaptic transmission is essentially blocked by a low Ca^{2+} concentration (Kaczmarek and Adey, 1975; Jefferys and Haas, 1982; Taylor and Dudek, 1982).

Two sufficiently fast means of neuronal interactions are well known. Gap-junction-mediated communication is available in the hippocampus, but it appears that the involved neurons are not sufficiently well connected for this to be an efficient means of signal propagation (Traub *et al.*, 1985). Ephaptic communication works through direct electric field interactions between touching, or nearby neurons – hence the name, εφαπτω = to touch (Jefferys, 1995). Related to ephaphsis is the term 'field effect': field influences on a group of neurons mediated by an electric field generated *locally*. The differences between these terms are ill defined, partially because they may essentially be the same phenomenon, only occurring on different scales. We may view ephaphsis as a process normally limited in space and time to a restricted domain of cells. In the pathophysiology of epilepsy, these constraints may be temporarily lost, allowing a far wider spread of nonsynaptic excitation. The ephaphtic mechanism for signaling is fairly subtle in terms of the electric fields involved. However, its speed, combined with structural pathways defined in the hippocampus, provides an alternative route to massive interconnectedness required for synchronization and dynamically interesting properties.

The phenomenon of periodic rhythms in the hippocampus are clearly, but not necessarily exclusively, electrical in nature. It is thus not surprising that sufficiently large electric fields are capable of influencing the system. The situation is very different for the magnetic field stimulus employed here, since there is no well-understood magnetic component to its function at present. However, some mechanistic insights may be gained from our studies that compared the effects of the magnetic field with that of NO on this system (see Section 9.2). Strikingly, all the experiments were consistent with the hypothesis that the magnetic field acts similarly to this chemical agent. From our work, a direct coupling to the short time scale mechanisms seen in RSA or epileptiform activity seems implausible. Our experiments designed on a dynamical premise (see Section 9.3.1) also argue against such an interpretation: only the interval *between* seizures was changed as a result of the magnetic field exposure – the exposure never significantly affected the trace amplitude, burst frequency or any other local measure.

In summary, our experimental evidence points to a field influence more akin to the diffusive action of NO than an acute perturbation at the level of the individual bursting processes. Furthermore, our results do not rule out that some complex dynamical state is subtly affected, but they imply an effect on a process that occurs on a longer time scale. While this process is still determined by neuronal activity, any dependence on the fast processes that produce the characteristic time series is entirely unknown at this point. When considering, for example, the existence of human consciousness, it may seem obvious that neuronal cooperative processes are capable of spanning life times of the order of minutes, yet more direct evidence from electrophysiological recordings is available (Adey, 1972).

Regardless of how hippocampal synchronization and communication is achieved, it is the *dynamics* at the neuronal level that must be influenced in order to control a disorder such as epilepsy. Our best option for attaining this goal may be to view functional hippocampal integration as a complex dynamical state, which is remarkably stable to many types of perturbation but still able to respond to subtle influences specifically designed to alter this state.

Acknowledgments

S. Engström is the recipient of a Fetzer Institute Fellowship in Theoretical Bioelectromagnetics. We thank W. Satmary, who was responsible for carrying out the electrophysiological recordings. This work was supported by the US Department of Energy under contract DE-AI01-95EE34020.

References

Able, K. (1994) Magnetic orientation and magnetoreception in birds. *Progr. Neurobiol.* **42**: 449–473.

Adey, W. R. (1972) Organization of brain tissue: is the brain a noisy processor? *Int. J. Neurosci.* **3**: 271–284.

Baker-Price, L. and Persinger, M. (1996) Weak, but complex pulsed magnetic fields may reduce depression following traumatic brain injury. *Percept. Motor Skills* **83**: 491–498.

Bamidis, P., Hellstrand, E., Lidholm, H., Abraham-Fuchs, K. and Ioannides, A. (1995) MFT in complex partial epilepsy: spatio-temporal estimates of interictal activity. *Neuroreport* **7**: 17–23.

Bawin, S., Satmary, W. and Adey, W. R. (1994) Nitric oxide modulates rhythmic slow activity in rat hippocampal slices. *Neuroreport* **5**: 1869–1872.

Bawin, S., Satmary, W., Jones, R., Adey, W. R. and Zimmermann, G. (1996) Extremely-low-frequency magnetic fields disrupt rhythmic slow activity in rat hippocampal slices. *Bioelectromagnetics* **17**: 388–395.

Bawin, S., Sheppard, A., Mahoney, M. and Adey, W. R. (1984) Influences of sinusoidal electric fields on excitability in the rat hippocampal slice. *Brain Res.* **323**: 227–237.

Bell, G., Marino, A. and Chesson, A. (1992a) Alterations in brain electrical activity caused by magnetic fields: detecting the detection process. *Electroencephalogr. Clin. Neurophysiol.* **83**: 389–397.

Bell, G., Marino, A., Chesson, A. and Struve, F. (1992b) Electrical states in the rabbit brain can be altered by light and electromagnetic fields. *Brain Res.* **570**: 307–315.

Berry, S. and Thompson, R. (1978) Prediction of learning rate from the hippocampal electroencephalogram. *Science* **200**: 1298–1300.

Bradshaw, D. and Simmons, M. (1995) Gamma-aminobutyric acid-A receptor function is modulated by cyclic GMP. *Brain Res. Bull.* **37**: 67–72.

Dunn, J., Fuller, M., Zoeger, J., Dobson, J., Heller, F., Hammann, J., Caine, E. and Moskowitz, B. (1995) Magnetic material in the human hippocampus. *Brain Res. Bull.* **36**: 149–153.

Durand, D. and Warman, E. (1994) Desynchronization of epileptiform activity by extracellular current pulses in rat hippocampal slices. *J. Physiol.* **480**: 527–537.

Eichwald, C. F. and Walleczek, J. (1996) Model for magnetic field effects on radical pair recombination in enzyme kinetics. *Biophys. J.* **71**: 623–631.

Eichwald, C. F. and Walleczek, J. (1997) Low-frequency-dependent effects of oscillating magnetic fields on radical pair recombination in enzyme kinetics. *J. Chem. Phys.* **107**: 4943–4950.

Eichwald, C. F. and Walleczek, J. (1998) Magnetic field perturbations as a tool for controlling enzyme-regulated and oscillatory biochemical reactions. *Biophys. Chem.* **74**: 209–224.

Jefferys, J. (1995) Nonsynaptic modulation of neuronal activity in the brain: electric currents and extracellular ions. *Physiol. Revs.* **75**: 689–723.

Jefferys, J. and Haas, H. (1982) Synchronized bursting of CA1 hippocampal pyramidal cells in the absence of synaptic transmission. *Nature* **300**: 448–450.

Kaczmarek, L. and Adey, W. R. (1975) Extracellular release of cerebral macromolecules during potassium- and low-calcium-induced seizures. *Epilepsia* **16**: 91–97.

Kalmijn, A. (1997) Electric and near-field acoustic detection, a comparative study. *Acta Physiol. Scand. Suppl.* **638**: 25–38.

Kavaliers, M., Ossenkopp, K., Prato, F., Innes, D., Galea, L., Kinsella, D. and Perrot-Sinal, T. (1996) Spatial learning in deer mice: sex differences and the effects of endogenous opioids and 60 Hz magnetic fields. *J. Comp. Physiol. A: Sensory, Neural, and Behavioral Physiology* **179**: 715–724.

Kawamichi, H., Kikuchi, Y., Endo, H., Takeda, T. and Yoshizawa, S. (1998) Temporal structure of implicit motor imagery in visual hand-shape discrimination as revealed by MEG. *Neuroreport* **9**: 1127–1132.

Kirkby, R., Carroll, D., Grossman, A. and Subramaniam, S. (1996) Factors determining proconvulsant and anticonvulsant effects of inhibitors of nitric oxide synthase in rodents. *Epilepsy Res.* **24**: 91–100.

Kirschvink, J., Kobayashi-Kirschvink, A., Diaz-Ricci, J. and Kirschvink, S. (1992a) Magnetite in human tissues: a mechanism for the biological effects of weak ELF magnetic fields. *Bioelectromagnetics Suppl.* **1**: 101–113.

Kirschvink, J., Kobayashi-Kirschvink, A. and Woodford, B. (1992b) Magnetite biomineralization in the human brain. *Proc. Natl. Acad. Sci. USA* **89**: 7683–7687.

Konopacki, J., MacIver, M., Bland, B. and Roth, S. (1987) Carbachol-induced EEG 'theta' activity in hippocampal brain slices. *Brain Res.* **405**: 196–198.

Korn, S., Giacchino, J., Chamberlin, N. and Dingledine, R. (1987) Epileptiform burst activity induced by potassium in the hippocampus and its regulation by GABA-mediated inhibition. *J. Neurophysiol.* **57**: 325–340.

Lai, H., Carino, M. and Ushijima, I. (1998) Acute exposure to a 60 Hz magnetic field affects rats' water-maze performance. *Bioelectromagnetics* **19**: 117–122.

Litovitz, T., Penafiel, M., Krause, D., Zhang, D. and Mullins, J. (1997) The role of temporal sensing in bioelectromagnetic effects. *Bioelectromagnetics* **18**: 388–395.

Lyskov, E., Juutilainen, J., Jousmeaki, V., Partanen, J., Medvedev, S. and Heanninen, O. (1993) Effects of 45-Hz magnetic fields on the functional state of the human brain. *Bioelectromagnetics* **14**: 87–95.

Markwort, S., Cordes, P. and Aldenhoff, J. (1997) Transcranial magnetic stimulation as an alternative to electroshock therapy in treatment resistant depressions: a literature review. *Fortsch. Neurologie-Psychiatrie* **65**: 540–549.

Pape, H. and Mager, R. (1992) Nitric oxide controls oscillatory activity in thalamocortical neurons. *Neuron* **9**: 441–448.

Persinger, M. (1997) Metaphors for the effects of weak, sequentially complex magnetic fields. *Percept. Motor Skills* **85**: 204–206.

Schiff, S., Jerger, K., Duong, D., Chang, T., Spano, M. and Ditto, W. (1994) Controlling chaos in the brain. *Nature* **370**: 615–620.

Sienkiewicz, Z., Haylock, R. and Saunders, R. (1998) Deficits in spatial learning after exposure of mice to a 50 Hz magnetic field. *Bioelectromagnetics* **19**: 79–84.

Stern, S. (1995) Do rats show a behavioral sensitivity to low-level magnetic fields? *Bioelectromagnetics* **16**: 335–336.

Taylor, C. and Dudek, F. (1982) Synchronous neural afterdischarges in rat hippocampal slices without active chemical synapses. *Science* **218**: 810–812.

Traub, R., Dudek, F., Taylor, C. and Knowles, W. (1985) Simulation of hippocampal afterdischarges synchronized by electrical interactions. *Neuroscience* **14**: 1033–1038.

Traub, R. D. and Richard, M. (1991) *Neuronal Networks of the Hippocampus*. New York: Cambridge University Press.

Vincent, S. and Hope, B. (1992) Neurons that say NO. *Trends Neurosci.* **15**: 108–113.

Walleczek, J. (1995) Magnetokinetic effects on radical pairs: a paradigm for magnetic field interactions with biological systems at lower than thermal energy. In *Electromagnetic Fields: Biological Interactions and Mechanisms* (ed. M. Blank), *Advances in Chemistry*, No. 250, pp. 395–420. Washington, DC: American Chemical Society.

Wexler, E., Stanton, P. and Nawy, S. (1998) Nitric oxide depresses GABA-A receptor function via coactivation of cGMP-dependent kinase and phosphodiesterase. *J. Neurosci.* **18**: 2342–2349.

Winson, J. (1978) Loss of hippocampal theta rhythm results in spatial memory deficit in the rat. *Science* **201**: 160–163.

Zorumski, C. and Izumi, Y. (1993) Nitric oxide and hippocampal synaptic plasticity. *Biochem. Pharmacol.* **46**: 777–785.

Part III

Stochastic noise-induced dynamics and transport in biological systems

The consideration of random fluctuations is central to any discussion of the limits of biological stimulus–response interactions. The conventional view holds that stochastic perturbations diminish the efficiency of biological signal detection and transduction pathways. Part III provides an overview of the recently found *constructive* role of noisy fluctuations in biochemical and biological processes. Two major concepts have emerged for describing such noise-facilitated processes, *stochastic resonance* and *fluctuation-driven transport*. In Chapter 10, Frank Moss opens with an overview of the history and of the physical principles of stochastic resonance. He then extends this discussion into the realm of biology and medicine, and offers experiments that demonstrate stochastic resonance in the detection by organisms of dynamical signals, including weak electric fields. The chapter concludes with new evidence for *spatiotemporal stochastic resonance* in two-dimensional systems. Sergey Bezrukov and Igor Vodyanoy take the discussion of stochastic resonance to the molecular level in Chapter 11. Their experiments demonstrate the noise-improved transduction of electrical signals by voltage-gated ion channels in a planar lipid bilayer. The authors also present a model that can theoretically account for the experimental observations. In Chapter 12, Dean Astumian introduces the theoretical foundations of *fluctuation-driven transport*. He explores the theoretical feasibility of different types of microscopic ratchet mechanisms that may harness random fluctuations to accomplish work such as directed transport. Experimental evidence in support of a biomolecular ratchet mechanism for free energy transduction is subsequently reviewed by Tian Tsong in Chapter 13. He found that *randomly fluctuating* electric fields were capable of increasing the activity of a membrane ion pump. This chapter also describes the sensitivity of membrane ion pumps to the frequency of *coherently oscillating* electric fields, and it concludes with a discussion of a potential role for frequency-specific electrical interactions in cell–cell communication.

10

Stochastic resonance: looking forward

FRANK MOSS

10.1 Introduction

Stochastic resonance (SR) has been a familiar topic to physicists since the early 1980s, to sensory neurobiologists since the early 1990s and is now becoming familiar to medical researchers interested in the possibilities it offers for new diagnostic and therapeutic techniques. It has been the subject of numerous editorial commentaries and three major reviews. In this chapter, I review very briefly the fundamental classical physical process and its realization or study in various substrates ranging from single elements to arrays of elements over a range of size scales. The emphasis is on the classical phenomenology and its applications in the biological or medical sciences. I then attempt to have a look into the future.

SR was first proposed as a theoretical explanation for the observed recurrences of the Earth's ice ages. In this view, dynamical orbital processes, which moved the Earth's orbit slightly closer to the sun periodically on an approximately 100 000-year period, were known to be too small to account for the reduced insolation necessary to trigger the onset of a glacial period in the Earth's climate. But if a temporal randomness in the insolation, or noise, was introduced into the dynamical motion of the climate state point, such quantities as the frequency spectrum of the ice age recurrences could be reproduced with some accuracy. For a review of SR applied to the climate problem and the early history of its introduction, see Nicolis (1993). Following demonstrations of the effect in an electronic circuit (Fauve and Heslot, 1983) and a ring laser (McNamara et al., 1988; Vermuri and Roy, 1989), SR was demonstrated or studied in a wide variety of physical contexts. This work, largely in physical systems, has been the subject of three extensive reviews (Moss, 1994; Moss et al., 1994; Gammaitoni et al., 1998), and hence will not be further outlined here. Instead, we shall look at its introduction into sensory biology (Douglass et al.,

1993; Wiesenfeld *et al.*, 1994; Wiesenfeld and Moss, 1995; Moss and Wiesenfeld, 1995) at the level of whole cells and at the molecular level (Petracchi *et al.*, 1994; Bezrukov and Vodyanoy, 1995), at array effects (Collins *et al.*, 1995; Lindner *et al.*, 1996), at the noise-enhanced propagation of coherent structures in chemistry (Kádár *et al.*, 1998), and in networks of cells and finally at what SR is beginning to mean to medical science. It should be understood that this sketch is certainly not the whole story. Currently, vigorous SR research on diverse systems is in progress in numerous laboratories scattered over the world.

10.1.1 *Basic principles underlying stochastic resonance*

The classical phenomenology of SR was for a long time thought to be dependent on the *dynamical motions* of a particle (or system state point) in some energy potential with two minima separated by a barrier; that is, a *bistable* potential (Gammaitoni *et al.*, 1989a,b; Jung and Hänggi, 1991). For quite some time, theoretical efforts, commencing with the seminal work of Wiesenfeld (McNamara and Wiesenfeld, 1989) were directed toward understanding the details of the stochastic motion of the state point in such energy potentials (for a recent review, see Gammaitoni *et al.*, 1998). A very nice exposition, using electronic circuits to illustrate the dynamics, has been written recently (Lanzara *et al.*, 1997). However, this picture changed abruptly in early 1995 with the advent of a *nondynamical*, purely statistical picture of SR (Gingl *et al.*, 1995). This considerably simplified view stimulated much further work, especially in biology where the picture seemed to fit the simplest and most general ideas about how neurons function. Indeed, the ingredients for SR in this view are generally found in a variety of settings in both natural and artificial systems and devices. For this reason SR has migrated into numerous fields in addition to the ones I discuss here. I introduce this simple picture in order to expose the basic principles of how SR works in a transparent way.

10.1.2 *The nondynamical picture of stochastic resonance*

In this view, only three ingredients are necessary for a system (or a device) to exhibit SR. They are (1) a *threshold*, (2) a *subthreshold signal*, which carries some information, and (3) *noise*, or some random process, which may be either *external* or *internal* to the system, but which adds to the subthreshold signal. This is illustrated in the diagram in Figure 1, wherein (a) the subthreshold signal (sine wave) and the threshold (horizontal line) are shown. We adopt a simple rule (which, in fact, is fundamental to the way neurons function): the

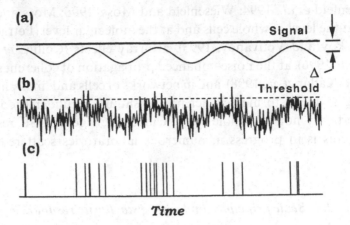

Figure 1. (a) A threshold (straight line) and a subthreshold signal (the sinusoidal wave) showing Δ, the distance between the threshold and the mean of the signal. (b) Noise has been added to the subthreshold signal. (c) Standard-shaped pulses mark the locations in time of the threshold crossings.

system or device can respond to the subthreshold signal only when it crosses the threshold, in which case it generates an output pulse of standard shape. In (a) no threshold crossings occur, since the signal is always subthreshold. However, if noise is added to the signal, as in (b), there are threshold crossings. These are marked in time by the train of standard shaped pulses shown in (c). Blurring out fine details, many devices or systems function in this basic way. As obvious examples, one can mention early model neurons and the thermal activation of electrons over a barrier, as in the room temperature operation of a back-biased electronic diode. In spite of the fact that the pulses seem to occur at random times, they carry a surprising amount of information about the subthreshold signal. In order to understand this, we need only turn to early electrical engineering. A formula for the mean *threshold-crossing rate* of a Gaussian process has been given (Rice, 1954),

$$\langle v \rangle = \frac{f_0}{\sqrt{3}} \exp\left(-\frac{\Delta^2}{2\sigma^2} \right), \tag{1}$$

where f_0 is the bandwidth of the noise (assumed to have a rectangular power spectrum to the cut-off frequency, f_0), Δ is the threshold as shown in Figure 1a, and σ is the standard deviation of the noise. The threshold is *time dependent*. Consider the case that the subthreshold signal is sinusoidal with amplitude ε and frequency ω, then,

$$\Delta(t) \rightarrow \Delta_0 + \varepsilon \sin \omega t, \tag{2}$$

where the signal is subthreshold for $\varepsilon < \Delta_0$. If $\Delta(t)$ is slowly varying compared to f_0, then the noise 'samples' the subthreshold signal at well above the Nyquist frequency, and the information content in the pulse train is determined by this 'sampling' rate; that is, the threshold-crossing rate. These two formulae show how the threshold-crossing rate $\langle v \rangle \rightarrow \langle v \rangle (t)$ can become a slow function of time. Moreover, with this the threshold-crossing rate has become *exponentially sensitive to the weak signal*. Under these conditions, one can calculate the power spectra of the output pulse train (Figure 1c) without the signal; that is, for the noise alone, P_n, and for the signal, P_s. The signal-to-noise ratio (SNR) is the ratio of the amplitudes of the first harmonic terms of these power spectra (for details, see Moss *et al.*, 1994; Gingl *et al.*, 1995),

$$\text{SNR} = \left[\frac{2f_0 \Delta_0^2 \varepsilon^2}{\sqrt{3}\sigma^4} \right] \exp\left[-\frac{\Delta_0^2}{2\sigma^2} \right]. \tag{3}$$

Note that this formula is remarkably similar to the one derived by McNamara and Wiesenfeld (1989) for a bistable energy potential, but with Δ_0^2 replacing the energy barrier, U_0, and σ^2 replacing the noise power density, D. Equation (3) together with the original data from the SR experiment with the crayfish mechanoreceptor is shown in Figure 2 (see also Section 10.2.1).

The foregoing picture of SR has also been called the *threshold* model for obvious reasons. It is *nondynamical* because it is the first model for SR that does not make use of the analogous 'particle' moving in an energy potential, for which dynamical equations of motion are necessary. The calculation sketched above is purely statistical, with the central question being 'What is the threshold-crossing rate?'. Thereafter, we are only asking questions related to signal processing, for example 'What is the signal-to-noise ratio?'.

This picture has recently been greatly expanded with the construction of a new theory for which the threshold is unnecessary (Bezrukov and Vodyanoy, 1995, 1997a,b). Biologists had long recognized that a 'hard' threshold, such as the one shown in Figure 1, was too simple an approximation for constructing neuron models. A more accurate description is obtained by replacing this threshold by some probability function. Bezrukov and Vodyanoy, by constructing their new theory with a probability function instead of a threshold function greatly expanded the list of candidate systems wherein SR can usefully be sought (see Bezrukov and Vodyanoy, Chapter 11, this volume).

10.2 Stochastic resonance moves into biology

The simple threshold model sketched above immediately calls to mind the functioning of a sensory neuron. The membrane potential governs the

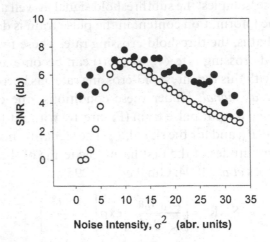

Figure 2. Signal-to-noise ratio (SNR) versus noise intensity. (Solid circles) Data from the crayfish mechanoreceptor experiment (for details see Section 10.2.1). (Open circles) Equation (3), $2f_0 \Delta_0^2 \varepsilon^2 / 3^{1/3} = 5.17 \times 10^3$ and $\Delta_0 = 6.33$.

dynamics of neurons. This potential has a resting value to which is added the sum of inputs to the neuron. When the potential exceeds some threshold, the neuron 'fires', or generates an action potential. All action potentials generated by the same neuron are virtually identical. Thus a train of action potentials traveling down an axon is a record of threshold crossings of the membrane potential. Moreover, the membrane potential is noisy in the millisecond time domain. This inherent noise has its origin in the random openings and closings of the membrane ion channels that are driven by thermal fluctuations (Gailey et al., 1997). Thus, even in the absence of a coherent signal on the membrane potential, the action potential train is noisy. In the presence of signals, the train contains a coherent component. Of course there are subtle details that we have ignored here, but this coarse-grained behavior is what we are interested in.

So where might one seek SR in the natural world? An obvious answer is where there are weak signals, noise and thresholds. Sensory neurons were an obvious first place to look.

10.2.1 Single neurons: the crayfish mechanoreceptor

The crayfish is a venerable animal in biology, having been the subject of the first experiments in modern neuronal physiology pioneered by Thomas Henry Huxley in the last century (Huxley, 1880). The animal has a no less estimable evolutionary history, tracing its ancestral origins to the Cambrian. Today it remains one of the most successful animals, populating the globe from the Arctic to the equator in many species. It may owe this remarkable success, at least in part, to an extraordinarily simple but remarkably effective predator avoidance system: the hydrodynamically sensitive hair mechanoreceptor (Wilkens and Douglass, 1994). The tail fan of the crayfish is covered with numerous small hairs, the longer of which (about 10 μm by 250 μm) move with low frequency (6 to 25 Hz) water motions (see Figure 3). Each of these hairs is connected to a pair of sensory neurons that fire action potentials when the hair moves. The sensory neurons converge on a small network of neurons called the sixth ganglion. This ganglion processes information from the hair receptors and sends it on to the higher nervous system. When necessary, the ganglion can excite a pair of motor neurons that trigger the animal's escape reflex. The animal escapes from predators, mainly fish, by spreading its tail fan, contracting its abdominal muscles and swimming backwards at a high speed.

Our experiments were performed with the most abundant species of crayfish found in the USA, the common 'red swamp' crayfish, *Procambarus clarkii*. Figure 3b shows the location of a suction electrode and preamplifier (A) for recording from a single sensory neuron. The stimulus was a periodic water motion of frequency in the range 8 to 100 Hz, plus a random component, the noise, with total amplitude in the range 10 to 100 nm. In these experiments, since the noise was added externally, it was necessary to choose a hair receptor with a small internal noise so that the effects of the noise added to the stimulus were maximized. Recordings were made extracellularly using completely standard techniques and apparatus. Trains of action potentials were recorded from identified single hair receptors. These were converted into standard-shaped rectangular pulses. The power spectra of these pulse trains were of the form of $\langle v \rangle^2 [\sin(at)/at]^2$ functions whose amplitude $\langle v \rangle^2$, was proportional to the noise intensity. Three example power spectra are shown in Figure 4 for cases of small, intermediate and large added noise. The signal due to the periodic component of the stimulus, which was held constant, is evident in every case as the sharp spike riding on the broad band noise background. The signal component is of maximum amplitude only for the intermediate noise intensity (Figure 4b). The SNR can be obtained by comparing the amplitude of the signal feature to the amplitude of the noise background at the signal frequency. The SNRs obtained from this experiment are plotted as the data points in Figure 2. This experiment was the first to demonstrate SR in any biological application (Douglass *et al.*, 1993).

(a)

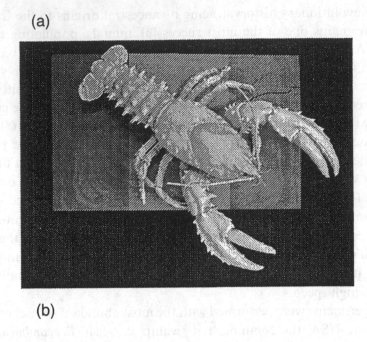

(b)

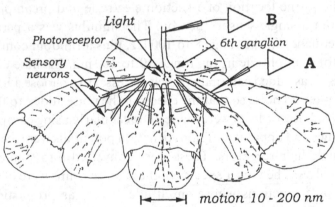

Light

Photoreceptor

Sensory
neurons

6th ganglion

B

A

motion 10 - 200 nm

Figure 3. (a) The Australian Murray River Crayfish, *Euastacus armatus* (Sandeman
and Wilkens, 1982). (b) Diagram of the mechanoreceptor: sixth ganglion and caudal
photoreceptor systems of the crayfish. Recording site A is at a sensory neuron, B is
from the caudal photoreceptor output neuron. The photosensitive area (about
$100\,\mu m \times 100\,\mu m$) is illuminated with dim steady light (arrow) during recording
sessions.

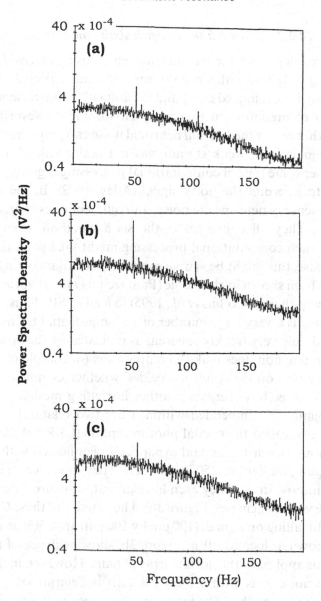

Figure 4. Three power spectra measured from a sensory neuron of the crayfish mechanoreceptor system, showing the broad-band noise background and the signal features as sharp peaks riding on the background. The fundamental is at about 50 Hz and the second harmonic shows in the top two panels. (a) Small noise, (b) optimal noise, and (c) large noise. The signal-to-noise ratio (SNR) is maximum for (b).

10.2.2 Networks: the crayfish sixth ganglion

Note that the foregoing experiment demonstrated SR using *external* noise. The results are biologically relevant, since the environment, especially the one in which the crayfish lives, is indeed noisy and the animal's survival depends upon timely detection of predators in this noisy environment. Nevertheless, the question of whether or not the *internal* neuronal noise can play some role in the processing of signals from weak stimuli was not addressed. Moreover, one might ask whether some type of computational processing is going on in order for the animal to interpret the noisy signal (Adey, 1972). In the absence of stimulation, the sensory neurons are noisy, and different ones exhibit different noise intensities. They all converge on the sixth ganglion, where the first opportunity for such computational processing might take place (Figure 3b). In the simplest case, this might be simple summation of all incoming signals, a process that has been shown in electronic (Pantazelou *et al.*, 1993) and numerical simulations and theory (Collins *et al.*, 1995) to lead to SR. Thus, as input to the ganglion, there is a very large number of action potential trains which are largely noise and only very weakly coherent as indicated by the power spectra in Figure 4. The ganglion does make a computation (of completely unknown nature) because based on the result it 'decides' whether or not to trigger the escape reflex. There is, however, yet another intriguing modality associated with the sixth ganglion. Embedded within it are two bilaterally symmetric photoreceptor cells, called the caudal photoreceptors (CPRs; Wilkens, 1988). These two cells are dendritically and synaptically connected with the interneurons of the ganglion (about 250 in number). They thus receive input from the network of interneurons. They each have an output neuron that connects to the higher nervous system (see Figure 3b). The activity of these CPR cells is mediated by light falling on a small (100 μm by 100 μm) area. Recordings at the CPR output, shown at location B in Figure 3b, show evidence of the hydrodynamic stimulus applied to the tail fan array of hairs. However, in the absence of stimulus, the noise level within the CPR cells is determined by the light intensity. And, the larger the light intensity, the larger is the resulting noise intensity.

In experiments with the CPR, while periodically stimulating the entire tail fan array of hair receptors with a weak periodic water motion, we observe that the *sensitivity* of the CPR to these processed signals is greatly enhanced by light (Pei *et al.*, 1996). In the dark, the CPR is only weakly sensitive to the hydrodynamic signal, but in the presence of light, its responses can be over one order of magnitude larger. Example data for two different cells, at two different stimulus amplitudes, are shown in Figure 5. One cell shows clear evidence of a

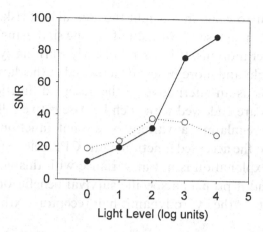

Figure 5. Signal-to-noise ratio (SNR) versus light level for two different photoreceptor cells in the crayfish sixth ganglion. (Open circles) One receptor shows the full stochastic resonance (SR) curve with a maximum SNR at an optimum light level. (Closed circles) The other has not yet reached its maximum at the largest light level. The light level 0 corresponds to complete darkness, and 4 corresponds to 1.1 $\mu W/cm^2$. The amplitude of the hydrodynamic forcing was 260 nm.

maximum, at optimal light (noise) intensities, while for the other, the optimal value was not achieved since the signal was still enhanced at the largest light intensity applied. The largest light intensities applied (about 1 $\mu W/cm^2$) were comparable to dim room light.

10.2.3 A survival trait?

The sixth ganglion together with the CPR cells function as a signal processor and decision-maker. The decision is whether or not to trigger the escape reflex, and it is based on the characteristics of the hydrodynamic signals received from the hair cell array on the tail fan. The 'danger' signal would be in the form of 8- to 12-Hz, weakly periodic water motions induced by swimming fish. But we must consider the whole life style of the crayfish. It is largely a nocturnal animal, remaining in its burrow near the bottom of streams or ponds during most of the day. During this time it is important that the escape reflex not be triggered 'accidentally', because then the animal might exit its burrow in the day and be subject to predators. I use the word 'accidentally', because the escape reflex trigger is essentially a stochastic process. If there are predator-type signals of some amplitude present together with noise, then there is some probability that the escape reflex will be triggered. Thus, when the animal is safely inside its dark burrow, the escape trigger probability must be greatly reduced. The animal, however, does sometimes emerge in daylight to forage for food. Out of the burrow and in the light the animal is at risk. In this situation the CPR drastically increases the sensitivity of the sixth ganglion to predator-induced weak water motions. It is a kind of early warning system that can be sensitized in the light and more or less deactivated in the dark.

We propose this as an alternative to the accepted function of CPR cells. Many crustaceans are endowed with such light-sensitive cells. They all show diurnal rhythms, regulating a variety of animal functions over the 24-h day–night cycle. So the accepted function of the CPR cells is to distinguish day from night. Our explanation is not at variance with this view, but does add essential detail, and it proffers a specific survival benefit for the light (noise) mediated sensitivity of the hydrodynamic hair receptor–sixth ganglion system.

10.2.4 But does the animal actually use stochastic resonance?

All experiments to date on SR in animals have relied on electrophysiological measurements of the responses of neurons to weak coherent signals and noise. Computer analysis of the data has shown increased information content in the neuronal responses at this level in the nervous system of the animal. But a serious weakness of this line of research is that such experiments can never test whether or not the living and functioning animal actually makes use of SR. In order to answer such questions, behavioral experiments are necessary. Psycho-physics experiments, using the human tactile sense, have been accomplished (Collins *et al.*, 1996) but no animal behavioral experiments showing SR have yet been reported. For a variety of reasons, the crayfish is not suitable for behavioral experiments. However, David Russell and Lon Wilkens of this laboratory have developed suitable behavioral research techniques using a different animal, the paddlefish, which uses an electroreceptor system.

10.2.4.1 The paddlefish electroreceptor

The animal model being developed for behavioral experiments is the pad-dlefish, *Polydon spathula*. This fish has a well-developed feeding apparatus based on the detection of weak electric fields from its customary prey, plankton such as daphnia or brine shrimp. These small animals (1 to 3 mm) emit weak electric fields (a few microvolts per centimeter) of 8 to 12 Hz associated with their characteristic swimming motions. The paddlefish uses an extensive array of electroreceptor cells spread over the upper and lower surfaces of its rostrum, a large paddle-shaped appendage at the front of the fish and extending out-ward above its mouth (see Figure 6a). Extensive experiments in this laboratory have shown that the paddle fish detects, locates and feeds on these small creatures by relying entirely on its electrical sense (Wilkens *et al.*, 1998).

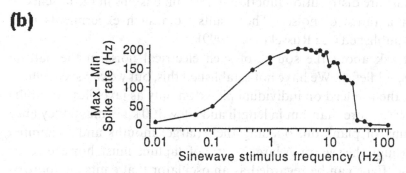

Figure 6. (a) The paddlefish with a rostrum supporting an array of electroreceptor organs. (b) Frequency response of a single cell showing a maximum response at the characteristic frequencies of the animal's planktonic prey ($\sim$ 10 to 20 Hz). The electric field in the vicinity of the receptor was about 1 μV/cm, which is also of the same order as the fields that are characteristic of the prey. (From Wilkens *et al.*, 1998.)

Electrophysiological experiments with the same type of electroreceptor cell in sharks have established that the cells are noise- and temperature-mediated oscillators (Braun *et al.*, 1994). Recent experiments in this laboratory have established that the frequency response and electric field sensitivity of the paddlefish's electroreceptors are closely matched to the characteristic fields emitted by swimming daphnia (Wilkens *et al.*, 1998).

10.2.4.2 A behavioral experiment

A swim mill has been constructed in this laboratory in such a way that the paddlefish can swim against a continuously flowing stream of water while remaining stationary in the laboratory frame of reference. The fish swims in a viewing chamber and is videotaped simultaneously from the side and below. The water stream is seeded with daphnia or brine shrimp that also show up on the videotape. The statistical feeding behavior of the swimming fish can thus be established by analyzing the videotapes. These data take the form of two-dimensional distributions of prey captures as a function of distance from a reference point on the fish's rostrum. The signal being detected is electrical, so that in order to study SR we introduce an *electrical noise* into the flowing stream. Our question is, can the fish detect prey at greater distances in the presence of an optimal level of electrical noise than in its absence? We have now established this (see below), and, for some fish, have found an increase in the overall feeding rate as a function of the amplitude of the applied electrical noise. These data show an enhanced feeding rate at an optimal noise intensity of around 1 μV/cm, that is, quite near the magnitude of the field emitted by the typical prey (see Figure 6b). However, what we wish to obtain is the detailed shape of the capture distribution function in two dimensions in the absence of noise and for a range of noises. The results from such experiments have recently been published (see Russel *et al.*, 1999).

One might ask about the source of such electrical noise in the natural habitat of the paddlefish. We have not established this, but we can speculate as follows. First, the fish feed on individual plankton only as juveniles. As adults (they can grow to more than 1 m in length and a few 100 kg in mass) they filter feed on swarms of plankton, opening their (large) mouths and swimming repeatedly through the swarm. A large swarm of daphnia must, however, emit electrical noise. Each can be regarded as an oscillator that emits an approximately dipole field pattern, but the daphnia in the swarm are oriented randomly in all possible directions. Moreover, their periodic swimming motions are not synchronized. Summing the electric field at any defined point in space over a large number of randomly oriented, asynchronous oscillators results in noise. A second possibility involves turbulently flowing water through which is passing a weak electric current. The current can be induced by Faraday induction as the conducting water flows in the Earth's magnetic field.

10.3 And into medical science

SR shows that weak signals can be enhanced in nonlinear systems characterized by thresholds. This suggests medical applications, especially in rehabilitative clinical applications. A variety of pathological conditions, arising through the natural aging process or through disease or accident, lead to reduced sensitivity of the nervous system to some sensory stimuli. A specific example might be the age-related, reduced sensitivity of the proprioceptive neurons. These sensory modalities detect and transmit information about joint angle and velocity to the central nervous system. In old age, the increased firing thresholds of these neurons cause the reduced sensitivity. With reduced information about joint position and limb velocity, arm and leg movements become difficult, leading to 'feebleness' in otherwise healthy elderly individuals. But the addition of noise may help to overcome the effects of the increased thresholds by restoring the normal firing rates to these degraded neurons. Efforts involving the basic research of such possibilities are currently in progress.

10.3.1 Electromyography of the median nerve

To begin this research, it is necessary to demonstrate that weak signals can indeed be enhanced by the noise inherent in typical sensory neurons of the human body. And it would be desirable to demonstrate this result in the simplest possible setting using apparatus familiar to every medical doctor. This has been done in a brilliant experiment designed by Dr Faye Chiou-Tan of Baylor College of Medicine (Chiou-Tan *et al.*, 1996) using electromyography of the median nerve. Transcutaneous electrical stimuli were introduced just above the elbow and detected on the middle index finger and thumb. The inherent noise in the median nerve was controlled by subject-controlled tension in the abductor pollicis brevis muscle, which was flexed against a force gauge (see Figure 7a). The noise in the median nerve is quite accurately Gaussian distributed. Preliminary experiments showed that the standard deviation of this distribution, i.e., the noise intensity, could be accurately controlled by the subject while exerting a fixed force against a force gauge. The experiment consisted of sending near-subthreshold electrical signals down the median nerve in the absence of noise. The internal noise was then increased by muscle flexure and noise-enhanced signals were recorded at the receiving electrodes. A SNR can be defined in a way similar to that outlined above. The measured SNRs versus the muscle tension-mediated noise levels are shown in Figure 7b. The data do not show a maximum at an optimal noise, but clearly demonstrate noise enhancement of the SNR in the sensory neurons of the median nerve. Moreover, we observed no SNR enhancement of signals traveling through the motor neurons in the median nerve. It should be noted that this was the first SR experiment to make use of well-controlled *inherent* noise. The experiment also demonstrated an important process: the transcutaneous introduction of electrical noise can enhance the sensitivity of sensory neurons. This experiment then opens the door for therapeutic and rehabilitative uses of noise in the human nervous system.

A series of further experiments have been reported by this group (Chiou-Tan *et al.*, 1997, 1998) aimed at demonstrating the effectiveness of the muscle tension-mediated inherent noise when the muscle is remotely located. Some of these experiments have included the use of drugs to block certain pathways in order to identify and localize the muscle-noise action.

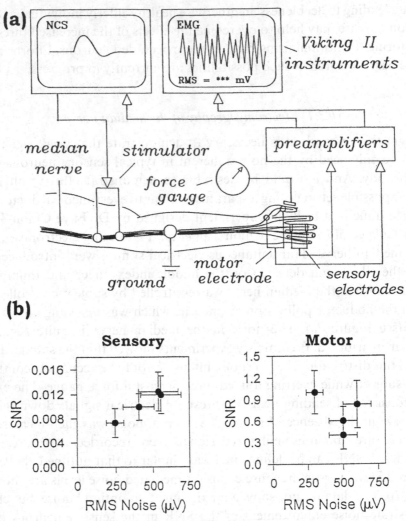

Figure 7. Electromyography of the human median nerve. (a) Set-up of apparatus and recording sites. NCS, neuronal cell system. (b) Sensory neurons in the median nerve show noise-enhanced signal transmission (left), but motor neurons during the same experiment do not (right). The noise was mediated by muscle tension and measured with the Viking II instrument shown in (a). SNR, signal-to-noise ratio; EMG, electromyograph; RMS, root mean square. (From Chiou-Tan *et al.*, 1996.)

10.3.2 Electrophysiology of proprioceptor neurons

Closely related to the experiments described in the foregoing section are the electrophysiological experiments on proprioceptive, or muscle-spindle neurons (Cordo *et al.*, 1996). These are neurons that sense the position and velocity of the movements of the joints. Without their proper functioning all muscle-mediated movement becomes impossible. In this experiment, electrodes were inserted into the median nerve above the elbow in humans. The nerve was searched until a signature of a proprioceptive neuron was found. The experiment from this point on entailed a classic SR procedure using external noise added to a weak signal. Moving the wrist periodically using a mechanical actuator provided a coherent signal. This motion was sensed through tendon stretch by the proprioceptive neuron and transmitted up the nerve to the recording site. The amplitude of wrist motion was reduced until the coherence in the neuronal recording was barely detectable. External noise was then added by mechanically vibrating the wrist tendon that couples into the stretch-sensing neuron. The SNRs of the responses were computed in the same way as in the original crayfish experiment. The SNRs were noise-enhanced at low levels of noise and degraded at high levels, thus demonstrating SR in this system.

10.3.3 Noise-mediated coherence in distributed systems

This topic is of great interest at present. Local random processes, or 'noise', probably affect many distributed natural phenomena. In one view, a number of interacting agents might be spread over a two-dimensional surface, each of which is being subjected to local noise. Certainly we expect the noise to affect the dynamics of such a system to some degree and, maybe, in some instances, even to a large degree. But could we imagine a system wherein the noise could enhance some collective, or coherent, dynamical property? If so, and if an optimal noise maximally enhances the property, the process is called *spatiotemporal* SR. Spatiotemporal SR has now been achieved experimentally for the first time in two very different media, which nevertheless may behave chemically in similar ways.

10.3.3.1 The subexcitable Belousov–Zhabotinsky reaction

In the first experiment the Belousov–Zhabotinsky nonequilibrium chemical reaction was operated in a subexcitable mode (Kádár *et al.*, 1998). Using a photosensitive version of this famous chemical reaction confined to a thin slab of silica gel, the group observed enhanced wave propagation in response to *spatiotemporal noise* applied to the gel slab as a two-dimensional optical image. The image was of a rectangular region divided into a large number of square subregions or cells. The intensity of the light falling on a given cell determined its state of excitability by controlling the rate of photoproduction of bromide ions (Br^-) which are inhibitors of autocatalysis in this reaction. A large light intensity maintains the region below the threshold of excitability, i.e., in the subthreshold state, with the result that disturbances are rapidly quenched. In the presence of little or no light, the disturbances grow into the familiar waves that propagate indefinitely. The average light intensity for all cells was adjusted to maintain the entire region in the subexcitable state, so that indefinitely sustained waves were impossible. To this average, a time-dependent noise was then added. The noise in each cell was generated independently of the noise in any other cell. It was noted that waves propagating into the region from an external source lived longer, and hence propagated further, in the presence of this spatially distributed, time-varying noise. An optimal value of the noise intensity resulted in sustained waves. Noise intensities larger than the optimal level resulted in degraded propagation and wave break-up into segments of random lengths.

10.3.3.2 Self-organized critical behavior in astrocyte syncytia

The second experiment takes place in a very different medium: a network of glial cells cultured from human brain tissue (Jung *et al.*, 1998). Long-range (a few centimeters), long-lived (many seconds), spiral chemical waves of calcium ions (Ca^{2+}) are observed in cultured networks of glial cells for normal concentrations of the neurotransmitter kainate (see Figure 8). A new method for quantitatively measuring the spatiotemporal size of the waves is described. This measure results in a power-law distribution of wave sizes, meaning that the process that creates the waves has no preferred spatial or temporal (size or lifetime) scale. This power law is one signature of *self-organized critical phenomena*, a class of behaviors found in many areas of science. The physiological results for glial networks are fully supported by numerical simulations of a simple network of noisy, communicating threshold elements. By contrast, waves observed in astrocytes cultured from human epileptic foci exhibited radically different behavior. The background random activity, or 'noise', of the network is controlled by the concentration of the neurotransmitter kainate.

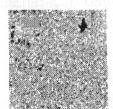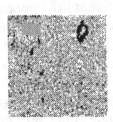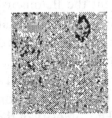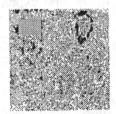

Figure 8. The growth and propagation of a spiral Ca^{2+} wave in a network of glial cells cultured from human brain tissue. Snapshot pictures of the wave with time elapsing to the right. The waves propagate over centimeter-scale distances for times of a few seconds. The images show the wave propagating against a background of network noise controlled by the concentration of neurotransmitter (kainate) in the bath solution.

The mean rate of wave nucleation is mediated by the network noise. However, the power-law distribution is invariant, within our experimental precision, over the range of noise intensities tested. These observations indicate that spatially and temporally coherent Ca^{2+} waves, mediated by network noise, may play an important role in generating correlated neuronal activity (waves) over long distances and times in the healthy vertebrate central nervous system (CNS).

Self-organized criticality (SOC) is an established dynamical behavior (Bak *et al.*, 1987) of numerous physical systems, including avalanches and earthquakes (Bak, 1996), magnetic noise and an economic index (Mantegna and Stanley, 1995) to name only a few. A common feature of these processes concerns power-law distributions of event sizes or lifetimes. By contrast, processes with characteristic times, for example the discharge of a capacitor through a resistor or the decay of a population of radioactive nuclides (and innumerable similar phenomena, for example those described by linear differential equations) proceed according to exponential laws. Systems showing SOC tend naturally toward a critical state where power-law scaling is the rule. Moreover, SOC arises in some complex systems that are far from equilibrium. Complexity means that an enormously large number of events are possible, the size or lifetime of any particular event being unpredictable. Thus a fundamental randomness, or noise, is characteristic of SOC. Far from equilibrium means that the systems are continuously driven from the external environment, for example the relentless build up of strain along a fault line, ultimately leading to an earthquake (of as yet unpredictable size to occur at an as yet unpredictable time).

The human brain is a complex object that operates far from equilibrium, owing to an incessant stream of stimuli to 'think about'. Thus one might speculate that brain function may also be an example of SOC, and, indeed, interesting and suggestive numerical simulations have been reported (Bak, 1996; see also Ding *et al.*, Chapter 4, this volume).

I report here the first experimental evidence supporting a role for SOC in a vertebrate CNS preparation. In a culture of glial cells, time-lapse images employing the Ca^{2+} fluorescent dye indicator Fluo-3, obtained during dose–response studies with kainate, revealed well-defined spiral and other waves. The waves were born in the background network noise, grew and propagated some distance before dying again in the noise (Figure 8). The spatiotemporal size, *s*, of a single wave comprises not only its physical size at any given time but also its growth during its lifetime and its propagation distance. Motivated by our physiological spiral wave observations, Peter Jung developed a novel statistical analysis suitable for quantitatively characterizing such waves. The analysis yields size distributions, $p(s) = s^{-a}$, which, for both numerical and biological data, are accurately described by power laws with quite similar exponents, *a*. In the healthy CNS, long-range signaling by Ca^{2+} probably occurs via such coherent waves.

10.4 Looking to the future

In this chapter I have outlined the movement of SR from the physical sciences into biology and briefly described only a few of the many experiments now in progress or recently completed. Moreover, with the experiments of Faye Chiou-Tan, we have seen the successful movement of SR into basic research in medical science. These experiments indicate that clinical applications may not be far off (e.g., see the recent commentary by Glanz, 1997). Certainly there are many medical applications at present where usually *periodic* stimulation is used. One can mention, for example, the electrical or mechanical vibratory stimulation to ease chronic pain currently used for many patients in rehabilitation programs. But periodic stimulation has disadvantages. It often leads to illusions or generally unpleasant sensations. By comparison, *random* stimulation may be advantageous. Moreover, many sensory modalities adapt rapidly to periodic stimuli but do not to random stimuli. So we can expect to see many applications of SR, or noise-supported excitatory behavior, in the near future in medical science.

Research on spatiotemporal SR has only just begun. If Bak's speculations (Bak, 1996) are even partially correct, self-organized, noise-supported and long-range signaling in the brain will develop into an exciting area of research in the near future. The experiments of K. Showalter (Kádár *et al.*, 1998) in the 'hard' science area of chemical dynamics, having demonstrated an almost identical dynamics, will surely stimulate interest in further observations of such waves in biological substrates.

Acknowledgment

I am grateful for past and continuing financial support for research on stochastic processes from the US Office of Naval Research.

References

Adey, W. R. (1972) Organization of brain tissue: is the brain a noisy processor? *Int. J. Neurosci.* **3**: 271–284.

Bak, P. (1996) *How Nature Works, the Science of Self-Organized Criticality*. New York: Copernicus.

Bak, P., Tang, C. and Wiesenfeld, K. (1987) Self-organized criticality: an explanation of 1/f noise. *Phys. Rev. Lett.* **59**: 381–384.

Bezrukov, S. M. and Vodyanoy, I. (1995) Noise-induced enhancement of signal transduction across voltage-dependent ion channels. *Nature* **378**: 362–364.

Bezrukov, S. M. and Vodyanoy, I. (1997a) Stochastic resonance in non-dynamical systems without response thresholds. *Nature* **385**: 319–321.

Bezrukov, S. M. and Vodyanoy, I. (1997b) Signal transduction across alamethicin ion channels in the presence of noise. *Biophys. J.* **73**: 2456–2464.

Braun, H. A., Wissing, H., Schäfer, K. and Hirsch, M. C. (1994) Oscillation and noise determine signal transduction in shark multimodal sensory cells. *Nature* **367**: 270–273.

Chiou-Tan, F. Y., Chuang, T-Y., Dinh, T., Robinson, L. R., Tuel, S. S. and Moss, F. (1998) Effect of nerve block on sural amplitude during remote muscle contraction. *Electromyogr. Clin. Neurophys.* **38**: 231–235.

Chiou-Tan, F. Y., Magee, K., Robinson, L., Nelson, M., Tuel, S., Krouskop, T. and Moss, F. (1996) Enhancement of subthreshold sensory nerve action potentials during muscle tension mediated noise. *Int. J. Bifurc. Chaos* **6**: 1389–1396.

Chiou-Tan, F. Y., Magee, K., Robinson, L. R., Nelson, M. R., Tuel, S. S., Krouskop, A. and Moss, F. (1997) Augmented sensory nerve action potentials during distant muscle contraction. *Am. J. Phys. Med. Rehabil.* **76**: 14–18.

Collins, J. J., Chow, C. C. and Imhoff, T. T. (1995) Stochastic resonance without tuning. *Nature* **376**: 236–238.

Collins, J. J., Imhoff, T. T. and Grigg, P. (1996) Noise-enhanced tactile sensation. *Nature* **383**: 770.

Cordo, P., Inglis, T., Verschueren, S., Collins, J., Merfeld, D., Rosenblum, S., Buckley, S. and Moss, F. (1996) Noise in human muscle spindles. *Nature* **383**: 769–770.

Douglass, J. K., Wilkens, L. A., Pantazelou, E. and Moss, F. (1993) Noise enhancement of information transfer in crayfish mechanoreceptors by stochastic resonance. *Nature* **365**: 337–340.

Fauve, S. and Heslot, F. (1983) Stochastic resonance in a bistable system. *Phys. Lett. A* **97**: 5–9.

Gailey, P. C., Neiman, A., Collins, J. J. and Moss, F. (1997) Stochastic resonance in ensembles of nondynamical elements: the role of internal noise. *Phys. Rev. Lett.* **79**: 4701–4704.

Gammaitoni, L., Hänggi, P., Jung, P. and Marchesoni, F. (1998) Stochastic resonance. *Rev. Mod. Phys.* **70**: 223–356.

Gammaitoni, L., Marchesoni, F., Menichaella-Saetta, E. and Santucci, S. (1989a) Stochastic resonance in bistable systems. *Phys. Rev. Lett.* **62**: 349–352.

Gammaitoni, L., Menichaella-Saetta, E., Santucci, S. and Marchesoni, F. (1989b) Extraction of periodic signals from a noise background. *Phys. Lett. A* **142**: 59–62.

Gingl, Z., Kiss, L. B. and Moss, F. (1995) Non-dynamical stochastic resonance: theory and experiments with white and arbitrarily coloured noise. *Europhys. Lett.* **29**: 191–196.

Glanz, J. (1997) Sharpening the senses with neural noise. *Science* **277**: 1759.

Huxley, T. H. (1880) *The Crayfish, An Introduction to the Study of Zoology.* New York: D. Appleton.

Jung, P., Cornell-Bell, A., Shaver Madden, K. and Moss, F. (1998) Noise-induced spiral waves in astrocyte syncytia show evidence of self-organized criticality. *J. Neurophysiol.* **79**: 1098–1101.

Jung, P. and Hänggi, P. (1991) Amplification of small signals via stochastic resonance. *Phys. Rev. A* **44**: 8032–8042.

Kádár, S., Wang, J. and Showalter, K. (1998) Noise-supported traveling waves in subexcitable media. *Nature* **391**: 770–772.

Lanzara, E., Mantegna, R., Spagnolo, B. and Zangara, R. (1997) Experimental study of a nonlinear system in the presence of noise: the stochastic resonance. *Am. J. Phys.* **65**: 341–349.

Lindner, J., Meadows, B., Ditto, W., Inchiosa, M. and Bulsara, A. (1996) Scaling laws for spatiotemporal synchronization and array enhanced stochastic resonance. *Phys. Rev. E* **53**: 2081–2086.

McNamara, B. and Wiesenfeld, K. (1989) Theory of stochastic resonance. *Phys. Rev. A* **39**: 4854–4868.

McNamara, B., Wiesenfeld, K. and Roy, R. (1988) Observation of stochastic resonance in a ring laser. *Phys. Rev. Lett.* **60**: 2626–2630.

Mantegna, R. N. and Stanley, H. E. (1995) Scaling behavior in the dynamics of an economic index. *Nature* **376**: 46–49.

Moss, F. (1994) Stochastic resonance: from the ice ages to the monkey's ear. In *Contemporary Problems in Statistical Physics* (ed. G. H. Weiss), pp. 205–253. Philadelphia: SIAM.

Moss, F., Pierson, D. and O'Gorman, D. (1994) Stochastic resonance: tutorial and update. *Int. J. Bifurc. Chaos* **6**: 1383–1397.

Moss, F. and Wiesenfeld, K. (1995) The benefits of background noise. *Sci. Am.* **273**: 66–69.

Nicolis, C. (1993) Long term climatic transitions and stochastic resonance. *J. Stat. Phys.* **70**: 3–14.

Pantazelou, E., Moss, F. and Chialvo, D. (1993) Noise-sampled signal transmission in an array of Schmitt triggers. In *Noise in Physical Systems and 1/f Fluctuations* (ed. P. H. Handel and A. L. Chung), pp. 549–552. New York: AIP Press.

Pei, X., Wilkens, L. A. and Moss, F. (1996) Light enhances hydrodynamic signaling in the multimodal caudal photoreceptor interneurons of the crayfish. *J. Neurophysiol.* **76**: 3002–3011.

Petracchi, D., Pellegrini, M., Pellegrino, M., Barbi, M. and Moss F. (1994) Periodic forcing of a K^+ channel at various temperatures. *Biophys. J.* **66**: 1844–1852.

Rice, S. O. (1954) Mathematical analysis of random noise. In *Selected Papers on Noise and Stochastic Processes* (ed. N. Wax), pp. 133–294. New York: Dover.

Russel, D. F., Wilkens, L. A. and Moss, F. (1999) Use of behavioural stochastic resonance by paddle fish for feeding. *Nature* **402**: 291–294.

Sandeman, D. C. and Wilkens, L. (1982) Sound production by abdominal stridulation in the Australian Murray River crayfish, *Euastacus armatus*. *J. Exp. Biol.* **99**: 469–472.

Vermuri, G. and Roy, R. (1989) Stochastic resonance in a bistable ring laser. *Phys. Rev. A* **39**: 4668–4674.

Wiesenfeld, K. and Moss, F. (1995) Stochastic resonance and the benefits of noise: from the ice ages to crayfish and SQUIDs. *Nature* **373**: 33–36.

Wiesenfeld, K., Pierson, D., Pantazelou, E., Dames, C. and Moss, F. (1994) Stochastic resonance on a circle. *Phys. Rev. Lett.* **72**: 2125–2129.

Wilkens, L. A. (1988) The crayfish caudal photoreceptor: advances and questions after the first half century. *Comp. Biochem. Physiol.* **91**: 61–68.

Wilkens, L. A. and Douglass, J. (1994) A stimulus paradigm for analysis of near-field hydrodynamic sensitivity in crustaceans. *J. Exp. Biol.* **189**: 263–272.

Wilkens, L. A., Russell, D. F., Pei, X. and Gurgens, C. (1998) Paddlefish rostrum functions and an electrosensory antenna. *Proc. Roy. Soc. (Lond.)* **264**: 1723–1729.

11

Stochastic resonance and small-amplitude signal transduction in voltage-gated ion channels

SERGEY M. BEZRUKOV AND IGOR VODYANOY

11.1 Introduction

Voltage-gated ion channels are crucial 'building blocks' in various systems of signal transduction and processing in living organisms. They are ultimately responsible for information flow at several hierarchical levels of biological complexity that include signal sensing (Lu and Fishman, 1994) and generation of nerve action potentials (Hille, 1992), and are crucially important in synaptic transmission and other intercellular communications (Alberts et al., 1994). Preceding biologically inspired work on the role of external noise in electrical signal transduction concentrated on rather complex objects such as neurons (Bulsara et al., 1994; Pei et al., 1996; Chapeau-Blondeau et al., 1996; Longtin, 1997; Plesser and Tanaka, 1997) and neuronal ensembles (Gluckman et al., 1996; Chialvo et al., 1997). It was demonstrated that addition of random fluctuations, or noise to the input of these systems could improve the transmission efficiency for small input signals. Yet even more elaborate physiological systems showing similar properties include isolated sciatic nerves of a toad (Morse and Evans, 1996; Moss et al., 1996), rat SA1 cutaneous mechanoreceptors (Collins et al., 1996a), mechanosensory transduction pathways in arthropods (Douglass et al., 1993; Levin and Miller, 1996), and human sensory perception (Cordo et al., 1996; Collins et al., 1996b; Chiou-Tan et al., 1996; Simonoto et al., 1997). The counterintuitive phenomenon of noise-improved signal transduction, called 'stochastic resonance' – first introduced as a possible explanation for the periodic recurrences of the Earth's ice ages (Benzi et al., 1981) – has now been established empirically for many macroscopic systems and, for some of them, is understood theoretically (Wiesenfeld and Moss, 1995; Gammaitoni et al., 1998; see also Moss, Chapter 10, this volume).

In a series of publications (Bezrukov and Vodyanoy, 1995, 1997a,b) that form the basis of this chapter, we have addressed the problem of electrical signal transfer in the presence of noise in a subcellular system consisting of voltage-gated ion channels reconstituted in a planar lipid bilayer. We explore the following: whether the phenomenon of noise-facilitated signal transduction described for a variety of different biological objects such as neurons or sensory systems can originate at the basic level of ion channels.

Theoretically, we discuss a model of the time-dependent Poisson wave (Cox and Lewis, 1966) where the average pulse generation rate is a function of an input signal. Our first finding shows that such a wave of identical current pulses with an exponential dependence on input voltage gives an adequate description of signal transfer properties in the absence of external noise. It explains the experimentally found threshold-free response of ion channels to arbitrarily small signals; it provides the right value of small-signal transduction coefficient and its dependence on membrane holding potential; it also determines the output noise value that reflects the stochastic behavior of ion channels.

Our second finding demonstrates that stochastic resonance is an inherent property of a *time-dependent Poisson process*. In the presence of an external input noise added to a small input signal, our model displays two features typical of a stochastic resonator. (1) It describes noise-facilitated signal trans- duction, i.e., output signal amplitude is increased by addition of noise to the system input. (2) It shows noise-induced improvement in the output signal quality; in particular, it yields the optimal noise amplitude corresponding to the maximal output signal-to-noise ratio (SNR).

Experimentally, we have studied 'model ion channels' formed by a 20-amino acid peptide alamethicin in planar lipid bilayer membranes (Hall *et al.*, 1984) from the point of view of electrical signal transduction. Alamethicin channels are highly voltage sensitive – while their open channel current–voltage charac- teristics are practically linear, the probability of finding the alamethicin chan- nel in an open state depends strongly on the applied transmembrane voltage. Thus, the equilibrium between the population of open alamethicin channels and the pool of 'inactive' alamethicin molecules can be regulated by an external voltage signal. When reconstituted into solvent-free membranes prepared from diphytanoyl phosphatidylcholine (DPhPC), the channel voltage sensitivity is defined by the gating charge of 5 to 7 elementary charges, which is remarkably close to that of voltage-gated channels in excitable membranes (Hille, 1992).

In the subsequent sections we show that the application of external noise facilitates the transfer of electrical signals through this system. Specifically, for small sine wave signals, the introduction of noise increases the output signal power by about 10^4-fold, conserving or even slightly improving the output SNR. For noise intensities around 10 mV (root mean square, r.m.s.), the output SNR, measured as the ratio of the output signal peak height to the output noise spectral density in the immediate vicinity of the signal, shows a 2-dB increase with respect to the initial output SNR in the absence of external noise. Thus the addition of noise improves signal transduction in this system.

11.2 Stochastic resonance in a time-dependent Poisson process

A time-dependent Poisson process represents a broad variety of reactions at the molecular level. The central issue of modern reaction-rate theory is the problem of escape from metastable states (Hänggi *et al.*, 1990). In a stationary system, thermally activated events usually occur with the escape rate r being exponentially dependent on reaction activation energy E_b according to the Van't Hoff–Arrhenius law

$$r = v \exp(-E_b/kT), \tag{1}$$

where k and T have their regular meanings. Prefactor v, giving an absolute reaction rate, is only weakly temperature-dependent and, in most models, does not depend on E_b at all. The absence of correlations between consecutive events – which accounts for the Poisson distribution – is caused by the independence of different reaction centers and/or by a significant separation between the time scale of a rare event of activation barrier crossing and the time scale characterizing a fast motion within a metastable state.

Empirically known for more than a century, Equation (1) has been the subject of numerous theoretical studies (e.g., for reviews, see Berezhkovskii *et al.*, 1988; Hänggi *et al.*, 1990). Eyring (1935) was the first to use quantum-statistical mechanics to calculate the prefactor v through the partition function of the metastable state and the activated complex. His approach is now commonly known as the transition state theory of escape rate and is regarded as an 'epoch-making' result, although at least several predecessors can be easily identified. For example, about 20 years before Eyring, a simplified one-dimensional version of this approach was used by Richardson (see e.g., Nye, 1976, pp. 236–240) in calculations of thermionic currents. Later, even simpler arguments, used to describe electron transfer across a semiconductor p–n junction (see e.g., Benedict, 1976, pp. 80–85), led to the same general result of the thermally activated escape rate expressed by Equation (1).

In many physico-chemical systems, reaction rates are controlled by activation barriers whose heights depend on an external parameter, e.g. voltage V, so that $E_b \equiv E_b(V)$. For small time-dependent voltage deviations $\delta V(t)$ from a stationary value V_0 we can write

$$r(V(t)) = r(V_0 + \delta V(t)) \cong r(V_0) + r(V_0)\beta\delta V(t), \tag{2}$$

where $\beta = -E'_b(V_0)/kT$ and $E'_b(V) \equiv \partial E'_b(V)/\partial V$. The module of factor β determines the reaction sensitivity to the external voltage. The output of such a system is a train of uniform, randomly arriving pulses, where information about the modulation by periodic, aperiodic or random signals is encoded in the statistical properties of pulse arrival times. A signal that increases the barrier height will decrease the pulse generation rate and vice versa. The detection of these rate changes (e.g., by appropriate filtering) permits the recovery of the input signal, although contaminated by a certain level of noise stemming from the randomness of the pulse arrival.

It is clear that the higher the average rate $r(V_0)$, the better is the signal transduction. The SNR is improved by increased statistics (similarly to decreasing the uncertainty in a coin throw experiment by increasing the number of throws). Indeed, the amplitude of the signal component prescribed by Equation (2) is proportional to $r(V_0)$, while the noise amplitude, according to general considerations, grows as the square root of the number of events; that is, it increases only as $\sqrt{r(V_0)}$. This means that the SNR is proportional to $\sqrt{r(V_0)}$ (or to $r(V_0)$, if one considers the ratio of signal and noise in terms of power spectral density components).

In the case of a superlinear dependence of the generation rate on the input signal, e.g., the activation energy in Equation (1) is proportional to an external parameter, the statistics can also be improved by adding zero-mean input noise. Due to the system's superlinear behavior, random voltages at the input that correspond to the activation energy decrease will add to the output rate more than equal voltages of the opposite sign will subtract. As a consequence, the average rate will go up. These improved statistics would result in an increased SNR at the output, if not for the contamination of the input signal by the added noise. This leads to the following central questions:

(1) Is it possible to increase the output signal quality in the time-dependent Poisson process by the addition of input noise and, if the answer to this question is yes, then:
(2) What are the noise properties that optimize signal transduction?

In a number of cases representing important phenomena in physics and biology, the dependence of activation energy on an external parameter is linear over a wide range of parameter changes. For semiconductor p–n junctions or ion channels of excitable membranes such a parameter is the applied voltage. In ion channels, voltage sensitivity is assigned to 'gating charge' (Hille, 1992). If channels spend most of their time in the closed conformation and channel openings are considered to be the events constituting the Poisson process, we find that

$$r(V(t)) = r(0)\exp(\beta V(t)), \qquad (3)$$

where

$$\beta = ne/kT, \tag{4}$$

ne is the channel gating charge, and e is the electron charge. In the case of the exponential dependence given by Equation (3), the influence of noise addition on signal transduction can be treated analytically. In this treatment we limit ourselves to the adiabatic small-signal regime by assuming that $V(t)$ is a sum of a constant V_0, a slow zero-mean Gaussian noise $V_N(t)$, and a slow small-amplitude sinewave signal

$$V(t) = V_0 + V_N(t) + V_S \sin(2\pi f_S t), \tag{5}$$

where $V_S \ll |\beta|^{-1}$. If the signal frequency, f_S, and noise 'corner frequency', f_c, are much smaller than all other characteristic frequencies in the system, we can characterize signal transduction by considering only the low-frequency part of the output spectrum. The shape of the output pulse gives a 'form-factor' that influences the spectrum only at high frequencies comparable to the inverse pulse duration. Indeed, at low frequencies, the power spectral density of a 'steady' Poisson wave of uniform pulses generated with a constant rate r is given by (Rice, 1954)

$$S(f)|_{f \to 0} = 2Q^2 r, \tag{6}$$

where Q is single pulse area. For a time-independent Poisson process the low-frequency spectrum is 'white'; that is, it does not depend on frequency.

If the generation rate r is time dependent, this is no longer true. Rate modulation by an external parameter $V(t)$ introduces correlations into pulse arrival times. At low frequencies around f_S, the power spectral density of such a time-dependent Poisson process (Cox and Lewis, 1966) can be calculated as

$$S(f) = 2Q^2 \langle r(V(t)) \rangle + 4(Qr(0))^2 \int_0^\infty \langle \exp(\beta V(t)) \exp(\beta V(t+\tau)) \rangle \cos(2\pi f \tau) \mathrm{d}\tau, \tag{7}$$

where angle brackets denote averaging over time t. The first term of Equation (7) represents the frequency-independent component expected from a 'stationary' Poisson process with a steady pulse generation rate $r = \langle r(V(t)) \rangle$. For the case of $V_S \ll \sigma$, where σ is the noise r.m.s. value, this average rate is easily obtained as

$$\langle r(V(t)) \rangle = r(V_0) \exp((\beta \sigma)^2 / 2). \tag{8}$$

The second term describes additional spectral components stemming from rate modulation by the input signal and noise. To calculate the correlator in the second term, we use an approach given by Rice (1954). We introduce a random vector $X \equiv \beta V(t)$, $Y \equiv \beta V(t + \tau)$ and employ relations obtained for two-dimensional normal distributions. As a result, we get

$$\langle \exp(\beta V(t)) \exp(\beta V(t + \tau)) \rangle = (\exp(\beta V_0))^2 \exp((\beta\sigma)^2(1 + \rho(\tau))), \quad (9)$$

where $\rho(\tau)$ is the normalized autocorrelation function of noise plus signal, represented by the last two terms in Equation (5). We consider here the input noise with a Lorentzian power spectrum,

$$S_N(f) = S_N(0)/[1 + (f/f_c)^2] = (2\sigma^2/\pi f_c)/[1 + (f/f_c)^2], \quad (10)$$

so that the spectrum is frequency independent, 'white', at low frequencies and, at frequencies higher than a 'corner frequency', f_c, decreases as $1/f^2$. The integral of $S_N(f)$ over all frequencies represents the total noise power and equals to the mean-square fluctuation, σ^2. The autocorrelation function for this noise is given by a single-exponential dependence with a characteristic decay time of $1/(2\pi f_c)$. At $f \ll f_c$, Equation (7) becomes

$$S(f) = 2Q^2 r(V_0) \exp\left(\frac{(\beta\sigma)^2}{2}\right) + \frac{2(Qr(V_0)\beta\sigma)^2}{\pi f_c} \exp((\beta\sigma)^2) \sum_1^\infty \frac{(\beta\sigma)^{2n-2}}{n!n}$$

$$+ \frac{(Qr(V_0)\beta V_S)^2}{2} \exp((\beta\sigma)^2)\delta(f - f_S). \tag{11}$$

The first term of this expression accounts for the noise of a 'steady' Poisson wave with time-independent pulse generation rate that exceeds the equilibrium rate $r(V_0)$ by a factor $\exp((\beta\sigma)^2/2)$. The second term yields the system response to the input noise and includes not only the small-amplitude transduction, but also the effects caused by frequency mixing. The mixing between different noise spectral components is described by the terms under the summation sign. For small noise intensities; that is, for $\sigma \to 0$, this sum approaches unity. The last term represents the transfer of a small signal indicating a finite power transduction coefficient of $(Qr(V_0)\beta)^2$ even in the absence of external input noise. The addition of noise enhances the signal component exponentially with the noise mean-square amplitude.

Encoding of the input voltage $V(t)$ into the time-dependent pulse generation rate $r(V(t))$ is described by the last two terms in Equation (11). Indeed, if σ and V_S were zeros, these terms would vanish and the output spectrum would be reduced to that of a steady Poisson wave given by Equation (6). The second term, representing the transduction of input noise to the system output, would also vanish if the noise were weakly correlated, that is, $f_c \to \infty$. This result is in accord with Equation (10), which shows a decrease in the low-frequency noise spectral density, $S_N(0)$, as the corner frequency is increased.

Experimentally, the intensity of the signal spectral component described by the last term in Equation (11) depends on the resolution of the spectrum analyzer used for measurements. To calculate this component (of dimensionality A^2/Hz), we divide the prefactor at the delta-function in Equation (11) by the spectrum analyzer resolution, Δf_A. The output SNR is now obtained as the ratio of the last term of Equation (11) to its two first terms

$$\text{SNR} = \frac{\dfrac{(\beta V_S)^2 r(V_0)}{2\Delta f_A} \exp\left(\dfrac{(\beta \sigma)^2}{2}\right)}{2 + \dfrac{2r(V_0)}{\pi f_c} \exp\left(\dfrac{(\beta \sigma)^2}{2}\right) \sum_1^\infty \dfrac{(\beta \sigma)^{2n}}{n!n}}. \tag{12}$$

It is clear that in addition to a trivial dependence on input signal amplitude, V_S, and spectrum analyzer resolution, Δf_A, the quality of the output signal can be regulated by the input noise parameters such as intensity, σ, and frequency bandwidth, f_c. A simple dependence of the SNR on pulse generation rate for a steady Poisson wave with adjustable rate, SNR $\propto r$, is modified by the second term of the denominator in Equation (12).

Figure 1 illustrates the main features of the SNR calculated according to Equation (12). The results show that it is *always* possible to select noise parameters in such a way that noise addition to the system input will improve the output signal quality. If the bandwidth of the input noise is large enough, then introduction of such small-intensity noise improves signal transduction. The output SNR first grows to a certain value defined by the initial rate and the ratio between the initial rate and the noise corner frequency. Then, as the noise intensity increases, the SNR starts to fall, eventually decreasing below its value for zero input noise. The optimal noise intensity corresponding to a maximum in the SNR is shown in Figure 2. The analysis demonstrates that for $f_c/r(V_0)$ ratios smaller than 10^5 it can be estimated using the following expression

$$\sigma_{\text{opt}} \cong \frac{1}{\beta} \sqrt{\ln \frac{\pi f_c}{2r(V_0)}}. \tag{13}$$

The introduction of input noise with sufficiently short correlation time (large f_c) and with intensity σ_{opt} or smaller *always* increases the output SNR value. It is apparent that in order to obtain any improvement in the output SNR by adding noise with a Lorentzian spectrum described by Equation (10) (single-pole

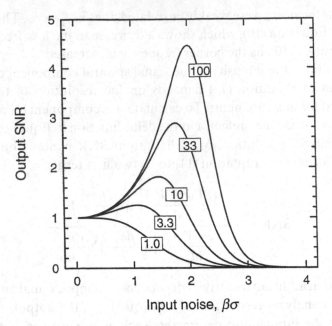

Figure 1. The output signal quality, measured as the signal-to-noise ratio (SNR), can be improved by addition of input noise if the ratio of the noise cut-off frequency, f_c, to the process initial rate, $r(V_0)$, is high enough. Numbers in the boxes show $\pi f_c/2r(V_0)$ values used in Equation (12). For the noise with a Lorentzian spectrum used in the calculations, an improvement in the output signal is obtained if $\pi f_c/2r(V_0) > 1$, otherwise, the introduction of noise only leads to deterioration of the output signal.

passive filtering), the condition $\pi f_c/2r(V_0) > 1$ has to be fulfilled. In the case of noise with a sharp spectral cut-off (rectangular spectrum) the $\pi/2$ factor here and in Equations (12) and (13) should be omitted (Bezrukov and Vodyanoy, 1997b).

Figure 3 reveals the dependence of the output SNR on the initial rate $r(V_0)$ at different input noise intensities. The trivial proportionality between the SNR and the statistics $(r(V_0))$ expected for a time-independent Poisson process is markedly changed by noise addition. At sufficiently small initial rates, the noise addition improves the output signal quality. This general conclusion holds for any parameter combination. Results are shown in Figure 3 for $\pi f_c/2 = 1$ and $(\beta V_S)^2/2\Delta f_A = 2$.

In summary, our model is very general as it describes noise-facilitated signal transduction in a variety of systems with exponential statistics ranging from 'kT-driven' molecular processes of electron transfer, ion channel conformational transitions, or biochemical reactions to more elaborate system with macroscopic sources of internal noise (Jung and Wiesenfeld, 1997).

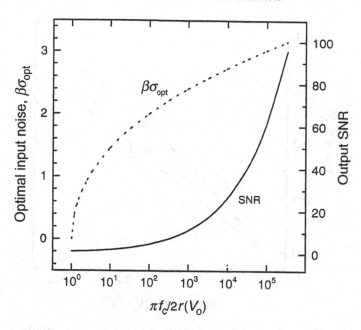

Figure 2. The maximal increase in the output signal-to-noise ratio (SNR) is achieved at an optimal noise intensity (vertical axis on the left) that depends on the ratio of the noise corner frequency, f_c, to the process initial rate, $r(V_0)$. The SNR value at the optimal noise intensity (vertical axis on the right) is given in units of the initial SNR, that is, the output SNR in the absence of input noise. Interestingly, the increase in the output SNR value can reach two orders of magnitude as the corresponding optimal noise changes by only three-fold.

11.3 Ion channels as molecular 'stochastic resonators'

One of the main parameters describing nonlinear properties of voltage-sensitive ion channels is *'gating charge'* (Hille, 1992). This parameter reflects the sensitivity of ion channel conformational dynamics to the electric potential drop across the cell membrane. Here, we consider a channel that has only two states, open and closed, with probabilities P_O and P_C, respectively. We also suppose that the conformational change underlying the closed–open transition moves n electron charges from one membrane surface to the other. Following Hodgkin and Huxley's (1952) and Hille's (1992) treatment, for the fraction of open channels we obtain

$$P_O = \frac{1}{1 + \exp((w - neV)/kT)}, \tag{14}$$

where w is the increase in the conformational energy in the closed–open transition when the transmembrane potential is zero. Equation (14) represents

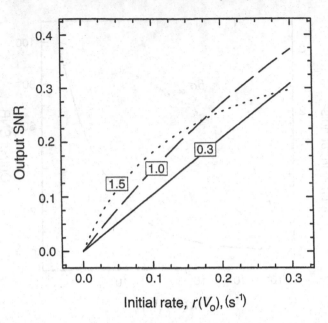

Figure 3. At small initial rates the output signal-to-noise ratio (SNR) is always increased by noise addition to the signal input. In the presence of noise the SNR deviates
from the straight solid line corresponding to an 'undisturbed' Poisson process (zero or
small input noise) giving larger SNR values. The higher the noise intensity ($\beta\sigma$,
numbers in boxes), the greater is the improvement in the output signal quality at small
$r(V_0)$. As the initial rate grows, the SNR starts to saturate and eventually declines to
values less than the SNR for the 'undisturbed' process.

the well-known Boltzmann equation rewritten for channel conformational
statistics. In the case when the closed channel population dominates, i.e.,
$P_O \ll P_C$, Equation (14) yields

$$P_O \cong \exp\left(\frac{-w}{kT}\right)\exp\left(\frac{neV}{kT}\right). \tag{15}$$

In this case it is easy to see that the fraction of the open channels and the rate of
single-channel openings, $r_1(V)$, are related to each other through the channel
lifetime in the open conformation, τ_O, according to $P_O \cong r_1(V)\tau_O$. Therefore, N
independent channels will produce a random process of ion current pulses with
the generation rate $r(V) = Nr_1(V)$ that is given by

$$r(V) \cong \frac{N}{\tau_O}\exp\left(\frac{-w}{kT}\right)\exp\left(\frac{neV}{kT}\right). \tag{16}$$

If the probability P_O of finding a channel in its open state is small enough, the
correlations between successive pulses generated by the same channel can be

ignored. Then, in the case of time-dependent voltage, $V \equiv V(t)$, that is slow enough for the channel gating system to respond, the current pulses represented by Equation (16) will constitute a time-dependent Poisson process described by Equation (3) with the 'equilibrium' generation rate being

$$r(0) = \frac{N}{\tau_0} \exp\left(\frac{-w}{kT}\right).$$ (17)

Experimentally, the gating charge is determined from the dependence of the fraction of open channels on the applied potential. The relative maximum conductance is measured during depolarizing voltage steps of different amplitudes (Hodgkin and Huxley, 1952). The 'limiting equivalent voltage sensitivities' are expressed in millivolts per e-fold increase in Na^+ or K^+ conductance and are calculated from the steepest part of the conductance versus voltage step amplitude curve at $P_O \ll P_C$. This procedure gives an equivalent sensitivity of 3.9 mV for Na^+ and 4.8 mV for K^+ channels that corresponds to about 6 and 4.5 electron charges of gating charge (Hille, 1992).

It should be noted that the gating charge determined in this way is just a convenient empirical parameter that probably underestimates the actual charge of the channel 'voltage sensor'. If the structural change that causes channel opening can move sensor charges only a fraction of the distance between membrane surfaces, the sensor charge has to be larger. Nevertheless, the concept of gating charge offers an useful quantitative description of channel conformational balance and also suggests a simple physical explanation for a complex biological phenomenon.

Biological ion channels have been successfully modeled with a number of channel-forming peptide compounds (Sansom, 1991; Woolley and Wallace, 1992; Cafiso, 1994). From the point of view of electrical signal transduction, the most interesting model of an ion channel is the antibiotic *alamethicin*. It forms channels with conformational dynamics that are highly voltage dependent. The 'equivalent voltage sensitivity' of these channels is remarkably close to that of Na^+ channels of excitable membranes. The mechanism of alamethicin channel sensitivity to transmembrane voltage is not fully understood (for a review, see, e.g., Cafiso, 1994). Most popular explanations refer to the interactions between the electric field within the membrane and the substantial dipole moment of the alamethicin molecule (about 75 debye). The voltage-dependent insertion of peptide molecules adsorbed to the membrane surface is the oldest model of alamethicin channel formation (Baumann and Mueller, 1974). We have chosen it from among the others for its simplicity and we use it here as an illustration only. Figure 4 shows how the electric field

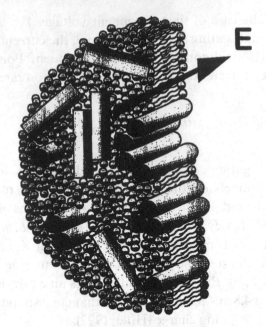

Figure 4. A model explaining alamethicin channel 'gating' by voltage-sensitive equilibrium between populations of peptide molecules adsorbed to the membrane surface and molecules traversing the membrane. The electric field (E) interacts with the alamethicin dipole moment, changing the equilibrium constant. Molecules within the membrane spontaneously and reversibly form ion-conducting clusters of different sizes. Bigger clusters correspond to higher current levels. (From Bezrukov and Vodyanoy, 1997b.)

across the membrane interacts with the alamethicin dipoles, inducing a change in the peptide orientation from the surface orientation to the transmembrane one. Molecules then reversibly aggregate into ion-conducting clusters of different sizes. The particular technique employed for alamethicin reconstitution into planar lipid bilayer membranes is described elsewhere (Bezrukov and Vodyanoy, 1993, 1997b). To obtain the steady-state conditions that are necessary for accurate measurements, the lipid bilayer membranes were equilibrated with alamethicin-containing aqueous solution for 2–3 h at a constant holding potential of 100–150 mV. The membrane capacitance was in the 30–50 pF range.

Ion channels of alamethicin show several distinctly different and well-defined conductance states that are probably related to peptide aggregates of different sizes (see Figure 4). They appear as spontaneous current 'bursts' where ionic current undergoes several random transitions between different current levels and then returns to the initial background level (Hall *et al.*, 1984). Figure 5 illustrates the phenomenon of ionic current bursts at two time

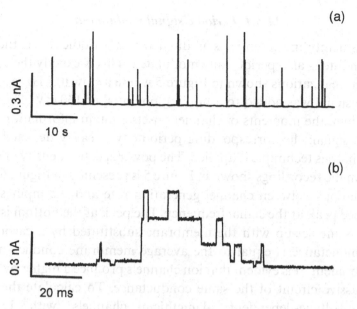

Figure 5. ((a) and (b)) Ionic currents through the membrane containing alamethicin channels are shown for two different time scales. A holding potential of 130 mV (positive from the side of peptide addition) and a 5-mV (r.m.s.), 0.5-Hz sinewave signal are applied to the membrane. Ion channels appear as random 'current bursts' with a fine structure revealed by the higher resolution recording. Different current levels within a single burst correspond to different cluster sizes (see Figure 4).

scales differing by a factor of 500. The probability of observing a channel burst event is voltage dependent. For the particular lipid (DPhPC), salt (1 M NaCl aqueous solution), range of transmembrane voltages (100–150 mV, positive from the side of peptide addition), and all other parameters held constant (alamethicin concentration, temperature etc.), the average alamethicin-induced conductance grows exponentially with voltage as $\exp(V/(4.1 \pm 0.6 \, \text{mV})$. Taking into account that the structure of a single burst does not depend on the applied voltage appreciably (Boheim, 1974), the rate of channel generation is described by Equation (16) with the gating charge ne being equal to 5 to 7 electron charges. Exponential dependence of channel open probability on voltages in this range means that alamethicin channels do not reach the saturation predicted by Equation (14) over more than five decades of probability change. The interpretation of this experimental observation is indeed very model dependent; however, formally, it suggests that the difference in the conformational energy between closed and open alamethicin channel states, w, is much larger than for Na^+ or K^+ channels of excitable membranes.

11.3.1 Periodic signal modulation

If the transmembrane potential is modulated by a periodic signal, then channel generation rate is also periodically modulated. This is exactly the case for the ion current realizations shown in Figure 5 where a slow (0.5 Hz) 5-mV (r.m.s.) sine wave signal was added to a holding potential of 130 mV. Though seemingly random, the moments of channel onsets contain information about the sine wave signal: the corresponding periodicity is easily detected when the spectral analysis technique is applied. The power spectral density of the output signal from the recordings shown in Figure 5 is presented in Figure 6. It reveals the correlations between channel generation rate and the input signal as a pronounced peak at the signal frequency. The peak at the bottom is measured using the same set-up with the membrane substituted by a carbon resistor whose conductance is equal to the average membrane conductance for the same experiment. It is evident that ion channels produce a higher output signal than a passive circuit of the same conductance. To calculate the 'gain', we compare voltage-dependent alamethicin channels with hypothetical

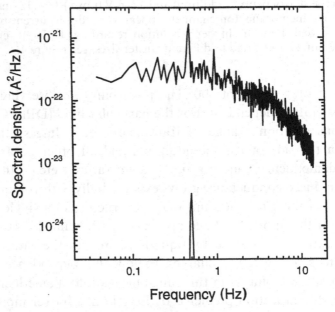

Figure 6. Spectral density of the membrane ionic current (upper trace) in comparison with the spectral density of the current through a resistor of equivalent conductance (peak at the bottom). No external noise is applied. It is evident that voltage-dependent ion channels give a much higher output signal than the equivalent resistor. Spectra were obtained as averages over 23 spectral estimates with a total duration of 20 min. (From Bezrukov and Vodyanoy, 1997b.)

'alamethicin channels' with the gating charge switched off, that is, $ne = 0$. Current through the parallel array of voltage-dependent channels is proportional to a product of applied votage, $V(t) = V_0 + \delta V(t)$, and channel generation rate. Using Equations (2) and (4) and keeping terms linear in $\delta V(t)$ only, we obtain

$$\delta[V(t)r(V(t))] \cong \delta V(t)r(V_0) + \delta V(t)r(V_0)neV_0/kT \qquad (18)$$

For voltage-insensitive channels n is zero and the corresponding expression for $\delta[V(t)r(V(t))]$ does not contain the second term in the right-hand side of Equation (18). The square ratio of these expressions yields the small-signal power transduction

$$\alpha^2 = (1 + neV_0/kT)^2. \qquad (19)$$

The obtained value is in good agreement with the data shown in Figure 6. The result, however, holds for small and slow signals only. If a signal is too fast for the channel 'gating system' (for alamethicin channels the gating time should include not only field-induced flipping of the molecules but also aggregation kinetics, as shown in Figure 4), frequency corrections should be added to Equations (18) and (19). Thus, while the measured transduction coefficient is in good agreement with the predictions ($n = 6.3$) of Equation (19) for a *slow* signal, Figure 7 illustrates that it deviates from the predictions for a *fast* signal. The output signal is calculated as $S(f_s) - N(f_s \pm \Delta f)$, where $S(f_s)$ is the height of the spectral component at the signal frequency, and $N_s(f_s \pm \Delta f)$ stands for the background noise spectrum that represents an average over spectral components in the immediate vicinity of the signal peak. The data show that transduction is proportional to the square of the holding potential and is nearly frequency-independent below 1 Hz. A 3-dB decline corresponds to about 2 Hz. Signals with higher frequencies are too fast for the alamethicin 'gating system' to respond, so that, with the signal frequency growing, the transduction coefficient approaches unity or 0 dB, corresponding to $ne = 0$ (see Equation (19)).

The data shown in Figures 6 and 7 refer to a *deterministic* input, i.e., a constant DC holding potential in the presence of a sine wave signal; no external noise was added at this stage. The smooth power spectra (an example of which is given in Figure 6) correspond to the intrinsic randomness of 'kT-driven' channel conformational transitions. Interestingly, the shapes of the noise spectrum in Figure 6 and of the transduction coefficient in Figure 7 are very similar. This similarity suggests that the mean duration of the single-channel current burst determining the noise spectrum off-set is also responsible for the frequency dispersion of the channel sensitivity to voltage.

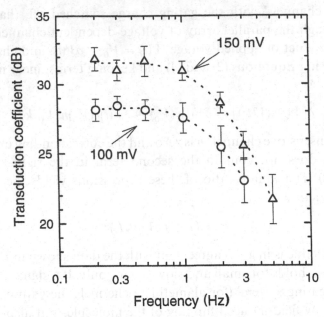

Figure 7. Transduction coefficient measured as the ratio of the channel output-signal peak to the signal peak of the equivalent resistors of holding potentials of 100 and 150 mV. The latter corresponds to the output signal expected for ion channels with their voltage sensitivity switched off, that is $ne = 0$ (see Equation 16). At low frequencies, the coefficient is in good agreement with the model prediction. As frequency increases above 1–2 Hz, the channel sensitivity to voltage decreases. The sinewave input signal becomes too fast for the channel-formation reactions (see Figure 4) to follow. (From Bezrukov and Vodyanoy, 1997b.)

11.3.2 *Noise enhancement of signal transduction*

The effect of external input noise on signal transduction was studied in terms of the output SNR and the output signal amplitude dependence on noise magnitude. The output SNR was calculated as the ratio of the signal spectral component (corrected for the background noise spectrum) to the background noise spectrum $SNR = (S(f_s) - N(f_s \pm \Delta f))/N(f_s \pm \Delta f)$. A white noise from a laboratory-made noise generator was filtered by a band-pass filter restricting its spectrum to the 3.2 mHz to 5.3 Hz frequency range (Bezrukov and Vodyanoy, 1997b). The generator output was varied within 0–20 mV (r.m.s.) limits. The upper curve in Figure 8 shows the power spectral density of ionic current through alamethicin channels in the presence of 8-mV external input noise. Under the particular conditions used in this experiments, a positive (from the side of peptide addition) holding potential of 140 mV supports an average initial rate of about 0.3 channels/s. Application of the 8-mV noise

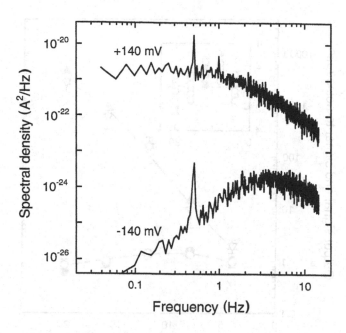

Figure 8. Spectral density of ionic current through the membrane in the presence of 8 mV (r.m.s.) input noise in comparison with the spectral density of membrane capacitive current. The membrane capacitive current was measured by inverting the initial holding potential of $+140\,mV$ to $-140\,mV$. Negative (from the side of peptide addition) voltage completely 'removed' ion channels from the membrane, restoring its resistance to about 10^{12} ohms and making all ohmic contributions negligible. An input signal of 0.5 Hz and 5 mV (r.m.s.) was applied in both cases. (From Bezrukov and Vodyanoy, 1997b.)

increases this rate by about 10-fold, increasing both the output signal and the background noise. The lower curve in Figure 8 corresponds to the bilayer capacitive currents. It was measured from the same bilayer but with ion channels removed by switching the polarity of the holding potential to negative. The input signal and noise are filtered by the membrane capacitance of about 35 pF in such a way that the frequency correction is proportional to the frequency squared. Indeed, the input voltage from noise and the signal generators is now applied to the operational amplifier input through a purely capacitive load (dielectric membrane devoid of ohmic channel conductance) whose impedance falls as the inverse of the frequency. Figure 8 illustrates that capacitive contributions to the output signal or noise in the signal vicinity are negligible. The decline of capacitive current spectral density starting at frequencies above 5 Hz is due to additional filtering of the noise generator output that was used to reduce possible overload of the electronics by wide-band noise.

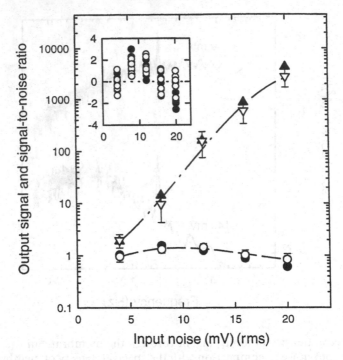

Figure 9. Introduction of input noise facilitates signal transduction. A major noise-induced increase in signal output is achieved at essentially constant, or even slightly improved, signal-to-noise ratio (SNR). The output signal (triangles) and the SNR (circles) are given as ratios of their values at zero input noise. Two signal frequencies, 0.2 Hz (filled symbols) and 0.5 Hz (open symbols), were used. The inset shows the SNR statistics at a finer scale. (From Bezrukov and Vodyanoy, 1997b.)

The addition of external input noise significantly increases the output signal and, for optimal noise amplitudes, even improves output signal quality as quantified by the SNR. Figure 9 presents results for two signal frequencies, 0.2 Hz (filled symbols) and 0.5 Hz (open symbols). One observes that a 20-mV input noise increases the output signal by about 3×10^3 times (or 35 dB), preserving the initial SNR value. At some intermediate noise intensities (around 10 mV) an increase in the SNR is observed, demonstrating a characteristic feature of systems with stochastic resonance: the small-intensity input noise increases both output signal and noise, but the signal grows somewhat faster. The statistics of the SNR measurements on a finer scale are displayed in the inset in Figure 9.

Before comparing our theoretical predictions with the experimental results we should note that our model has been formulated for idealized inertia-free, nondynamical systems with an instantaneous response to the applied stimulus. Alamethicin channels, as any other real object, however, have inertia in their response. As it follows from Figure 7, signals with frequencies exceeding 2 Hz are progressively less effective in influencing alamethicin channel behavior, implying that the response cut-off frequency of 2 Hz should be used as the noise corner frequency, f_c. Taking $r(V_0)$ equal to $0.3\,\mathrm{s}^{-1}$ and n in the range of 5 to 7, from Equation (13) we obtain $\sigma_{\mathrm{opt}} \approx 6$ to 8 mV. This value is in good agreement with the experimental findings presented in Figure 9. The same is true for the output SNR (Equation (12)) given by our model for these parameters. The predicted increase is about 3 dB and is very close to the measured SNR value.

The agreement between the experimental results and the predictions from the time-dependent Poisson model is surprisingly good, taking into consideration that several of the theoretical assumptions do not hold in our experiments. While the assumption about the exponential dependence of pulse generation rate on applied voltage has been verified by our measurements of the average alamethicin-induced conductance versus applied DC holding potential, the absence of pulse–pulse correlations is not so well established. In fact, autocorrelation analysis of alamethicin channel currents at high open-channel densities shows an increase in correlation times, suggesting interactions between different pulses (Kolb and Boheim, 1978). This behavior may be important at high noise intensities that induce a significant increase in alamethicin conductance and, therefore, 'switch on' interpulse interactions. The assumption of the weakness of the input signal strength, namely $V_S \ll kT/ne$ and $V_S \ll \sigma$, is also obviously violated. In our experiments we used signals with amplitudes comparable to the equivalent channel voltage sensitivity (3.5–4.7 mV) and to the input noise amplitude (0–20 mV). These 'high' signal amplitudes were necessary for the reliable differentiation of the signal from the noise background. Nevertheless, the theoretical predictions are accurate enough to allow us to state that our model not only gives a solid *qualitative* picture of the phenomenon, but that it also provides a robust *quantitative* tool for the analysis of stochastic resonance at the molecular level.

11.4 Concluding remarks on small-signal stochastic resonance

Our theoretical and experimental results allow us to discuss a number of generic and frequently discussed questions concerning noise-facilitated small-signal transduction.

(1) *Addition of input noise can increase output signal amplitude.* Our model shows (the last term in Equation (11)) that the signal component in the output spectrum grows with noise (r.m.s.) value σ as

$$S_{\mathrm{signal}} \propto \exp((\beta\sigma)^2), \tag{20}$$

and does not depend on the noise frequency band. The increase in signal transduction does not saturate with σ as in other models (e.g. Kiss, 1996) and does not depend on signal frequency.

(2) *Addition of noise with optimized parameters enables the improvement of the output signal quality.* Our model predicts an infinite growth in the output SNR value, if additive noise is correlated poorly enough, i.e.,

$$f_c/r(V_0) \to \infty, \tag{21}$$

and its amplitude is kept at the level prescribed by Equation (13). As it follows from Equation (12), the noise-induced SNR improvement does not depend on signal frequency. It is necessary to note that our calculations were performed for a *periodic* signal (Equation (5)), whereas several authors argue, quite reasonably, that practically interesting cases should refer to *aperiodic* signals occupying a finite frequency range (Neiman and Schimansky-Geier, 1994; Collins *et al.*, 1996; Kiss, 1996). It is exactly the frequency-*independent* behavior of the signal amplification and output SNR values obtained from our model that makes it applicable also to aperiodic signals. Indeed, any aperiodic signal occupying a finite frequency range can be represented by a superposition of periodic signals. As long as the highest frequency of this range is much smaller than all the other characteristic frequencies in the model, all our conclusions hold for aperiodic stimuli as well.

(3) *Small-signal stochastic resonators and the input–output SNR issue.* This is, probably, the most controversial issue in stochastic resonance research. First of all, a discussion of this issue depends on the particular definition of the SNR. If, following standard textbooks on random data analysis (e.g., Bendat and Piersol, 1986), we define the SNR as the ratio of the signal amplitude to that of noise in some *finite* frequency range, then the output SNR value can exceed the input SNR value. Indeed, if the input noise is distributed over a finite frequency range, Δf, and the signal frequencies are limited to a sufficiently smaller range, $\Delta f_s < \Delta f$, then it is always possible to 'filter out' noise contributions outside Δf_s. At Δf_s small enough, even a deteriorated SNR at the output of a stochastic resonator can be substantially improved to exceed that at the input, if the resonator bandwidth is properly adjusted.

To distinguish between optimal filtering and some alternative mechanism that could be responsible for the improvement in the SNR, researchers have defined the SNR as a ratio of the squared Fourier signal component to the power spectral density of the output noise at the signal frequency (e.g., Douglass *et al.*, 1993; Gingl *et al.*, 1995; Jung, 1995). Then, in the linear-response limit, it is easy to show that the output SNR never exceeds the input SNR (DeWeese and Bialek, 1995; Dykman *et al.*, 1995). Indeed, noise components that, by their frequencies, are infinitely close to the input signal are indistinguishable from the signal itself and are amplified by the stochastic resonator with the same transduction coefficient. Frequency mixing only adds additional noise components to the system output. In the case of large signals, which is not addressed by our work, the situation is not that simple (Kiss,

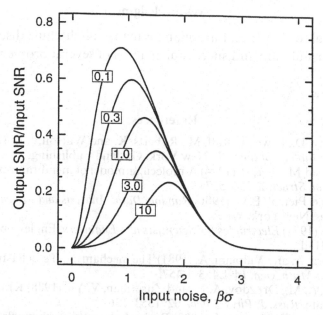

Figure 10. In the case of small-signal stochastic resonance, the output signal-to-noise ratio (SNR) never exceeds the input SNR. The model, based on a time-dependent Poisson process, shows that the output SNR can approach its input value only for small ratios of the noise corner frequency, f_c, to the initial process rate, $r(V_0)$. The numbers in boxes show different $\pi f_c / 2r(V_0)$ values used in the calculations.

1996; Loerincz *et al.*, 1996).

In our model the output SNR versus input SNR ratio can be calculated from Equations (10) and (12) as

$$\chi = \frac{(\beta\sigma)^2 \exp\left(\dfrac{(\beta\sigma)^2}{2}\right)}{\dfrac{\pi f_c}{r(V_0)} + \exp\left(\dfrac{(\beta\sigma)^2}{2}\right) \sum_1^\infty \dfrac{(\beta\sigma)^{2n}}{n!\,n}}. \tag{22}$$

This ratio is plotted in Figure 10 as a function of $\beta\sigma$ for different $f_c/r(V_0)$ values. As can be observed, the output SNR never exceeds the input SNR but approaches its value for small noise amplitudes when the initial rate is high, $r(V_0) \to \infty$. Thus, stochastic resonance is a mechanism for noise-induced signal amplification (Equations (11) and (20)); it is also a mechanism for noise-induced improvement of the performance of a signal transducer (Equations (12) and (13)). However, with respect to the *input* SNR, the small-signal stochastic resonance does not yield any additional benefits in comparison with optimal filtering techniques.

Acknowledgment

We are grateful to V. Adrian Parsegian for many enlightening discussions that have led to formulation and successful analysis of several problems presented here.

References

Alberts, B., Bray, D., Lewis, J., Raff, M., Roberts, K. and Watson, J. D. (1994) *Molecular Biology of the Cell*. New York: Garland Publishing.

Baumann, G. and Mueller, P. (1974) A molecular model of membrane excitability. *J. Supramol. Struct.* **2**: 538–557.

Bendat, J. S. and Piersol, P. G. (1986) *Random Data: Analysis and Measurement Procedures*. New York: Wiley.

Benedict, R. R. (1976) *Electronics for Scientists and Engineers*. Englewood Cliffs, NJ: Prentice-Hall.

Benzi, R., Sutera, A. and Vulpiani, A. (1981) The mechanism of stochastic resonance. *J. Phys. A: Math. Gen.* **14**: L453–L457.

Berezhkovskii, A. M., Drozdov, A. N. and Zitserman, V. Yu. (1988) Kramers kinetics: present state. *Russ. J. Phys. Chem.* **62**: 1353–1361.

Bezrukov, S. M. and Vodyanoy, I. (1993) Probing alamethicin channels with water soluble polymers. Effect on conductance of channel states. *Biophys. J.* **64**: 16–25.

Bezrukov, S. M. and Vodyanoy, I. (1995) Noise-induced enhancement of signal transduction across voltage-dependent ion channels. *Nature* **378**: 362–364.

Bezrukov, S. M. and Vodyanoy, I. (1997a) Stochastic resonance in non-dynamical systems without response thresholds. *Nature* **385**: 319–321.

Bezrukov, S. M. and Vodyanoy, I. (1997b) Signal transduction across alamethicin ion channels in the presence of noise. *Biophys. J.* **73**: 2456–2464.

Boheim, G. (1974) Statistical analysis of alamethicin channels in black lipid membranes. *J. Membrane Biol.* **19**: 277–303.

Bulsara, A. R., Lowen, S. B. and Rees, C. D. (1994) Cooperative behavior in the periodically modulated Wiener process: noise-induced complexity in a model neuron. *Phys. Rev. E* **49**: 4989–5000.

Cafiso, D. S. (1994) Alamethicin: a peptide model for voltage gating and protein-membrane interactions. *Annu. Rev. Biophys. Biomol. Struct.* **23**: 141–165.

Chapeau-Blondeau, F., Godivier, X. and Chambet, N. (1996) Stochastic resonance in a neuron model that transmits spike trains. *Phys. Rev. E* **53**: 1273–1275.

Chialvo, D. R., Longtin, A. and Mullergerking, J. (1997) Stochastic resonance in models of neuronal ensembles. *Phys. Rev. A.* **55**: 1798–1808.

Chiou-Tan, F. Y., Magee, K., Robinson, L., Nelson, M., Tuel, S., Krouskop, T. and Moss, F. (1996) Enhancement of sub-threshold sensory nerve action potentials during muscle tension mediated noise. *Int. J. Bifurc. Chaos* **6**: 1389–1396.

Collins, J. J., Imhoff, T. T. and Grigg, P. (1996a) Noise-enhanced information transmission in rat SA1 cutaneous mechanoreceptors via aperiodic stochastic resonance. *J. Neurophysiol.* **76**: 642–645.

Collins, J. J., Imhoff, T. T. and Grigg, P. (1996b) Noise-enhanced tactile sensation. *Nature* **383**: 770.

Cordo, P., Inglis, T., Verschueren, S., Collins, J. J., Merfeld, D. M., Rosenblum, S., Buckley, S. and Moss, F. (1996) Noise in human muscle spindles. *Nature* **383**:

769–770.

Cox, D. R. and Lewis, P. A. W. (1966) *The Statistical Analysis of Series of Events*. London: Methuen.

DeWeese, M. and Bialek, W. (1995) Information flow in sensory neurons. *Nuovo Cimento D* **17**: 733–741.

Douglass, J. L., Wilkens, L., Pantazelou, E. and Moss, F. (1993) Noise enhancement of information transfer in crayfish mechanoreceptors by stochastic resonance. *Nature* **365**: 337–340.

Dykman, M. I., Luchinsky, D. G., Manella, R., McClintock, P. V. E., Stein, N. D. and Stocks, N. G. (1995) Stochastic resonance in perspective. *Nuovo Cimento D* **17**: 661–683.

Eyring, H. (1935) The activated complex in chemical reactions. *J. Chem. Phys.* **3**: 107–115.

Gammaitoni, L., Hänggi, P., Jung, P. and Marchesoni, F. (1998) Stochastic resonance. *Rev. Mod. Phys.* **70**: 223–287.

Gingl, Z., Kiss, L. B. and Moss, F. (1995) Non-dynamical stochastic resonance: theory and experiments with white and arbitrary colored noise. *Europhys. Lett.* **29**: 191–196.

Gluckman, B. J., Netoff, T. I., Neel, E. J., Ditto, W. L., Spano, M. L. and Schiff, S. J. (1996) Stochastic resonance in a neuronal network from mammalian brain. *Phys. Rev. Lett.* **77**: 4098–4101.

Hall, J. E., Vodyanoy, I., Balasubramanian, T. M. and Marshall, G. R. (1984) Alamethicin: a rich model for channel behavior. *Biophys. J.* **45**: 233–247.

Hänggi, P., Talkner, P. and Borkovec, M. (1990) Reaction-rate theory: fifty years after Kramers. *Rev. Mod. Phys.* **62**: 251–342.

Hille, B. (1992) *Ionic Channels of Excitable Membranes*. Sunderland, MA: Sinauer.

Hodgkin, A. L. and Huxley, A. F. (1952) A quantitative description of membrane current and its application to conduction and excitation in nerve. *J. Physiol.* **117**: 500–544.

Jung, P. (1995) Stochastic resonance and optimal design of threshold detectors. *Phys. Lett. A* **207**: 93–104.

Jung, P. and Wiesenfeld, K. (1997) Too quiet to hear a whisper. *Nature* **385**: 291.

Kiss, L. B. (1996) Possible breakthrough: significant improvement of signal to noise ratio by stochastic resonance. In *Chaotic, Fractal, and Nonlinear Signal Processing* (ed. R. Katz), pp. 382–396. New York: American Institute of Physics.

Kolb, H.-A. and Boheim, G. (1978) Analysis of the multipore system of alamethicin in a lipid membrane. II. Autocorrelation analysis and power spectral density. *J. Membrane Biol.* **38**: 151–191.

Levin, J. E. and Miller, J. P. (1996) Broadband neural encoding in the cricket cercal sensory system is enhanced by stochastic resonance. *Nature* **380**: 165–168.

Loerincz, K., Gingl, Z. and Kiss, L. B. (1996) A stochastic resonator is able to greatly improve signal-to-noise ratio. *Phys. Lett. A* **224**: 63–67.

Longtin, A. (1997) Autonomous stochastic resonance in bursting neurons. *Phys. Rev. A* **55**: 868–876.

Lu, J. and Fishman, H. M. (1994) Interaction of apical and basal membrane ion channels underlies electroreception in ampulary epithelia of skates. *Biophys. J.* **67**: 1525–1533.

Morse, R. P. and Evans, E. F. (1996) Enhancement of vowel coding for cochlear implants by addition of noise. *Nature Med.* **2**: 928–932.

Moss, F., Chiou-Tan, F. and Klinke, R. (1996) Will there be noise in their ears? *Nature Med.* **2**: 860–862.

Neiman, A. and Schimansky-Geier, L. (1994) Stochastic resonance in bistable systems driven by harmonic noise. *Phys. Rev. Lett.* **72**: 2988–2991.

Nye, J. F. (1976) *Physical Properties of Crystals.* Oxford: Clarendon.

Pei, X., Wilkens, L. and Moss, F. (1996) Noise-mediated spike timing precision from aperiodic stimuli in an array of Hodgkin–Huxley-type neurons. *Phys. Rev. Lett.* **77**: 4679–4682.

Plesser, H. E. and Tanaka, S. (1997) Stochastic resonance in a model neuron with reset. *Phys. Lett. A* **225**: 228–234.

Rice, S. O. (1954) Mathematical analysis of random noise. In *Selected Papers on Noise and Stochastic Processes* (ed. N. Wax), pp. 133–294. New York: Dover.

Sansom, M. S. P. (1991) The biophysics of peptide models of ion channels. *Progr. Biophys. Mol. Biol.* **55**: 139–235.

Simonoto, E., Riani, M., Seife, C., Roberts, M., Twitty, J. and Moss, F. (1997) Visual perception of stochastic resonance. *Phys. Rev. Lett.* **78**: 1186–1189.

Wiesenfeld, K. and Moss, F. (1995) Stochastic resonance and the benefits of noise: from ice ages to crayfish and SQUIDs. *Nature* **373**: 33–36.

Woolley, G. A. and Wallace, B. A. (1992) Model ion channels: gramicidin and alamethicin. *J. Membrane Biol.* **129**: 109–136.

12

Ratchets, rectifiers, and demons: the constructive role of noise in free energy and signal transduction

12.1 Introduction

A particle in solution is subject to random collisions with solvent molecules, giving rise to the erratic, 'Brownian', motion first observed and reported by Robert Brown in 1826. This dynamic behavior was described theoretically by Langevin (1908), who hypothesized that the forces on the particle due to the solvent can be split into two components: (1) a fluctuating force that changes magnitude and direction very frequently compared to any other time scale of the system, and (2) a viscous drag force that always acts to slow the motion induced by the fluctuation term. Einstein (1906) derived a (fluctuation–dissipation) relationship between the magnitude of the fluctuation term and the viscous drag coefficient that dampens its effect. Since the strength of the fluctuation increases with temperature, the fluctuating force is often called *thermal noise*. If the particle is a molecule, bombardment by the solvent also allows exploration of the different molecular configurations, i.e., the arrangements of the atoms of the molecule relative to each other. Biological (and many other) macromolecules often have only a few stable configurations, called conformations, with large energy barriers separating them. Thermal noise 'activates' transitions over these barriers from one conformation to another. Kramers (1940) formulated a theory for thermal noise-activated transitions between different conformations based on diffusion over energy barriers (e.g., Hänggi *et al.*, 1990). Indeed almost all chemical reaction pathways are described in terms of rate constants that specify the probability that thermal noise will provide sufficient energy to surmount barriers separating chemical states. Since the development of a statistical description of thermodynamics, where heat continuously sloshes back and forth between different parts of any system and homogeneity is maintained only on average, scientists have attempted to devise ways to harness the fluctuations inherent even in equilibrium systems to accomplish work. If, without expending energy, we could trap with any certainty a system in a state where one part is hotter than another, it would be possible to harness the heat flow resulting from removal of the constraints on the system to do useful work. The thought experiments involved in understanding why this is impossible have provided excellent theoretical laboratories in which to investigate the deeper implications of the second law of thermodynamics.

12.2 Maxwell's demon

Perhaps the first and most influential such construct is that due to Maxwell, who proposed a device (often whimsically described as an intelligent being or demon) that could open or close a gate separating two containers of a gas depending on a measurement of the velocity of an oncoming molecule (Leff and Rex, 1990; von Baeyer, 1998). If the demon would open the gate only to fast molecules coming from one side and to slow molecules coming from the other, a thermal gradient would be formed without expenditure of external energy – a clear violation of the second law. Imagine two identical chambers filled with gas that are open to one another. At some time the two chambers are isolated from one another by sliding a partition between them. Because of fluctuations in density and average kinetic energy, the pressure and temperature in the two chambers will not be identical. This is most easily recognized if there is only one or a few molecules, but it is true in general. With an appropriate apparatus, the reequilibration between the two chambers when the partition is removed could be used to do work, e.g., the lifting of a weight attached to a piston that moves when the gas in the chamber that is hotter and denser expands into the chamber where the gas is cooler and less dense. If we simply guess in which direction the expansion will occur and arrange our apparatus and weight accordingly, half the time the weight will be lifted, but half the time we will be wrong and the weight will descend. On average no net ork results. There are two approaches to attempt to improve on this result. The first, discussed by most authors writing on Maxwell's demon, is to determine in which chamber the product of density and temperature is greater and use this information to guide the set-up of an apparatus to extract the energy released from reequilibration. If such a measurement could be done without expenditure of energy, a perpetual motion machine could be constructed. Early attempts to resolve the paradox focused on the measurement process itself, arguing that interacting with a system to make a measurement intrinsically involves a finite dissipation that overcompensates any work output of the device. Landauer (1991), however, demonstrated several convincing physical set-ups by which a measurement can be made without any dissipation. This stimulated Bennet (1987) to consider how much energy is dissipated in the process of returning the measuring device to its initial state by erasing the acquired information. He found that the minimum energy required is more than enough to make up for any work done by lifting weights. So, the principle of 'you can't get something for nothing' remains, yet the investigations leading to exorcism of Maxwell's demon have been of great importance in the development of modern ideas of computing and information theory.

12.3 Ratchets and rectifiers

A second possible approach to extract work from thermal fluctuations is to design an apparatus involving weights and pulleys where, if we guess correctly, the reequilibration raises a weight, but, if we guess wrongly, the weight is prevented from descending. Such a device is a ratchet. Macroscopic ratchets and rectifiers passively convert external zero-average fluctuating forces into net motion (current). What happens if such devices are shrunk to microscopic size? This was the question posed by von Smoluchowski (1912) and later by Feynman in his *Lectures on Physics* (Feynman et al., 1966), where he analyzed a microscopic ratchet (a cog with asymmetric teeth) and pawl (Figure 1a). The ratchet's teeth are arranged so that it is impossible to drag the pawl up one face. At first glance it would seem that a paddlewheel attached to such a ratchet should convert thermal fluctuations to unidirectional motion. If the wheel cannot go backward, molecules hitting the paddle would cause an irregular but relentless rotation of the wheel that could lift a weight. This would be a perpetual motion machine of the second kind, a contradiction to the second law. Feynman resolved this paradox, showing that consistent application of the laws of statistical mechanics to all parts of the device restores the result that work cannot be extracted from thermal noise in an isothermal system. In order to function, the pawl must be attached to the ratchet by some elastic element, say a spring, which is also influenced by thermal noise. When the spring is extended the pawl is down and the device works as anticipated. However, occasionally, by thermal noise the spring contracts, lifting the pawl and disengaging the ratchet mechanism. When this happens, molecular collisions with the paddlewheel cause forward and backward motion with equal probability. Because of the asymmetry, however, it takes only a tiny motion backward to set the device back far enough that when the pawl reengages the device moves counterclockwise by one tooth, but it takes a much larger movement forward to advance the device by one tooth in the clockwise direction. If the paddlewheel and pawl are at the same temperature, the

a)

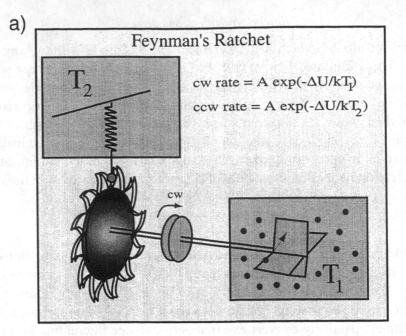

Feynman's Ratchet

T_2

cw rate = A exp(-ΔU/kT$_1$)

ccw rate = A exp(-ΔU/kT$_2$)

cw

T_1

b)

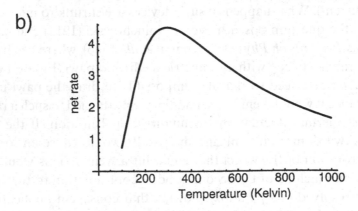

Figure 1. (a) Depiction of Feynman's ratchet. Because of the asymmetry of the ratchet's teeth, it might seem that even thermal fluctuations acting on a paddlewheel attached to the ratchet could be used to do work. However, the behavior of the spring is also influenced by collisions with molecules, which cause it to vibrate. When the spring is down, molecular collisions with the paddlewheel indeed tend to cause the cog to turn only in the planned direction. However, in the rare event that the spring is up, disengaging the one-way mechanism, the random molecular forces on the paddle cause forward and backward motion with equal probability. It only takes a very tiny movement of the wheel backwards to set the device back one tooth, whereas to send it forward by a tooth requires a much greater motion. If the paddle and the pawl are at the same temperature so that the fluctuations on the pawl are as strong as those driving the paddlewheel, the ratchet, despite appearances, will not turn. The equations are approximate formulae for the frequency of moving a step in the clockwise (cw) and counterclockwise (ccw) direction. $\Delta U = \kappa h^2$, where κ is the spring constant and h is the height of a tooth on the cog. (b) Plot of net rate versus T from Equation (1), with $A = 10^4$/s and $\Delta U = 50$ meV.

tendencies to move forward, owing to the fluctuating force acting at the paddlewheel, and to move backward, owing to the fluctuating engagement and disengagement of the pawl, exactly cancel; despite our macroscopic intuition, the ratchet will not rotate at equilibrium.

Recently, Kelly and colleagues at Boston College have synthesized an organomolecular ratchet with a triptycene as the paddlewheel, linked to a four-ring helicene as the pawl and spring (see Figure 2; Kelly *et al.*, 1997, 1998; Davis, 1998). Because the helicene has a twist to it, the force necessary to turn the triptycene paddlewheel clockwise is less than is necessary to turn it counterclockwise. This is illustrated by manipulation of 'tinker-toy' molecular models and by the calculated plot of ΔH versus the dihedral angle. The energy profile is strongly anisotropic, as expected for a ratchet, but of course is periodic, with a repeat every 120°. Using the nuclear magnetic resonance

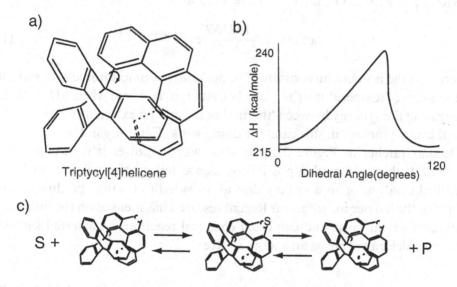

Figure 2. (a) A molecular ratchet constructed of a triptycyl paddlewheel attached to a helix-forming helicene pawl by a single bond around which rotation is possible. (b) Calculated ΔH as a function of the dihedral angle about this bond. Despite the anisotropy in the structural design, at equilibrium, clockwise and counterclockwise transitions over the barrier are equally likely, for structural reasons analogous to those given by Feynman for his ratchet and pawl. (c) Hypothetical scheme for converting this ratchet to a chemically driven molecular motor by linking the 'pawl' to a catalyzed reaction (conversion of S to P). When the active site is empty, the pawl is flexible and the barrier to rotation small. Binding substrate at the active site reduces the flexibility of the pawl, and raises the barrier to rotation. If the catalyzed reaction is far from equilibrium, the fluctuations of the barrier height will not obey detailed balance and can lead to unidirectional rotation of the triptycyl paddlewheel – a chemically driven molecular motor.

(NMR) technique of spin polarization transfer, it was shown that despite the anisotropy, consistent with the second law, the frequencies of clockwise and counterclockwise turns are exactly equal – a neat experimental verification of Feynman's theoretical analysis.

Feynman also considered a modification of his ratchet. The temperatures of the pawl and of the gas in the box containing the paddlewheel need not be equal. Feynman showed that when the temperatures are different the imbalance in the strength of the thermal noise acting on the paddlewheel and on the spring does indeed cause the cog to rotate. Astonishingly, the device works both ways – when the temperature of the paddlewheel is greater than that of the spring, the ratchet rotates clockwise as our intuition would suggest. But when the temperature of the spring is greater than that of the paddlewheel, the ratchet's rotation is counterclockwise! A simple equation illustrating this can be derived from the two relations for the unidirectional rates shown in Figure 1, with $T_1 - T_2 = 2\Delta T$ and $T_1 + T_2 = 2T$, and

$$\text{net rate} = \frac{A\Delta U \Delta T}{2k_b T^2} e^{-\Delta U/k_b T}, \tag{1}$$

where k_b is the Boltzmann constant. The net rate is nonmonotonic with respect to the average temperature T (Figure 1b), and is maximized when $\Delta U = 2k_b T$, illustrating the synergy between 'thermal noise' and an external energy supply (the thermal gradient in this case) for doing work. Certainly in the case of the molecular ratchet in Figure 2 there is no way to power it with a thermal gradient. Is there some other possibility, such as linking the helicene pawl to a chemical catalyst, or to a moiety that absorbs light and thus modulates the height of the barrier for rotation? Recent research has focused on the interplay between thermal noise and catalyzed chemical reactions or external signals, both of which can serve as an energy source.

12.4 Biasing Brownian motion

To get a better insight into how Brownian motion can be biased by a fluctuating input to cause directed motion, let us consider two simple examples based on diffusion of a particle in solution in the presence of electric potentials that are turned on and off at random. The first is based on the ideas behind Maxwell's demon, and the second on the ideas behind Feynman's ratchet and pawl.

12.4.1 Information ratchet

Imagine a negatively charged particle diffusing on a glass slide where there is a periodic array of electrodes. The negative electrode of each pair is much closer to the solution than is the positive electrode. Thus, when the potential is on, the particle feels a series of repulsive barriers centered at the electrodes (see Figure 3a). If the ionic strength of the solution is large, the Debye distance is very small and these barriers can be very sharp. Now, let us add a mechanism for turning the potential on or off depending on where the particle is located at a particular instant in time. We place a series of light beams, one just to the left of each electrode pair (dashed line in Figure 3a), and one to the right (dotted line in Figure 3a), and arrange it so that when the particle trips the dotted light beam immediately to the left of an electrode pair a switch that turns the potential off is flipped, and tripping the dashed beam to the right of the electrode pair turns the potential on. Thus, when a particle approaches the electrodes from the left, the potential barriers are turned off. The potential will turn back on when the particle diffuses either the short distance to the next dashed beam to the right, or the longer distance to the dashed beam on the left. Because diffusion the shorter distance is more probable the particle moves on average to the right. If the two beams are very close to the electrodes, the net velocity to the right approaches D/L, where L is the spacing between the electrodes and D is the diffusion coefficient. If the voltage is very large, motion to the right occurs even against a strong opposing force, but the velocity decreases exponentially with the force

$$\text{velocity} = \frac{k_b T}{\gamma L} e^{-FL/k_b T}, \tag{2}$$

where we have used the Einstein relation $D = k_b T/\gamma$, and where γ is the coefficient of viscous drag. It is worth noting that at no point is elastic energy stored in this mechanism – useful motion occurs solely by biasing the Brownian motion of the particle.

12.4.2 Energy ratchet

The information ratchet described above works very well for driving directed motion of a single particle, but it involves a rather complicated set-up. This is because the mechanism requires actively obtaining and using information about the particle's whereabouts to decide when to turn the potential on and off, much like Maxwell's demon. It turns out that we can circumvent this

a)

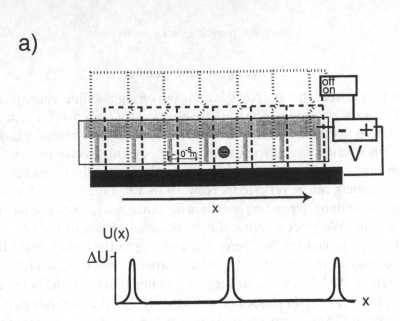

b)

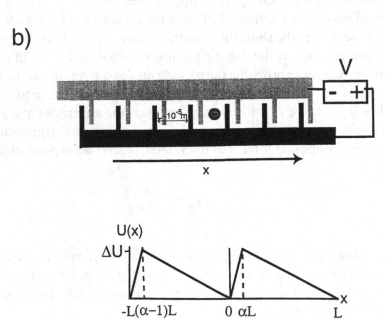

Figure 3. Illustration of electronic implementation of biasing Brownian motion. (a) Information ratchet. When a particle passes the dotted line, a circuit is tripped that turns off the potential. It is more likely that the particle will trip the closer dashed line to the right turning the potential back on than it is that the particle will diffuse to the dashed line on its left. In this way, net left-to-right motion is ensured even though the particle never experiences a 'force' in any direction. (b) Energy ratchet. When the potential is on, the particle is tightly pinned in one of the wells at positions iL. Turning off the potential allows free diffusion, and then turning the potential back on forces the particle back to one of the wells. Because of the anisotropy, net motion from left to right occurs as explained in Figure 4.

requirement by setting up the system so a particle 'feels' an anisotropic force when the potential is on. Consider a simple example where fluctuations are imposed from the outside by turning a potential on and off in the device shown in Figure 3b. The electric potential felt by a charged particle fluctuates as shown in Figure 4, where a macroscopic force is superimposed as a net tilt to show that the turning the potential on and off can actually drive the particle uphill and thereby do work. If the external force is not too large the particle is pinned near the bottom of one of the wells (for example at $x = 0$) when the potential is on. Without noise, the particle would move to the left with a velocity F_{ext}/γ, when the potential is off, where γ is the coefficient of viscous drag. Thermal noise changes the situation dramatically. Because of Brownian motion, a random walk is superimposed on the deterministic drift, and the position of the particle can be determined only in terms of a probability distribution. While the potential is off, the probability distribution simultaneously drifts downhill with a velocity F_{ext}/γ, and spreads out like a Gaussian function. After a time t_{off}, the distribution (Figure 4) is

$$P(0\,|\,x;t_{off}) = \frac{e^{-\frac{(x-t_{off}F_{ext}/\gamma)^2}{4Dt_{off}}}}{\sqrt{4\pi Dt_{off}}}. \tag{3}$$

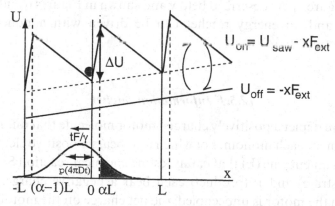

Figure 4. Schematic picture of how turning a ratchet potential (U) on and off can lead to motion against an applied force. When the sawtooth potential (U_{saw}) is on, the Brownian particle is tightly pinned at the bottom of an energy well. Then, when the sawtooth potential is turned off, the particle begins to move downhill (to the left) with velocity F/γ, but also begins to undergo Brownian motion, which superimposes a random walk on the deterministic drift to the left. If the potential remains off long enough for the particle to diffuse the short distance αL, but not the longer distance $(1 - \alpha)L$, net motion to the right occurs even though the applied force F_{ext} tends to move the particle to the left.

When the potential is turned on again, the particle is trapped in the well at iL if it is between $L(i - 1 + \alpha)$ and $L(i + \alpha)$. The probability of finding the particle in each well, P_{iL}, is calculated by integrating the probability density (Equation (3)) between these limits. Because of the anisotropy of the potential, a particle starting at iL is more likely to be trapped in the well at $(i + 1)L$ than in the well at $(i - 1)L$ for small enough F_{ext} and intermediate values of t_{off}. The average number of steps R in a cycle of turning the potential on for a time t_{on} long enough that the particle reaches the bottom of a well $(t_{on} > (L^2/\Delta U))$, and then off for a time t_{off}, and then back on again is $R = \sum_{i=-\infty}^{\infty} iP_{i}L$, and the average velocity is $\langle v \rangle = RL/(t_{off} + t_{on})$. Motion to the right occurs even though the macroscopic force acts to push the particle to the left.

12.5 Chemically driven motion

The above descriptions show how external fluctuations can lead to directed transport, but what about a nonequilibrium chemical reaction? Figure 5 shows how a chemical reaction can drive directed motion by coupling a nonequilibrium-catalyzed reaction to the diffusion of a particle along a polymer chain. A key requirement that is necessary (but not sufficient) is that detailed balance for the transitions between one 'state' and another is broken (for details, see caption to Figure 5). As described below and shown in Figures 6 and 7, both an information and an energy ratchet can be driven with a nonequilibrium chemical reaction.

12.5.1 Information ratchet

In Figure 6 we depict a positively charged motor molecule that diffuses along a polymer filament each monomer of which also bears a positive electric charge. The motor is an enzyme (E) that catalyzes the chemical reaction $S \rightleftharpoons P$, where both S (substrate) and P (product) each bear a negative charge. When the active site on the motor is unoccupied, the net charge on the motor is positive and there is an energy barrier for the motor to pass over the positive charge on the filament backbone, as shown on the curve labeled U_{free}. When the active site is occupied by either S or P the motor is electrically neutral, and in this chemical state there is no barrier for the motor to diffuse over the positive charges on the filament, as reflected in the flat potential profile labeled U_{bound}.

If it could be arranged that the active site of the motor would most probably be occupied when the motor is just to the left of a charge on the filament (the

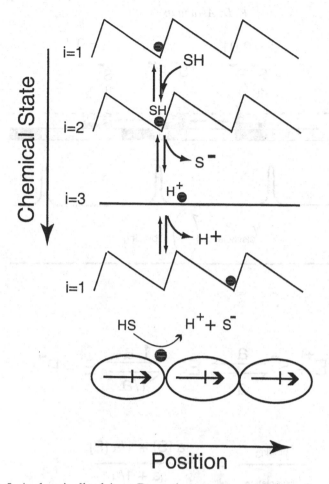

Figure 5. A chemically driven Brownian motor consisting of an electrically charged catalytic particle undergoing one-dimensional diffusion along a linear polymer of electric dipoles. As the particle catalyzes the chemical reaction $SH \rightleftharpoons S^- + H^+$, its net charge fluctuates depending on its chemical state (i.e., to what it is bound), and so the interaction with the dipole fluctuates. In states $i = 1$ and 2, the particle is negatively charged and so is pinned in the region near the positive end of a dipole. When the particle is bound to proton (state $i = 3$), the net charge is zero and the particle is free to diffuse along the backbone of the dipole with equal probability to the left and right. Because of the asymmetry of the dipole potential, a short excursion to the right results in the particle being one step to the right when proton dissociates. A much longer and less probable excursion to the left is necessary to trap the particle a step to the left when proton dissociates. Thus it might seem that the charge fluctuations caused by the chemical reaction could drive net motion even when the reaction is at equilibrium. Just as in Feynman's ratchet, a more careful consideration shows that this is not true. Since S^- and H^+ are electrically charged, their concentrations depend on the position along the dipole axis. The excess probability to bind H^+ near the δ^- end and to bind S^- near the δ^+ end results in a detailed balance for all transitions, and zero net flow. However, when the reaction is away from equilibrium, directed motion does occur. For $L = 10^{-8}$ m, $\gamma = 2 \times 10^{-8}$ N s/m and reasonable values of the rate coefficients (Astumian, 1997a), the average velocity is greater than 3 μm/s with $\Delta G \approx 1$ kJ/mol.

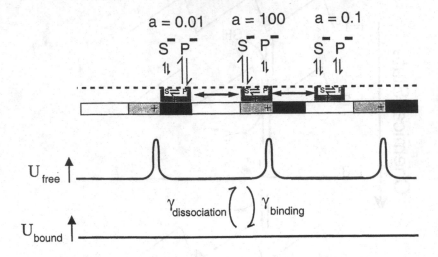

$$E^+ + S^- \underset{a}{\overset{a}{\rightleftharpoons}} ES \underset{1/a}{\overset{1/a}{\rightleftharpoons}} E^+ + P^-$$

$$\frac{\gamma_{binding}}{\gamma_{dissociation}} = \frac{a\,[S] + 1/a\,[P]}{a + 1/a}$$

Figure 6. Schematic illustration of a chemically driven 'information ratchet' in which the anisotropy necessary to break the symmetry of the system and allow net motion is contained within the position dependence of the transition constants between the chemical states. For simplicity we have taken the equilibrium constant for the chemical reaction to be unity, but the results do not depend on this scaling.

gray regions), and unoccupied when the motor is just to the right of a charge on the filament (the black regions), left-to-right motion arises trivially, as described in the case of an information ratchet driven by external fluctuations. The barriers act as a series of gates. Every time a motor approaches a barrier from the left, S or P most likely binds, thus causing the 'gate' to open. As soon as the motor crosses the threshold, S or P dissociates, causing the gate to close and preventing backward diffusion.

This condition is achieved if, for example, the association/dissociation of S is fast and association/dissociation of P is slow in the gray region, and vice versa in the black region, but only when the $S \rightleftharpoons P$ reaction is far from equilibrium. A

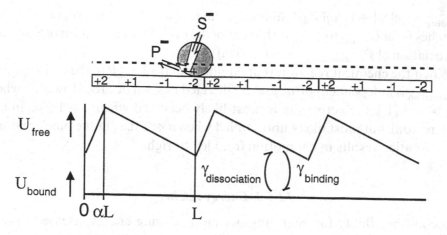

$$E^+ + S^- \underset{\phi}{\overset{1}{\rightleftharpoons}} ES \underset{10^5/\phi}{\overset{10^5}{\rightleftharpoons}} E^+ + P^-$$

$$\phi = \exp\left(\frac{U_{bound} - U_{free}}{RT}\right)$$

$$\frac{\gamma_{binding}}{\gamma_{dissociation}} = \frac{\phi^{-1}\{\phi\,[S] + 10^5\,[P]\}}{\phi + 10^5}$$

Figure 7. Illustration of a chemically driven energy ratchet, where the anisotropy is found in the function describing the potential energy of interaction between the motor and its track. The transition constants can be isotropic. These illustrative examples focus on electrostatic models for simplicity, but the same symmetry and energetic principles could equally well be implemented with models involving protein conformational change.

simple parametrization in terms of a factor 'a' is shown in Figure 6, where in the units used here, the dissociation constant $K_{D,S} = K_{D,P} = 1$ for both S and P. Note that the equilibrium constant of the overall reaction is independent of 'a', as it must be. The rate at which the system switches from the potential U_{free} to U_{bound} is the sum of the rates for binding S and for binding P,

$\gamma_{\text{binding}} = \{a[S] + (1/a)[P]\}$. Similarly, the rate at which the potential profile switches from U_{bound} to U_{free} is the sum of the rates for dissociation of S and for dissociation of P, $\gamma_{\text{dissociation}} = [a + (1/a)]$.

When the chemical reaction is at equilibrium ($[S] = [P]$), the ratio $\gamma_{\text{binding}}/\gamma_{\text{dissociation}}$ is independent of 'a', and no directed motion occurs. However, when $[S] > 1 > [P]$, the active site is most likely occupied when $a \gg 1$ (i.e., in the gray region), but most likely unoccupied when $a \ll 1$ (i.e., in the black region). This situation results in net motion from left to right.

12.5.2 Energy ratchet

A second possibility for achieving net motion using energy released from a chemical ratchet is modeled after an 'energy' ratchet, where the required anisotropy is found in the interaction between the motor and the track. This possibility is illustrated in Figure 7. Once again the motor bears a net charge, but now the filament is depicted as an array of dipoles lined up head to tail. We have shown several charges along the axis of each monomer of the filament so that the motor is never more than a Debye distance (represented by the dashed line) from one of the charges on the filament. When the active site of the motor is unoccupied, the interaction between the charge on the motor and the charges on the dipoles gives rise to a potential energy profile U_{free} that is well approximated by a saw tooth function as shown. When either S or P occupies the enzyme active site, the charge on the motor is effectively neutralized and the potential energy is an approximately flat function of position.

Here, the dissociation constants $K_{\text{D,S}}$ and $K_{\text{D,P}}$ must each depend on position, since the difference in electrostatic potential energy between the bound and free states depends on position. In general, this position dependence is not equally apportioned between the association and dissociation rate constants, since the transition state may 'look' either more like the free or more like the bound state. In Figure 7, we have illustrated a specific case loosely inspired by the structure of the myosin molecule. In this picture, S enters and leaves the binding cleft from the 'top', where we imagine the transition state is at the mouth of the cleft, far from the filament surface. Thus, almost all of the position dependence of $K_{\text{D,S}}$ appears in the rate constant for *dissociation* of S, since the transition state looks more like the unbound state. On the other hand, P enters and leaves the binding cleft through an opening that is quite close to the binding site itself, and so the transition state 'looks' more like the bound state. Thus almost all of the position dependence of $K_{\text{D,P}}$ is expressed in the *association* rate constant for binding P. At equilibrium, ($[S] = [P]$), a Boltzmann relation between the γ_{binding} and $\gamma_{\text{dissociation}}$ holds and no flow occurs (Astumian and Bier, 1996). Far from equilibrium, however (where $[S] > 10^5[P]/\phi$ (see Figure 5)), the transitions between 'bound' and 'free' states are approximately independent of position, and flow results as described for the energy ratchet driven by external fluctuations.

12.6 Biased diffusion in practice

The phenomenon of fluctuation-driven transport was anticipated in some sense by many authors. Even before publication of Feynman's ratchet, A. F. Huxley (1957) proposed a model for muscle movement based on biased Brownian motion. His idea in some ways is very similar to Maxwell's demon because it requires the rate constants for binding and release of adenosine triphosphate (ATP), adenosine diphosphate (ADP), and inorganic phosphate (P_i) to depend on the position of a motor molecule (myosin) along a bio-polymer (actin). Passive devices that can convert external random fluctuations into directed motion have also been investigated. Risken (1989) showed that the mobility of a particle in an anisotropic periodic potential is not symmetric with respect to an applied force. Wonnenberger, Breymayer and colleagues investigated the effect of harmonic forcing on particles in a spatially periodic pinning potential (Breymayer *et al.*, 1982; Wonnenberger and Breymayer, 1984). Closer in spirit to Feynman's work, Buttiker (1987) showed that a symmetric spatially periodic temperature could drive transport of a Brownian particle along a spatially periodic potential, with a velocity and sign that depends on the phase relation between the temperature and potential periods.

A very significant recent development was the demonstration by Ajdari and Prost (1992) that directed motion of a particle in a viscous medium could be induced simply by turning a periodic anisotropic potential on and off. Orig-inally proposed as a possible method for particle separation, this represents a true Brownian motor because the mechanism utterly fails in the absence of thermal noise, irrespective of the amplitude of the potential. Astumian and Bier (1994) and Prost *et al.* (1994) showed that the mechanism (now known as a flashing ratchet) works even if the potential (or equivalently the interaction) is turned on and off randomly, but only within a frequency window for the inverse correlation time of the fluctuation.

There have been several recent reviews (Hänggi and Bartussek, 1996; Astumian, 1997a; Jülicher *et al.*, 1997; Astumian and Moss, 1998) and popular articles (Astumian, 1997b; Bier, 1997) dealing with fluctuation-driven trans-port, to which we refer the interested reader. Recent work has dramatically extended the domain of fluctuation-driven transport, with two specific topics coming to the fore – applications to particle separation, and possible ties to mechanism of biomolecular motors and pumps. And perhaps a third topic should now be added to this list – chemically synthesized nanomolecular ratchets, brakes and molecular versions of other macromechanical devices.

Rousellet et al. (1994) demonstrated experimentally that diffusion of small particles can be biased by turning an anisotropic dielectrophoretic potential cyclically on and off. Shortly after, Faucheaux et al. (1995) used an optical trap to show the same behavior at the level of a single particle. This work has begun an avalanche of proposals concerning ways to use fluctuation-driven transport for particle separation. Several authors (Chauwin et al., 1995; Mielke, 1995; Bier and Astumian, 1996) have shown that with appropriate potentials there can be a threshold value of the diffusion coefficient where particles with a diffusion coefficient greater than the threshold move in a direction opposite to those with a diffusion coefficient smaller than the threshold. This opens up the possibility of a continuous separation, where particles are continuously fed into the middle of the device and large particles collect on one side and small particles collected on the other side. These experimental set-ups have used a flashing ratchet approach. Duke and Austin (1998) and Ertas (1998) have recently proposed a novel two-dimensional separation method based on constructing a system with anisotropic barriers using microlithography. If the system has a nondiagonal mobility tensor, a DC force in one direction will cause transport perpendicular to the force. They have argued that this can also be used as the basis for a continuous separation technique. Derenyi and Astumian (1998) have recently shown that a nondiagonal mobility tensor is not necessary, and in fact a continuous method for separation can be based on a simple two-dimensional system where the linear coupling coefficients are identically zero. The biased diffusion in this case results from a quadratic non-Onsager effect and allows a great reduction in the size and increase in the throughput of the device as compared with the proposal of Duke and Austin (1998).

One of the main reasons for the tremendous recent interest in fluctuation-driven transport has been the potential application to biological systems (Astumian and Bier, 1996), particularly in light of the fact that it is now possible to study biomolecular motors at the level of a single molecule (Svoboda et al., 1993). Molecular motors use chemical energy to move along a biopolymeric track constructed of identical monomeric subunits. Because of this, the motors see an energy landscape that is spatially periodic at every instant in time and the flashing ratchet models are thus more relevant for biology than the fluctuating force ratchets. Analysis of the simplest flashing ratchet models shows that with a spatial period of 10^{-8} m, energy barriers of roughly $10 k_b T$ (hydrolysis of one molecule of ATP releases about $20 k_b T$) and a viscosity somewhat greater than that of water, the velocity induced by turning the potential on and off is around 10^{-6} m/s and a force of about 1–10 pN is necessary to bring the particle to a halt. These values are in good agreement with what is observed experimentally. Needless to say, such simple models do not reflect all aspects of the very complicated motor proteins, and a number of authors have described more complicated models that take into account more degrees of freedom for the system (Derenyi and Vicsek, 1996; Cilla and Floria, 1998). Perhaps the strongest direct evidence for a ratchet mechanism for free energy transduction by a biomolecule comes from recent experiments showing that the Na^+,K^+-ATPase, a biomolecular ion pump, can use an external oscillating (Tsong and Astumian, 1986; Liu *et al.*, 1990) or randomly fluctuating (Xie *et al.*, 1994, 1997) electric field to drive unidirectional transport (Astumian *et al.*, 1987, 1989; Astumian and Robertson, 1989, 1993). These results were interpreted theoretically in terms of a four-state kinetic ratchet (Robertson and Astumian, 1991). For the detailed experimental results and their interpretation consult Tsong (Chapter 13, this volume).

12.7 Perspective

Typically, approaches to designing molecular motors and pumps have focused on transposing and scaling down our macroscopic inventions to the microscopic world. Such attempts have been rewarded with only limited success, partly because of the ubiquitous presence of thermal noise. In the microscopic world, moving deterministically is like trying to walk in a hurricane – the forces prescribing the desired path are puny in comparison with the random forces exerted by the environment.

A more promising approach has emerged over the last several years where, instead of working against Brownian motion, thermal noise is harnessed to provide the energy for motion and to surmount energy barriers, and the external or chemical energy source provides the energy to bias the diffusive motion. Returning for example to consideration of the molecular ratchet in Figure 2, we begin to see a way that chemical or light energy can be used to power a simple molecular motor. Kelly and colleagues (1994) previously synthesized a molecular brake, with a polycyclic metal chelating agent in place of the helicene pawl. When a metal ion is bound to the chelator, the polycycle is rigid, presenting a large barrier to rotation of the triptycyl paddlewheel – the brake is 'on'; but when the chelator is not bound to a metal ion, the barrier to rotation is low and the brake is off. A similar strategy could be adopted in the case of the ratchet, but with a catalyst in place of the simple chelator. Then, if the catalyzed reaction were far from equilibrium, chemical energy would drive a nonequilibrium fluctuation between a high barrier and low barrier – a flashing ratchet, as shown in Figure 4. Coupled with the intrinsic anisotropy of the system, energy from the chemical reaction would drive (counterclockwise!) unidirectional rotation of the triptycyl paddlewheel. With an appropriate photoactive moiety, it might also be possible to modulate the barrier height between a photochemically excited state and a ground state, leading to a light-driven molecular motor.

Analogies between biological motors and systems have often been formulated in terms of mechanisms familiar from standard, macroscopic deterministic physics. Molecular motors designed as described above, based on the physics of chemical reactions and stochastic processes, may instead take cues from recent studies at the single molecule level of actual biological engines. This may be the start of a new chapter in the quest for making microscopic machines.

Acknowledgments

I am grateful to NIH for funding and to Martin Bier and Imre Derenyi for very helpful discussions.

References

Ajdari, A. and Prost, J. (1992) Mouvement induit par un potential de basse symétrie: dielectrophorèse pulsée. *C. R. Acad. Sci. Paris* **315**: 1635.

Astumian, R. D. (1997a) Thermodynamics and kinetics of a Brownian motor. *Science* **276**: 917–922.

Astumian, R. D. (1997b) Body works. *New Scientist* **156**: 38–41.

Astumian, R. D. and Bier, M. (1994) Fluctuation driven ratchets: molecular motors? *Phys. Rev. Lett.* **72**: 1766–1769.

Astumian, R. D. and Bier, M. (1996) Mechanochemical coupling of the motion of molecular motors to ATP hydrolysis. *Biophys. J.* **70**: 689–711.

Astumian, R. D., Chock, P. B., Tsong, T. Y., Chen, Y. D. and Westerhoff, H. V. (1987) Can free energy be transduced from electric noise? *Proc. Natl. Acad. Sci. USA* **84**: 434–438.

Astumian, R.D., Chock, P.B., Tsong, T.Y. and Westerhoff, H.V. (1989) Effects of oscillations and fluctuations on the dynamics of enzyme catalysis and free-energy transduction. *Phys. Rev. A* **39**: 6416–6435.

Astumian, R. D. and Moss, F. (1998) Overview: the construction role of noise in fluctuation driven transport and stochastic resonance. *Chaos* **8**: 533–538.

Astumian, R. D. and Robertson, B. (1989) Nonlinear effect of an oscillating electric field on membrane proteins. *J. Chem. Phys.* **72**: 4891–4899.

Astumian, R. D. and Robertson, B. (1993) Imposed oscillations of kinetic barriers can cause an enzyme to drive a chemical reaction away from equilibrium. *J. Am. Chem. Soc.* **115**: 11063–11068.

Bennet, C. (1987) Demons, engines, and the second law. *Sci. Am.* **257**: 108–116.

Bier, M. (1997) Brownian ratchets in physics and biology. *Cont. Phys.* **38**: 371–379.

Bier, M. and Astumian, R. D. (1996) Biasing Brownian motion in different directions in a 3-state fluctuating potential and an application for the separation of small particles. *Phys. Rev. Lett.* **76**: 4277–4280.

Breymayer, H. J., Risken, H., Vollmer, H. D. and Wonneberger, W. (1982) Harmonic mixing in a cosine potential for large damping and arbitrary field strengths. *Appl. Phys. B* **28**: 335–339.

Buttiker, M. (1987) Transport as a consequence of state dependent diffusion. *Z. Phys. B* **68**: 161–167.

Chauwin, J. F., Ajdari, A. and Prost, J. (1995) Current reversal in asymmetric pumping. *Eur. Phys. Lett.* **32**: 373–376.

Cilla, S. and Floria, L. M. (1998) Internal degrees of freedom in a thermodynamical model for intracell biological transport. *Physica D* **113**: 157–161.

Davis, A. P. (1998) Tilting at windmills? The second law survives. *Angew. Chem. Int. Ed. Engl.* **37**: 909–910.

Derenyi, I. and Astumian, R. D. (1998) AC separation of particles by biased Brownian motion in a two-dimensional sieve. *Phys. Rev. E* **58**: 7781–7784.

Derenyi, I. and Vicsek, T. (1996) The kinesin walk: a dynamic model with elastically coupled heads. *Proc. Natl. Acad. Sci. USA* **93**: 6775–6779.

Duke, T. A. J. and Austin, R. H. (1998) Microfabricated sieve for the continuous sorting of macromolecules. *Phys. Rev. Lett.* **80**: 1552–1555.

Einstein, A. (1906) Theory of Brownian motion. *Ann. Phys.* **19**: 371–381.

Ertas, D. (1998) Lateral separation of macromolecules and polyelectrolytes in microlithographic arrays. *Phys. Rev. Lett.* **80**: 1548–1551.

Faucheux, L. P., Bourdieu, L. S., Kaplan, P. D. and Libchaber, A. J. (1995) Optical thermal ratchets. *Phys. Rev. Lett.* **74**: 1504–1507.

Feynman, R. P., Leighton, R. B. and Sands, M. (1966) *The Feynman Lectures on Physics*. Reading, MA: Addison-Wesley.

Hänggi, P. and Bartussek, R. (1996) Brownian rectifiers: how to convert Brownian motion into directed transport. In *Nonlinear Physics of Complex Systems: Lecture Notes in Physics* (ed. J. Parisi), pp. 1–7. Berlin: Springer-Verlag.

Hänggi, P., Talkner, P., Borkevec, M. (1990) Reaction rate theory: 50 years after Kramers. *Rev. Mod. Phys.* **62**: 251–341.

Huxley, A. F. (1957) Muscle structure and theories of contraction. *Progr. Biophys. Biophys. Chem.* **7**: 255–318.

Jülicher, F., Ajdari, A. and Prost, J. (1997) Modeling molecular motors. *Rev. Mod. Phys.* **69**: 1269–1281.

Kelly, T. R., Bower, M. C., Bhaskar, K. V., Bebbington, D., Garcia, A., Lang, F., Kim, M. H. and Jette, M. P. (1994) A molecular brake. *J. Am. Chem. Soc.* **116**: 3657–3658.

Kelly, T. R., Tellitu, I. and Sestelo, J. P. (1997) In search of molecular ratchets. *Angew. Chem. Int. Ed. Engl.* **36**: 1866–1868.

Kelly, T. R., Tellitu, I. and Sestelo, J. P. (1998) New molecular devices: in search of molecular ratchets. *J. Org. Chem.* **63**: 3655–3665.

Kramers, H. A. (1940) Brownian motion in a field of force and diffusion model of chemical reaction. *Physica* **7**: 284–305.

Landauer, R. (1991) Information is physical. *Phys. Today* **44**(5): 23–29.

Langevin, P. (1908) Sur la théorie du mouvement brownien. *Comptes Rendue* **146**: 530–533.

Leff, H. S. and Rex, A. F. (1990) *Maxwell's Demon: Entropy, Information, Computing*. Princeton, NJ: Princeton University Press.

Liu, D. S., Astumian, R. D. and Tsong, T. Y. (1990) Activation of Na^+ and K^+ pumping mode of (Na,K)-ATPase by an oscillating electric field. *J. Biol. Chem.* **265**: 2760–2767.

Mielke, A. (1995) Transport in a fluctuating potential. *Ann. Phys. (Leipzig)* **4**: 721–723.

Prost, J., Chawin, J. F., Peliti, L. and Ajdari, A. (1994) Asymmetric pumping of particles. *Phys. Rev. Lett.* **72**: 2652–2655.

Risken, H. (1989) *The Fokker–Planck Equation.* Berlin: Springer-Verlag.

Robertson, B. and Astumian, R. D. (1991) Frequency dependence of catalyzed reactions in a weak oscillating field. *J. Chem. Phys.* **94**: 7414–7419.

Rousselet, J., Salome, L., Ajdari, A. and Prost, J. (1994) Directional motion of Brownian particles induced by a periodic asymmetric potential. *Nature* **370**: 446–448.

Svoboda, K., Schmidt, C. F., Schnapp, B. J. and Block, S. M. (1993) Direct observation of kinesin stepping by optical trapping interferometry. *Nature* **365**: 721–727.

Tsong, T. Y. and Astumian, R. D. (1986) Absorption and conversion of electric field energy by membrane bound ATPases. *Bioelectrochem. Bioenerg.* **15**: 457–476.

von Baeyer, H. C. (1998) *Maxwell's Demon: Why Warmth Disperses and Time Passes.* New York, NY: Random House.

von Smoluchowski, M. (1912) Experimentell nachweisbare der üblichen Thermodynamik wiedersprechende Molekularphänomene. *Physik. Z.* **13**: 1069–1074.

Wonnenberger, W. and Breymayer, H.-J. (1984) Broadband current noise and ac induced current steps by a moving charge density wave domain. *Zeit. Phys. B* **56**: 241–246.

Xie, T. D., Chen, Y. D., Marszalek, P. and Tsong, T. Y. (1997) Fluctuation-driven directional flow in biochemical cycles: further study of electric activation of Na,K pumps. *Biophys. J.* **72**: 2495–2502.

Xie, T. D., Marszalek, P., Chen, Y. D. and Tsong, T. Y. (1994) Recognition and processing of randomly fluctuating electric signals by Na,K-ATPase. *Biophys. J.* **67**: 1247–1251.

13

Cellular transduction of periodic and stochastic energy signals by electroconformational coupling

TIAN Y. TSONG

13.1 Introduction

The transport of life-sustaining materials across the cell membrane consumes large fractions of the total energy expenditures of cells. As such, membrane transport is tightly regulated and controlled. Membrane proteins are mostly responsible for the transport of molecules and ions in and out of the cell. There are many classes of membrane transporters. A simple classification lists three types: *channels, carriers* and *pumps*. Channels and carriers are *passive downhill* transporters, whereas pumps are *active uphill* transporters. However, a channel may also be able to fulfill the role of an uphill transporter by coupling a downhill transport reaction to an uphill transport reaction: the transporter, however, will still be a downhill transporter overall. Beside proteins, a small number of transporters come from heterocyclic organic compounds. These small molecules can serve as ion channels or ion carriers, but only a protein with *enzymatic activity* can function as a pump. All transporters are embedded in lipid bilayers and, for a pump to function, some degrees of freedom must be hindered or restricted. As I shall discuss later, this feature of a pump is crucial for capturing energy from a *periodically oscillating* or *randomly fluctuating force field* (Tsong and Astumian, 1986, 1988; Tsong, 1990, 1992).

In order for a pump to capture energy another feature is essential: it must be able to interact effectively with a driving force, e.g., an *electric* field or an *acoustic* field (Tsong and Astumian, 1986, 1988). This chapter focuses on mechanisms that employ these two properties of a membrane transporter (1) for harvesting energy from a periodic or a randomly fluctuating electric field and (2) for using the energy to perform chemical work. The ability to transduce energy from a fluctuating force is a remarkable design of Nature: *a cell must be able to harvest energy from seemingly chaotic fluxes of energy in its environment in order to survive.* By what mechanisms can a cell perform such task? I

301

examine certain properties of biochemical reactions that are relevant to the main theme of this volume, namely what roles do the dynamics, especially the nonlinear dynamics of molecules, play in biological signal transduction and sensory detection. And, what mechanisms may a cell or an organism adopt to improve its sensitivity and efficiency? Are there limits below which the detection of signals would be considered unattainable by our current state of knowledge? I begin by examining some unique features of membrane proteins and chemical behavior or reactivity arising from these features. Proteins that make up part of supramolecular assemblies, such as myosin and actin in a myofibril or sonic receptors of the cochlea, which may have similar properties, are mentioned briefly as well.

13.2 Chemical reactions in isotropic and anisotropic media

A chemical reaction occurs by bringing two or more reactant molecules to proximity or to collide, thus producing the end-products. The rate and the behavior of a reaction depend not only on the reactivity of the reactants but also on the properties of the medium. In a low-viscosity, noninteracting medium of a homogeneous phase, the rate of an elementary step is either diffusion limited or proportional to the diffusion rate. No directionality is involved. Hence, the medium, and so the reaction, is *isotropic*. The driving force for such a reaction is simply the free energy of the reaction. The reaction proceeds downhill with respect to the free energy. However, in an organism, only small fractions of biochemical reactions take place in homogeneous solution phase. Most reactants in a cell constitute parts of a supramolecular structure, tissue, or organ and these reactants have limited mobility. Chemical reactions of these molecules depend on how reactants are brought together and how products are dissipated. The media of these chemical processes are, therefore, *anisotropic*. Directionality becomes an inherent property of such biochemical reactions. Most biochemical reactions of cells and tissues belong to this class and they manifest many interesting properties, some of which are only slowly unfolding.

Adding to this complex nature of a reaction environment, the *driving force* for a reaction may also be complex and anisotropic. Furthermore, a driving force may not be constant. It may exhibit both spatial and temporal fluctuations. One of the most commonly occurring fluctuating force fields is temperature, which reflects the *thermal energy* of the system, or the kinetic energy of the molecules. If a reaction has a nonzero enthalpy that is not compensated for by the entropy of the reaction, thermal energy will be able to shift the chemical equilibrium of the system. The extent of change is described by the van't Hoff equation, $\partial \ln K / \partial(1/T) = -\Delta H_R / R$, where K, T, R, and ΔH_R are, respectively, the equilibrium constant, the absolute temperature in Kelvin, the gas constant, and the enthalpy of the reaction. The system, in this case, will have absorbed energy, shifting the chemical reaction equilibrium, and the thermal energy will have done chemical work on the system. A nonzero reaction enthalpy, ΔH_R, is, thus, crucial for a system's interaction with thermal energy. In such a case, the reaction is isotropic. This means that the reactivity of molecules is *scalar* or *nonvectorial*. As a consequence, the reaction cannot produce directional flux of molecules or energy. Hence, mass moves by simple diffusion. Take a ligand binding reaction as an example. Macromolecules and ligands can approach each other from any direction with equal probability. Their paths of diffusion are not obstructed in any direction. In other words, all reactant molecules are rapidly tumbling, and chemical reactivity of these molecules is, for all practical purposes, considered uniform in all directions. As long as a ligand approaches a macromolecule within a certain distance, they will either react or not react, depending solely upon whether the impact energy (kinetic energy) of the two reactants is sufficient to overcome the *activation barrier*. If the impact energy exceeds the barrier, they will react to produce products; if not, they will bypass each other. Case A in Figure 1 illustrates such a system. The macromolecule might be a protein or DNA, and the ligand a substrate, an ion, or a small molecule. Case B illustrates a system in which the active site of the macromolecule is located at a specific locale on the molecule. Only a ligand approaching from the right direction can react; others, from incorrect directions, cannot. Strictly speaking, the latter system would behave differently from case A. However, if the molecule and the ligand can tumble much faster than the frequency of the molecular collisions, the system can still be regarded as quasi-isotropic, and thermodynamic and kinetic equations developed for case A should also be applicable to case B in Figure 1. In the real world, case B is more common than case A despite the fact that most thermodynamic and kinetic equations with which we are familiar have been derived on the basis of systems represented by case A. For example, recent advances in structural biology have taught us that the active site of a protein is typically located in a small, highly localized area within its three-dimensional structure.

The situation would be different if the macromolecule were too big, or carried with it a piece of membrane fragment. This might be the case when one prepares a sample, which retains much of the supramolecular assembly, or disrupts a cell membrane to prepare a partially purified protein. In this case, the rotation of the protein–membrane complex would not be uniform, and its rotational relaxation time could conceivably be longer than the mean

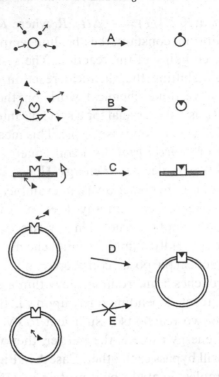

Figure 1. Isotropic and anisotropic chemical reactions. Ligand binding to a macro-molecule is used as an example to illustrate the two basic types of chemical reaction. (A) A ligand can bind to any site of the macromolecule and both ligands and macromolecule are free floating and the medium of the reaction is homogeneous and of low viscosity. The reaction is *isotropic*. (B) The ligand can bind only to a specific site on the macromolecule. If both ligands and macromolecules are free floating and tumbling much faster than the collision rate, and the medium is homogeneous and of low viscosity, the reaction is *quasi-isotropic*. In the laboratory, most reactions occur-ring in the homogeneous phase belong to this class. (C) The rotation of the macro-molecule is slower than the collision time and there exists also only one specific site on the macromolecule at which ligand binding can take place. The reaction will display *anisotropic* characteristics. (D,E) The reaction is *vectorial*; only ligands approaching from the outside of the vesicle may react. The reaction is anisotropic (for details, see text in Section 13.2).

time constant for collisions. Then, binding of the ligand to the protein is no longer an isotropic event (see case C in Figure 1). Further, mass action in case B would behave differently from that in case A, although, in kinetic treatments, effects of nonproductive collisions can easily be corrected by inserting a coefficient into the rate equation, e.g., a cross-section of collisions. As we shall see later in our analysis, when such a system interacts with a periodically oscillating or randomly fluctuating driving force, there may oc-cur many interesting effects that do not occur in cases A and B (Tsong, 1992,

1994). The anisotropic nature of a chemical reaction becomes quite obvious in case D in Figure 1, where membrane reactions are mostly *vectorial* or anisotropic. The difference between an isotropic and an anisotropic reaction forms the basis of our discussion of an 'animate' chemical system versus an 'inanimate' chemical system.

13.2.1 *Periodic driving force and randomly fluctuating driving force*

Our system is an anisotropic one: a membrane ATPase, an immobile enzyme, or a protein, which is a part of a supramolecular complex, for example myosin in a myofibril. Chemical reactions at surfaces, for example adsorption or catalysis, belong to this class of reactions. But, in catalysis and reactions at surfaces, the potential or the driving force of a chemical reaction is still treated as constant, for example the chemical potentials of reactants, or the free energy of a reaction. What if a driving force is a nonuniform, nonconstant force: an oscillatory, fluctuating, or noise-function driving force, with a spatial dynamic? In other words, a driving force that is temporally and spatially fluctuating with time. In a cell, a tissue, or an organ, this situation should be the prevalent situation. The contraction of a muscle is a good example of a directional flow of energy. In electron transport, reducing equivalents follow a unique path, and, in membrane transport, a driving force must be directed normal to the membrane to be effective; exerting a force in any wrong direction would not be productive.

We consider the simple situation illustrated in Figure 2. Here, a membrane-embedded protein is assumed to have a permanent or an induced electric moment in its structure. This property will enable the protein to interact with an externally applied electric field. We assume that the protein does have different conformational states with different values of electric moments, and that two of these conformational states are functional states of the enzyme in the catalytic cycle. When an applied field points in the direction of the electric moment, a conformational state (P_1) with a smaller net moment will be favored, and when a field points in the opposite direction, a conformational state (P_2) with a larger electric moment will be favored. If an applied field is oscillatory, it will cause the protein to undergo a conformational oscillation between P_1 and P_2. This phenomenon has been termed '*enforced conformational oscillation*' (Tsong, 1990). Without an applied field, there are fluctuations in protein conformation but the amplitude of fluctuations will be small. If an electric field is randomly fluctuating, the conformation of the molecule will also be fluctuating, not randomly, i.e., in a spontaneous manner, but in response to the field. Next, a simple estimate of the interaction energy is in

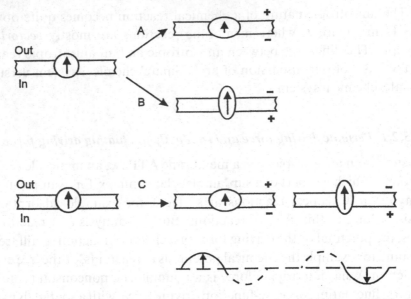

Figure 2. Enforced conformational oscillation by a periodic electric field. (A) The membrane is polarized, which forces the protein with an outward molar electric moment (vertical arrow) to assume a conformational state with smaller moment. (B) When the membrane is depolarized, the protein assumes another conformational state with a greater outward molar electric moment. (C) An oscillating electric field is used to drive the conformational oscillation of the protein. The phenomenon is termed *enforced conformational oscillation* by an alternating electric field (Tsong, 1990, 1992).

order. The thermodynamic relationship governing an electric field-induced conformational change of a protein is,

$$(\partial K_E/\partial E)_{P,T,V\ldots} = (\Delta M/RT),\tag{1}$$

where K_E, E, R and T are, respectively, the equilibrium constant (although K_E would not be constant in an oscillating field) for the $P_1 \to P_2$ conversion, the electric field intensity, the gas constant, and the absolute temperature in Kelvin. Subscripts P,T,V, denote pressure, absolute temperature and molar volume. $\Delta M = M_2 - M_1$ represents the difference in the molar electric moments for P_1 and P_2. The shift in equilibrium is best illustrated in Figure 3 where the relative energy levels of P_1 and P_2 are shown. When the field is off, P_1 and P_2 exhibit the same energy level, meaning that they are equally populated. When the membrane is polarized, the energy level favors P_1, but when the membrane is depolarized, the energy level favors P_2. The energy difference between the two different polarization conditions is ΔG_E, where E is the peak-to-peak field strength of an oscillating electric (AC) field. Simply, $\Delta G_E = \Delta M \cdot \Delta E$, where ΔM of a protein can be a large number; hundreds to

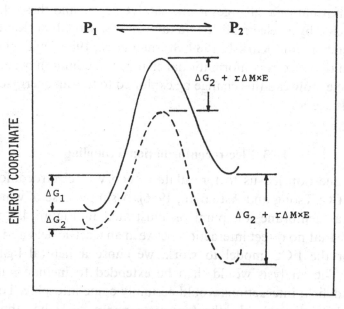

Figure 3. Energy diagram of a membrane protein under the influence of an oscillatory electric field. The relative energy levels of P_1 and P_2 in the absence and presence of the field are shown. At zero electric field, the reaction diagram assumes the shape given by the solid curve. The conformational state P_1 is favored over P_2 by ΔG_1. When an electric field is present, the energy diagram shifts to the one represented by the dashed curve. The overall energy level declines because of the interaction between the molar electric moment (M) of the protein and the electric field. The electric field interaction energy with P_1 is ΔG_2. However, the field interaction favors P_2 over P_1 by $\Delta M \times E$, shifting the chemical equilibrium in favor of the conformational state P_2. If the electric field oscillates, the protein will also oscillate between P_1 and P_2. As mentioned, this phenomenon is termed enforced conformational oscillation (Tsong, 1990, 1992).

thousands of debye (D) have been reported in some proteins. If P_1 and P_2 differ by one gating charge and membrane thickness is 50 Å (1 Å = 0.1 nm), ΔM will be roughly equal to 240 D. The dipole moment of a transmembrane helical segment is 120 D. A transmembrane potential oscillating at ± 50 mV will generate a transmembrane ΔE of 200 kV/cm (ΔG_E, in this case, would be a few kilocalories per mole). In a cell, field fluctuations are likely to be much greater than the ΔE value given here, when both temporal and spatial resolutions are high, and the ΔM values for membrane proteins are also likely to be greater than 240 D. Thus, electroconformational changes of membrane proteins cannot be ignored. The thermodynamics of electric field-induced conformational changes has been discussed in great detail elsewhere (Tsong and Astumian, 1986; Tsong, 1990, 1992). From experiments, Na^+,K^+-ATPase is known to undergo conformational changes in the catalytic cycle and these changes can also be triggered by an electric field (Nakano and Gadsby, 1986; Rephaeli *et al.*, 1986a,b; Steinberg and Karlish, 1989; Stürmer *et al.*, 1989; Wuddel and Apell, 1995). Given that electroconformational change is a common occurrence in a cell membrane, how can this change be exploited to fuel an energy-consuming biochemical reaction?

13.3 Electroconformational coupling

The above question led us to formulate a theory of electroconformational coupling (ECC; Tsong and Astumian, 1986). A simple case in which ECC is applied to a membrane ion pump is illustrated in Figure 4. In order to demonstrate that no direct interaction between an electric field and a ligand is required for the ECC model to work, we chose a neutral ligand in the simulation. The analysis would then be extended to include ionic ligands where ligand–field interactions would certainly come into play. To make the pump simple and workable, the four-state pump cycle has the following characteristics. P_1 has its ligand-binding site facing out and P_2 has its binding site facing in. The affinity of P_1 for ligand (S) in the external medium (S_{out}) is weak compared with the affinity of P_2 for ligand in the cytoplasm (S_{in}). First, the system is allowed to equilibrate. Then, a sinusoidal AC field is applied. Several events will happen in sequence. (1) Before the AC field, E, is turned on, $[P_1] \gg [P_1 \cdot S_{out}]$, $[P_2] \ll [P_2 \cdot S_{in}]$, $[P_1] \gg [P_2]$, and $[P_1 \cdot S_{out}] \ll [P_2 \cdot S_{in}]$. (2) An AC field, E, is applied. The first phase of the AC field depolarizes the membrane, which favors the formation of P_2 and $P_2 \cdot S_{in}$. This triggers a flux from $P_1 \rightarrow P_2$ and another one from $P_1 \cdot S_{out} \rightarrow P_2 \cdot S_{in}$. Because $[P_1] \gg [P_1 \cdot S_{out}]$, the first flux is greater than the second flux; thus a net clockwise flux is produced. The accumulation of P_2 leads to association with S_{in} because of its high affinity for S_{in}, and $[P_2 \cdot S_{in}]$ increases further. (3) The field reaches the second half of the sinusoidal wave and the membrane is now polarized. The reversal of membrane polarization will now favor the formation of P_1 and $P_1 \cdot S_{out}$ over P_2 and $P_2 \cdot S_{in}$. (4) There now will be two types of flux, $P_1 \rightarrow P_1$ and $P_2 \cdot S_{in} \rightarrow P_1 \cdot S_{out}$. Since at this moment, $[P_2 \cdot S_{in}] \gg [P_2]$, the second flux will be greater than the first flux, and the net flux is again a clockwise one. (5) Since the wheel is turned clockwise no matter which way the membrane is polarized, a ligand is pumped out of the cytoplasm for each complete turn of the four-state catalytic wheel. (6) The next cycle of the sinusoidal wave will repeat the processes of steps 1 to 4. In summary, the ECC model successfully describes the action of a molecular pump. The direction of the wheel spin can be easily reversed if the affinity of S for P_1 and P_2 is

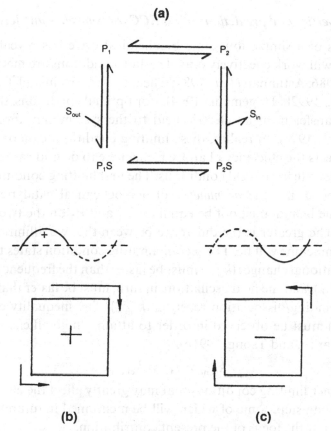

(a)

(b)   **(c)**

Figure 4. Electroconformational coupling (ECC) model of a protein pump. A model of a protein pump can be constructed with the enforced electroconformational change coupled to a ligand binding reaction. In the four-state cyclic transport model shown here, membrane depolarization (+) favors P_2 (with the substrate binding site facing the extracellular medium) and membrane polarization (−) favors P_1 (with the substrate binding site facing the cytoplasm). If substrate (S) is neutral, then membrane depolarization will also favor P_2S and polarization will favor P_1S. With a suitable design of binding affinity for S, e.g., affinity of P_1 for S is smaller than P_2 for S, the pump will spin clockwise under the influence of an alternating electric field. The substrate will be pumped out of the cytoplasm when the applied field drives the catalytic wheel to spin clockwise (Tsong and Astumian, 1986; Tsong, 1989).

reversed. In such a case, the transporter will pump S inward instead of outward. Subsequently, more detailed analyses have been carried out and the ECC model has been shown to exhibit some interesting properties that are discussed below. Next, experimental data relevant to the discussion of the ECC membrane transport model are presented and compared with the model's predictions.

13.3.1 Properties and predictions of an ECC transporter – an electric ratchet

Our analysis of a simple four-state biochemical cycle has revealed that the ECC model will work effectively only if certain conditions are met (Tsong and Astumian, 1986; Astumian *et al.*, 1987; Chen, 1987; Markin and Tsong, 1991b; Markin *et al.*, 1992b; Hilgemann, 1994). For optimal conditions, the efficiency of energy transfer from an electric field to the system can approach unity (Markin *et al.*, 1990). In reality, these limiting conditions cannot be attained and one expects the efficiency of an ECC machine to degrade as experimental conditions deviate from ideal conditions. The first limiting condition has been discussed above, i.e., the *asymmetry* of a biochemical catalytic cycle. The affinity for the ligand must not be equal for P_1 and P_2 on the two sites of the membrane. The greater is the difference between the two affinities, the more efficient the machine will be. The second limiting condition states that the rate for conformational changes (k_{conf}) must be faster than the frequency of the field oscillation (f_{AC}), and the field oscillation, in turn, must be faster than the rate of the ligand binding/dissociation reaction (k_{chem}). The inequality expressed in Equation (2) must be observed in order to attain a high efficiency in energy coupling (Markin and Tsong, 1991b),

$$k_{conf} \gg f_{AC} \gg k_{chem}. \tag{2}$$

There are other limiting conditions that may greatly affect the efficiency of the energy-coupling step, some of which will be mentioned in future discussions, but they are not the focus of the present contribution.

We next ask the question, what does the ECC model predict that can be put to experimental test? Some of the more prominent features, and in many cases also the obvious features of the model, are listed below. Many of these features were observed prior to the formulation of the ECC model (Serpersu and Tsong, 1983, 1984; Liu *et al.*, 1990). Indeed, they formed the basis of the construction of the model:

(1) *Windows of field strength and frequency to achieve maximal efficiency in energy coupling*: This is a rate-dependent phenomenon and was obvious intuitively and checked with experiments before the conception of the ECC model (see below).

(2) *Window for ligand concentration*: After this prediction, a search was undertaken and, for example, a window for $[Ca^{2+}]$ was identified for Ca^{2+}-pump activity in human erythrocytes (Tsong, 1992).

(3) *Energy coupling with randomly fluctuating electric fields*: The intuition was born out by theoretical analysis and later verified by experiments. The theory allowed a quantitative fitting of the data (Xie *et al.*, 1994, 1997).

(4) *Energy coupling for systems with various combinations and ratios of charges between a ligand and a transporter protein*: Charges in the two species may have identical or

opposite signs. Transports due to charge rectification and to the ECC mechanism have been separated and their contributions to energy coupling quantitatively evaluated. Several predictions as laid out by Markin and Tsong (1991a,b) await experimental tests.

(5) *Effects of white noise on the ECC coupling efficiency*: Experiments have shown that when electric fields of suboptimal strengths are used, the superposition of a broad-band electric noise enhances the pumping efficiency. But, in no cases did the observed activities exceed the maximal activity using an optimal electric signal. In other words, there is a maximal enzyme activity, which cannot be exceeded by the imposition of electric energy in the forms of either signal or white noise. The theory for this observation remains to be developed. The enhancement of the ion pump activity by the white noise presumably is by *stochastic resonance*; however, the detailed mechanisms remain unknown and are currently under study.

(6) *An ECC pump is an electric ratchet*: Equations and conditions are identical with an electric ratchet (for details, see Astumian, Chapter 12, this volume).

(7) *The ECC concept is equally applicable to other driving forces*: for example, acoustic, concentration (chemical potential), osmotic pressure, mechanical, and temperature fluctuations (Tsong, 1989, 1990, 1992; Markin and Tsong, 1991a). Few modifications of the basic thermodynamic and kinetic equations are necessary. The ECC system is in fact a Brownian ratchet (Astumian and Bier, 1994; Astumian, 1997).

13.3.2 Energy coupling of periodic electric fields

We selected membrane ATPases for electric activation experiments, because the proton electrochemical potential had been postulated to be an intermediate state in the biosynthesis of ATP. The proton electrochemical potential consists of two components, the proton gradient (ΔpH across the energy-transducing membrane) and the transmembrane electric potential. As most experiments have failed to detect a sizable ΔpH across any energy-transducing membrane, such as the inner membrane of the mitochondrion or the photosynthetic membrane, the electric potential would appear to be the dominant force of ATP synthesis. In mitochondria, electron transport from the consumption of foodstuffs can produce a transmembrane potential of the order of 200 mV. For Na^+,K^+-ATPase, the electrogenic nature of the Na^+-pump and the K^+-pump can also generate a transmembrane potential. It was reasoned that these two enzymes should also respond to electric stimulation, and this turned out to be indeed the case. Submitochondrial particles, when exposed to an electric pulse of about 20 kV/cm with a decay time constant of 60 µs, produced ATP, and this activity was inhibited by oligomycin, an inhibitor of mitochondrial F_0F_1-ATPase (Teissie *et al.*, 1981). The high field strength was required to input sufficient energy, and the short electric pulse avoided Joule heating. It was found that the ATP yield increased with increasing concentration of dithiothreitol (DTT; F. Chauvin, R. D. Astumian and T. Y. Tsong, unpublished results). The reason for this dependence remains unclear. Conceivably, the energy captured from the electric field could be temporarily stored in a reduced form of the enzyme complex. As is evident, mitochondrial ATPase was not the best system for our electrical activation experiment because of the high field strength needed to generate a sufficient amount of ATP for the assay. Furthermore, the size distribution of submitochondrial particles, as they were prepared, was broad, making the estimation of the $\Delta\psi_{memb}$ generated by the external field difficult (Marszalek et al., 1990). The Na^+, K^+-pump of human erythrocytes turned out to be a simpler, more suitable system. This enzyme specifically responds to the inhibitor ouabain, which would allow us to monitor pump activity specifically due to Na^+, K^+-ATPase. Because of this possibility, we chose to perform experiments with intact erythrocytes, where the enzyme exists in its natural environment (Teissie and Tsong, 1980, 1981). Radioactive tracers were used to measure the electric field-induced flux of Na^+, K^+ and Rb^+. It turned out that only the two ATP-dependent pumping modes of the enzyme were activated by the external electric field. The electric field did not affect the passive flux of these ions. Furthermore, the specific activity of the electric field-stimulated pump activity never exceeded the maximal ATP-dependent pump activity. Because of this property, experiments were either done at 2–4 °C, at which ATP hydrolysis activity was negligible, or in samples in which ATP was depleted by biochemical treatment. Many experiments have been performed during the past years and a great number of results are now available. Those obtained prior to the use of randomly fluctuating electric fields are summarized below. These experiments were conducted with sinusoidal or equally spaced, square-waved AC fields (Serpersu and Tsong, 1983, 1984; Liu et al., 1990).

Ouabain-sensitive AC field-induced pump activities were also detected. These activities exhibited the same Michaelis–Menten constant (K_m) and maximum velocity (V_{max}) as the ATP-dependent pump activities. Thus, electric field-activated pump pathways are likely to be similar to the molecular pathways involved in ATP-dependent enzyme catalysis. Furthermore, we observed windows of field strength and frequency for AC field-stimulated pump activity. The optimal field strength was $40 V_{p-p}$/cm (p–p denotes peak–peak), or amplitude 20 V/cm, for both Na^+- and K^+-pumps, which generates a maximum transmembrane electric potential of about 25 mV. This value corresponds roughly to a transmembrane electric field of 50 kV/cm for this system, a value substantial enough to enable the use of the ECC model for analysis. Our experiments also led to the discovery of optimal AC field frequencies for Rb^+-

and K^+-pump activity ($\sim 1.0\,kHz$) and for Na^+-pump activity ($\sim 1.0\,MHz$). This large difference in effective frequencies seemed to suggest that the two pumps could be uncoupled under our experimental conditions; at $1.0\,kHz$, when the K^+-pump was actively pumping, the Na^+-pump was completely inactive, and vice versa, at $1.0\,MHz$. These observations were considered unusual but not at all implausible. Many publications report the uncoupling of the two pumps under special solvent conditions.

Our work exploring the dependence of Rb^+-pump activity on AC field frequency also revealed a second type of frequency dependence; that is, the activity versus frequency diagram exhibited a shoulder around $10\,kHz$. This shoulder was reproducible but not always reliably measured. However, a similar shoulder was also detected by Wei Chen of the University of Chicago, who used a whole-cell clamp method to measure Na^+,K^+-pump activity in muscle cells (W. Chen, personal communication). The experimental method in this case was very different from ours, and his observation of a similar frequency dependence for the K^+-pump was a remarkable affirmation of our own data.

The experiments described so far for the Na^+,K^+-pump have led us to propose that membrane proteins are capable of *receiving* and *deciphering* electric signals and signals of other physical origins such as pressure, sound waves, concentrations or mechanical forces, as was mentioned above.

13.3.3 *Energy coupling of randomly fluctuating or stochastic electric fields*

With respect to the question of the physiological significance of the experiments, there appear to be several shortcomings when only *regularly* oscillating electric fields are employed. First, electric fields on, around, or across a cell membrane are unlikely to be regularly oscillatory. Because of the many simultaneous actions of ion channels, pumps, redox enzymes, and ATPases, the electric potential in the vicinity of a cell membrane should display considerable fluctuations. The magnitude of the fluctuations would depend on how a signal is measured; measuring at a local level, the fluctuations could be quite substantial. Theoretical analysis suggested that the four-state cyclic ECC transporter, which was shown to respond to AC field stimulation, should also respond to stimulation with *random telegraph fluctuating electric noise* (RTN; Astumian *et al.*, 1987; Chen, 1987; Zhou and Chen, 1996). RTN consists of alternating square pulses, with life times randomly distributed according to the exponential function. The pulse width of RTN is $t = -\tau^* \ln(R)$, where τ^* is the mean life time and R is a random number, e.g., of a value between 0.01 and 1. A representation of typical RTN with an amplitude of $20\,V/cm$ and a τ^* of $1\,ms$ is shown in the lower panel of Figure 5a as compared to a sinusoidal wave

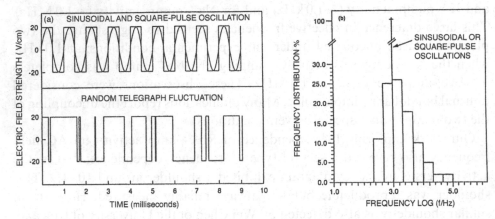

Figure 5. Different waveforms of applied electric fields. In the upper panel of (a), oscillatory electric fields, one with sinusoidal waveform and the other with squared waveform, are shown. Both have an amplitude of 20 V/cm and a frequency of 1.0 kHz. In the lower panel, the life time of the electric field varies randomly with a distribution according to the random telegraph function (RTF). The distribution of the RTF is given in (b); the frequency distributions of the sinusoidal and the squared electric fields are shown as delta-functions at 1.0 kHz (Xie *et al.*, 1994).

and a square wave of the same mean life time (or periodicity) in the upper panel. If one plots the histogram of frequency distribution versus frequency, a sinusoidal wave, or a square wave of constant periodicity, would show a delta-function at 1 kHz, while RTN would have the distribution shown in Figure 5b.

For our experiments, RTN-triggering signals of varied mean life times were generated by computer and used to drive a function generator. The resulting RTN electric pulses were then employed in the electric stimulation experiment summarized next. Not surprisingly, we observed that the Rb^+-pumping mode of the Na^+,K^+-pump was activated at a mean frequency of 1.0 kHz. As shown in Figure 6, the optimal amplitude was 20 V/cm, as was the case for the above-described experiments with sinusoidal electric fields (Xie *et al.*, 1994). In the figure the solid lines drawn through the data points are simulations based on the four-state cyclic ECC model. The fact that a simple four-state ECC model could reproduce these observations was gratifying because of (1) the complexity of the involved experimental conditions and (2) the simplicity of the ECC concept and the employed membrane transport model.

Our work was then extended to include fluctuations in the electric field amplitude. The field amplitude was allowed to fluctuate according to the Gaussian distribution function, with a varied standard deviation (σ): $A(x) = A_0 \exp(-2x^2/\sigma)$, where A_0 is the amplitude at $x = x_0$, and x is a

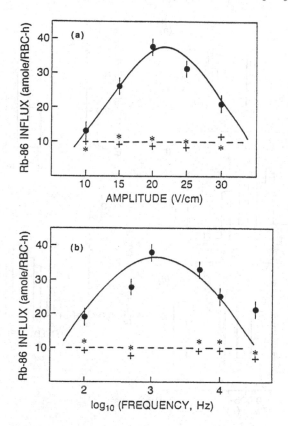

Figure 6. Electric activation of Rb^+ pumping mode of Na^+,K^+-ATPase in human erythrocytes (red blood cells, RBC). (a) Rb^+-influx into human red cells stimulated with a random telegraph function (RTF) field of mean frequency 1.0 kHz, at different amplitude values. Filled circles are data for electric field-stimulated samples and stars for stimulated samples in the presence of 0.2 mM ouabain. Crosses represent the controls (no stimulation, with or without ouabain). The experiment was done at $3 \pm 1\,°C$, and no significant ATP-dependent pump activity was measured. (b) Rb^+-influx at constant amplitude (20 V/cm) as dependence on the mean frequency of the RTF (Xie *et al.*, 1994). amole, attomole.

random number between 1 and 0. A graphical representation of typical RTN of mean amplitude 20 V/cm is shown in Figure 7c. Several different experiments were performed: one was to apply a square wave of constant life time, with fluctuating Gaussian amplitude that centered around 20 V/cm and with σ varying. The result showed that when $\sigma = 0$, the Rb^+-pump activity was about 25% higher compared with the application of a regular sine wave of identical amplitude. However, when a field with finite σ was imposed, the activity was found to decline linearly (Xie *et al.*, 1997). Simulations with a four-state ECC model were capable of reproducing these observations. In another experiment,

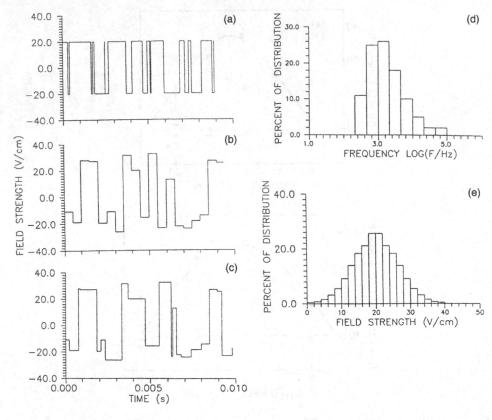

Figure 7. Different waveforms of randomly fluctuating electric fields. (a) A square random telegraph function (RTF) electric field (amplitude distribution with a mean frequency of 1.0 kHz as given in (d) is shown). (b) The square waveform has a constant life time of 1.0 kHz, but with the amplitude fluctuating according to the Gaussian distribution shown in (e). (c) Both the life time and the amplitude are fluctuating. The frequency follows the RTF in (d), and the amplitude follows the Gaussian distribution in (e) (Xie *et al.*, 1997).

we employed a RTN-Gaussian field. In this case, the dependence of pump activity on σ was monotonic for certain conditions but nonmonotonic for others. Typical experimental results are summarized in Figure 8. The non-monotonic feature observed at $E = 10$ and 15 V/cm could not be reproduced by the ECC model simulations. However, in the simulations, we had assumed that the Rb^+-pump had only one frequency optimum at 1 kHz, but previous experiments had also suggested a shoulder around 10 kHz. Thus the question remains to be investigated whether the nonmonotonic feature apparent from the experimental study could be accounted for by including in the model the additional dependence at 10 kHz.

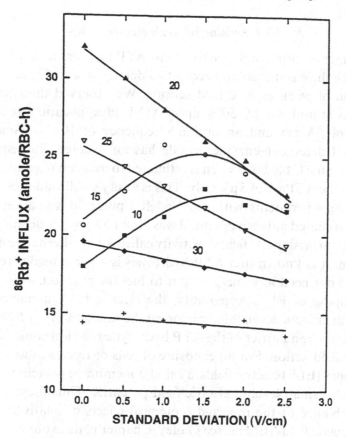

Figure 8. Random telegraph function (RTF)-electric field induced, ouabain-sensitive Rb$^+$ pump activity. Electrical stimulation was done with a RTF of mean frequency 1.0 kHz, with the peak amplitude value given for each curve. The amplitude of the field fluctuated according to a Gaussian distribution of varying values of standard deviation. Data represented by crosses are for the controls (no electrical stimulation). A nonmonotonic behavior of pump activity is seen for experiments with peak amplitudes of 10 and 15 V/cm (Xie *et al.*, 1997). amole, attomole.

Ca^{2+}-ATPase from human erythrocyte was also tested for its ability to respond to an electric field. It was found that when a sinusoidal electric field was used, the optimal amplitude for the coupling efficiency was 30 V/cm and the optimal frequency was 100 kHz. In this case, however, a ligand (Ca^{2+}) concentration window was also found at 0.8 mM. Indeed, the ECC model predicted the existence of a window for the ligand concentration before the experiment. We also tested other ATPases, but, at present, the results are not detailed enough for a report.

13.4 Sensing of weak electric fields

Our experiments with a highly active Ecto-ATPase preparation from chicken oviduct, with the enzyme solubilized with a detergent, are of special interest to the question of weak electric field sensing. We observed that Ecto-ATPase activity was stimulated by 50% upon AC field application at an optimal amplitude of 5 V/cm and an optimal frequency of 10 kHz. Since the AC field-exposed detergent–enzyme micelle has an average dimension of only about 100 Å, an AC field of 5 V/cm is estimated to induce a maximum potential drop across the ATPase of 5 μV only. This is a very small value compared with the exposure experiments with the Na^+,K^+-pump in human erythrocytes, where the estimated induced potential was about 50 mV (for details see Section 13.2.1). This surprising AC field sensitivity called for an alternative mechanistic explanation. It is known that ATP hydrolysis is a spontaneous reaction and that it does not need any energy input to fuel the reaction, as in the case of uphill pumping of Rb^+. Apparently, the electric field interacted with the enzyme at the kinetic level, thus enhancing the rate of ATP hydrolysis. In this view, the activation barrier of the ATP hydrolytic reaction could be the site of the electric field action. For an exposure of cells or tissues to weak extremely low-frequency (ELF) electric fields, a cellular membrane protein would experience a similar small level of electric field potential. Thus these observations bear resemblance to the reported biological effects of relatively weak ELF electromagnetic fields (EMFs; see Gailey, Chapter 6, this volume).

As was described previously, we interpreted these results with our oscillatory activation barrier (OAB) model (Markin *et al.*, 1992a,b). In this case, the conformational oscillator should have a characteristic frequency of 10 kHz. Apparently, the transition state of the ATP hydrolysis involves a large change in the molar electric moment; as a result, the enzyme–substrate complex was able to interact with the weak electric field. When the applied field matched the intrinsic frequency of the enzyme oscillator, a *resonance* would occur, and hence the amplitude of the oscillation would increase. A greater barrier oscillation could lead to either an enhancement or a suppression of the catalytic rate, depending on the detailed mechanisms of interactions. When a simple harmonic oscillator was adopted as a model of an activation barrier oscillator, however, the AC field was found to enhance the rate. The numerical simulation based on the OAB model reproduced both the magnitude and the general features of the observed AC field effects on ATP hydrolysis.

The OAB concept was simple, however, and our analysis was cryptic. Thus, in the future, the model needs to be further refined and tested more rigorously and thoroughly with other enzyme systems. Despite these limitations, our analysis suggested that the rate enhancement originated from nonlinear effects due to enhanced barrier oscillations. For example, one may reason that when the field strength is low, any nonlinear effects may vanish. In other words, all nonlinear systems may become linearly behaving systems at a lower bound. The question is, of course, at what level of field intensity should potential nonlinear effects be omitted and at what levels they should not. In any case, in order to be able to theoretically reproduce the above experimental findings, our calculations needed to include *nonlinear* effects. We take this as evidence that, at the tested electric field intensity, nonlinear effects play a significant role. Obviously, an OAB system should also respond to a fluctuating force field, as was the case for the ECC system, but a detailed study of such interactions remains to be done.

13.5 The effects of broad-band (white) electric noise: stochastic resonance

From the very beginning of our experiments, we asked the questions 'what would be the effect of white noise on the ability of a molecule to recognize a signal?' 'Would white noise enhance or mask a signal, or would it have no effect on signal recognition?' This question lingered and other experiments always took precedence. During that time, many laboratories reported that low-level noise could greatly enhance the signal-to-noise ratio (SNR) in biological rate processes such as the impulse firing rate of crayfish neurons and other neurons, ion conductance of gramicidin channels, and other biological processes (Douglass *et al.*, 1993; Wiesenfeld and Moss, 1995; Moss, 1997; Bezrukov and Vodyanoy, 1995, 1997). These enhancements in the SNR-manifested effects and properties of *stochastic resonance* (for an introduction, see Moss, Chapter 10, this volume). With our system, a suboptimal field strength (10 V/cm) was used to stimulate Rb^+-pump activity, and white noise of varied power levels was superimposed on the stimulating signal. With a sinusoidal electric field, 10 V/cm is at the onset of the stimulation activity, and by itself can produce only little pump activity, but when white noise of broad bandwidth was also imposed, Rb^+-pump activity was *amplified*. We do not know whether the phenomenon reported here meets the criteria of stochastic resonance, but until a better explanation is found, the noise enhancement of Rb^+-pumping will be treated as a case of stochastic resonance. This observation may be of special importance, since the Na^+,K^+-pump is a membrane transport protein of universal occurrence and it is an essential energy-transducing protein. Furthermore, our experiments did not simply demonstrate the improvement in the SNR of Na^+ or Rb^+ currents, but they also showed that the activity of an uphill transport process could be aided by white noise. In essence, the energy contained in white noise is used by the stochastic resonance mechanism to boost the efficiency of energy coupling of an ECC transport system. Our data to be published indicate that while stochastic resonance may enhance cation-pump activity, the highest activity we have obtained so far never exceeded the optimal activity of the enzyme in the absence of noise.

13.6 Michaelis–Menten enzyme models

Although we have discussed only the effects of electric fields on the activity of several membrane ATPases, the four-state membrane transport model we discussed here is considered a special case of a more general Michaelis–Menten enzyme model (Tsong, 1990, 1992; Robertson and Astumian, 1990). Figure 9 illustrates this idea. A simple Michaelis–Menten enzyme mechanism can be rewritten into a cyclic biochemical reaction, step by step, species by species, the same as the four-state membrane transport reaction. As was mentioned in the beginning of this chapter, the concept and the equations of the ECC model are equally applicable to other types of physical force, such as pressure, chemical potential, thermal energy, etc., with minimal modification. This means that mechanisms similar to the ECC model may play an important role in the normal functioning of cells. For example, a thermal ratchet model has been employed to explain muscle contraction and motor functions in cell locomotion and division (Astumian, 1997).

It would appear that molecules of cell membranes, over millions of years of evolution, have acquired the ability to sense, decipher and respond to low-level electric fields, in the form of either periodic or randomly fluctuating signals (Kalmjin, 1982; Weaver and Astumian, 1990). We propose that a high-level background noise may not be sufficient to mask such effects. Furthermore, since nonstationary magnetic fields can induce electric fields, it is reasonable to assume that organisms also have the ability to sense and respond to magnetic fields on this basis. A current concern of bioelectrochemists is to understand the mechanisms by which cells interact with EMFs (e.g., Walker et al., 1997). We have shown that interactions may start at the level of the cell membrane, and that ATPases comprise a class of membrane proteins capable of recognizing and deciphering an electric signal. Whether the biological effects of EMFs are harmful or beneficial to human health is a fundamental question of great public concern. Our experiments so far do not directly address this issue. It is, however, understood that once mechanisms of interaction are clarified, their effects can be assessed with higher confidence. Effects on enzyme mechanisms also serve as a reference point for understanding the effects observed with cells, organisms and in humans. Concentration oscillations of metabolites, ions,

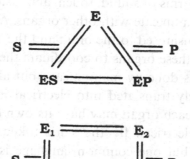

Figure 9. Comparison of the electroconformational coupling (ECC) model with Michaelis–Menten enzyme models. The upper scheme is the four-state ECC model discussed in this chapter. The next scheme was proposed to interpret the electric field induced ATP synthesis by mitochondrial ATPase (Tsong and Astumian, 1986); it is also an ECC model. The linear scheme that follows is a typical Michaelis–Menten enzyme catalytic model. Since enzyme (E) is recycled in each turnover, it can be rewritten as a cyclic model. Furthermore, since the enzyme conformation that favors binding of substrate (S) should be different from the conformation that favors binding of product (P), the catalytic reaction is better represented by the four-state mechanism shown at the bottom. If the enzyme is immobilized, as is the case for a membrane transporter, it can become a transducer of signal or energy in the form of fluctuating electric fields, chemical potentials, acoustic fields, magnetic fields, or other fluctuating potentials (see Tsong, 1990, 1992).

oxygen, etc. are common occurrences in cells. In effect, these events produce oscillations of chemical potentials, or oscillations of chemical driving forces.

13.7 The language of cells

What can we learn from these experiments and model analyses? We have found that a molecule with restricted motion can interact with an oscillatory driving force of particular bandwidth and amplitude. Fluctuating electric fields of similar mean frequency and distribution give equivalent results. Most molecules in cells or organisms are parts of greater supramolecular structures; few molecules are truly free floating. Their chemistry is bound to be different from that of the homogeneous phase, for example chemical reactions in an aqueous solution. One of the differences with which we were particularly fascinated, and have been exploring here, is their ability to decipher a signal, process it, and transmit the information contained therein into a more permanent record of a cell or an organism.

We all accept that an individual communicates with other individuals by a language, either in the forms of sound, touch, and vision, or by other means. An organ can also communicate with other organs. All sensory and perceptual signals must be transmitted to the brain and the brain must, conversely, transmit commands to these organs to coordinate their activities. Organ-to-organ communication is done by many means but all these different signal types must be ultimately translated into electrical impulses of neurons to reach the brain. While each organ may have its own distinct languages, one common language is electrical. By the same token, cells have their own diversity of languages, but one common language is again electrical. Ultimately, all 'cellular languages' must be received and processed by molecules (Tsong, 1989).

How do cells converse in different kinds of languages? What are the essential elements of a language? Amplitude, frequency, and waveform, and different combinations of these properties of force fields may be some critical elements. For example, our experiments described here have shown that the Na^+,K^+-pump may recognize an electric field of a particular amplitude, frequency, and waveform; these characteristics may thus constitute essential elements of a cellular language. As I mentioned before, inside a cell, there are molecules that can recognize different types of force and potentials beside electric potentials, e.g., pressure (Hudspeth, 1989), magnetic fields (Walker et al., 1997; see Wallec-zek and Eichwald, Chapter 8, this volume), concentration, temperature, etc. Even for membrane ATPases, different ATPases responded to electric fields of different amplitudes and frequencies. In other words, a cell may contain all the machinery needed for deciphering and processing a basic type of language. Tissues and organs are much more sophisticated in their handling of complex signals of communication. Still, I propose that the basic unit of communication is the cell and that molecules carry out the initial deciphering of signals. Thus, to understand the mechanisms of signal transduction, one must start with molecules.

In summary, the experiments discussed in this chapter indicate that each enzyme or ion pump has a characteristic frequency by which it can most effectively interact with an oscillating or a fluctuating potential of particular amplitude. Therefore, different enzymes and pumps can recognize and process different frequencies and amplitudes. In this view, these oscillatory and fluctuating potentials aimed at particular cellular enzymes and transporters make up the language of cells. The transmission of acoustic, or vocal signals in organisms, and of electric signals in organs and cells, are well-known examples. Can an organism transmit an EMF signal, and if so by what mechanisms? These are questions we would like to address in our future work. In an earlier publication I proposed that cells do have their own languages for communication (Tsong, 1989). Now our task is to decipher these languages. Once we can understand the languages of cells, we should be able to communicate with them, or even to control their activities and command their actions. It would appear that the time is ripe for a rigorous pursuit of this concept and to develop it into a viable new area of biomedical research.

References

Astumian, R. D. (1997) Thermodynamics and kinetics of a Brownian motor. *Science* **276**: 917–922.

Astumian, R. D. and Bier, M. (1994) Fluctuation-driven ratchets: molecular motors. *Phys. Rev. Lett.* **72**: 1766–1769.

Astumian, R. D., Chock, P. B., Tsong, T. Y., Chen, Y.-D. and Westerhoff, H. (1987) Can free energy be transduced from electric noise? *Proc. Natl. Acad. Sci. USA* **84**: 434–438.

Bezrukov, S. M. and Vodyanoy, I. (1995) Noise-induced enhancement of signal transduction across voltage-dependent ion channels. *Nature* **378**: 362–364.

Bezrukov, S. M. and Vodyanoy, I. (1997) Signal transduction across alamethicin ion channels in the presence of noise. *Biophys. J.* **73**: 2456–2464.

Chen, Y. D. (1987) Asymmetry and external noise-induced free energy transduction. *Proc. Natl. Acad. Sci. USA* **84**: 729–733.

Douglass, J. L., Wilkens, L., Pantazelou, E. and Moss, F. (1993) Noise enhancement of information transfer in crayfish mechanoreceptors by stochastic resonance. *Nature* **365**: 337–340.

Hilgemann, D. W. (1994) Channel-like function of the Na,K pump probed at microsecond resolution in giant membrane patches. *Science* **263**: 1429–1432.

Hudspeth, A. J. (1989) How the ear works work? *Nature* **341**: 397–404.

Kalmjin, A. J. (1982) Electric and magnetic field detection in elasmobranch fishes. *Science* **218**: 916–918.

Liu, D. S., Astumian, R. D. and Tsong, T. Y. (1990) Activation of the Na$^+$-pumping and the Rb$^+$-pumping modes of Na,K-ATPase by oscillating electric field. *J. Biol. Chem.* **265**: 7260–7267.

Markin, V. S., Liu, D. S., Rosenberg, M. D. and Tsong, T. Y. (1992a) Resonance transduction of low level periodic signals by an enzyme: an oscillatory activation barrier model. *Biophys. J.* **61**: 1045–1049.

Markin, V. S., Liu, D. S., Gimsa, J., Strobel, R., Rosenberg, M. D. and Tsong, T. Y. (1992b) Ion channel enzyme in an oscillating electric field. *J. Membrane Biol.* **126**: 137–145.

Markin, V. S. and Tsong, T. Y. (1991a) Reversible mechanosensitive ion pumping as a part of mechanoelectric transduction. *Biophys. J.* **59**: 1317–1324.

Markin, V. S. and Tsong, T. Y. (1991b) Electroconformational coupling for ion transport in an oscillating electric field: rectification versus active pumping. *Bioelectrochem. Bioenerg.* **26**: 251–276.

Markin, V. S., Tsong, T. Y., Astumian, R. D. and Robertson, B. (1990) Energy transduction between a concentration gradient and an alternating electric field. *J. Chem. Phys.* **93**: 5062–5066.

Marszalek, P., Liu, D. S. and Tsong, T. Y. (1990) Schwan equation and transmembrane potential induced by oscillating electric field. *Biophys. J.* **58**: 1053–1058.

Moss, F. (1997) Stochastic resonance at the molecular level. *Biophys. J.* **73**: 2249–2250.

Nakano, M. and Gadsby, D. C. (1986) Voltage dependence of Na$^+$ translocation by the Na,K-pump. *Nature* **323**: 628–630.

Rephaeli, A., Richards, D. E. and Karlish, S. J. D. (1986a) Conformational transitions in fluorescence-labeled (Na,K)-ATPase reconstituted into phospholipid vesicles. *J. Biol. Chem.* **261**: 6248–6254.

Rephaeli, A., Richards, D. E. and Karlish, S. J. D. (1986b). Electrical potential accelerates the E1P(Na)–E2P conformational transition of (Na,K)-ATPase in reconstituted vesicles. *J. Biol. Chem.* **261**: 12437–12440.

Robertson, B. and Astumian, R. D. (1990) Generalized Michaelis–Menten equation for an enzyme in an oscillating electric field. *Biophys. J.* **57**: 689–695.

Serpersu, E. H. and Tsong, T. Y. (1983) Stimulation of Rb$^+$ pumping activity of (Na,K)-ATPase in human erythrocytes with an external electric field. *J. Membrane Biol.* **74**: 191–201.

Serpersu, E. H. and Tsong, T. Y. (1984) Activation of electrogenic Rb$^+$ transport of Na,K-ATPase by electric field. *J. Biol. Chem.* **259**: 7155–7162.

Steinberg, M. and Karlish, S. J. (1989) Studies on conformational changes in Na,K-ATPase labeled with 5-iodoactamido-fluorescein. *J. Biol. Chem.* **264**: 2726–2734.

Stürmer, W., Apell, H. J., Wudell, I. and Läuger, P. (1989) Conformational transitions and charge translocation by the Na,K pumps: comparison of optical and electrical transitions elicited by ATP-concentration jumps. *J. Membrane Biol.* **110**: 67–86.

Teissie, J., Knox, B. E., Tsong, T. Y. and Wehrle, J. (1981) Synthesis of adenosine triphosphate in respiration inhibited submitochondrial particles induced by microsecond electric pulses. *Proc. Natl. Acad. Sci. USA* **78**: 7473–7477.

Teissie, J. and Tsong, T. Y. (1980) Evidence of voltage-induced channel opening in (Na,K)-ATPase of human erythrocyte membranes. *J. Membrane Biol.* **55**:

133–140.

Teissie, J. and Tsong, T. Y. (1981) Voltage modulation of Na^+/K^+ transport in human erythrocytes. *J. Physiol. (Paris)* **77**: 1043–1053.

Tsong, T. Y. (1989) Deciphering the language of cells. *Trends Biochem. Sci.* **14**: 89–92.

Tsong, T. Y. (1990) Electric modulation of membrane proteins: enforced conformational oscillation and cellular energy and signal transductions. *Annu. Rev. Biophys. Biophys. Chem.* **19**: 83–106.

Tsong, T. Y. (1992) Molecular recognition and processing of periodic signals in cells: study of activation of membrane ATPases by alternating electric fields. *Biochim. Biophys. Acta* **1113**: 53–70.

Tsong, T. Y. (1994) Exquisite sensitivity of electroreceptor in skates. *Biophys. J.* **67**: 1367–1368.

Tsong, T. Y. and Astumian, R. D. (1986) Absorption and conversion of electric field energy by membrane-bound ATPases. *Bioelectrochem. Bioenerg.* **15**: 457–476.

Tsong, T. Y. and Astumian, R. D. (1988). Electroconformational coupling: an efficient mechanism for energy transduction by membrane-bound ATPases. *Annu. Rev. Physiol.* **50**: 273–290.

Walker, M. M., Diebel, C. E., Haugh, C. V., Pankhurst, P. M. and Montgomery, J. C. (1997) Structure and function of the vertebrate magnetic sense. *Nature* **390**: 371–376.

Weaver, J. C. and Astumian, R. D. (1990) The response of living cells to very weak electric fields: the thermal noise limit. *Science* **247**: 459–462.

Wiesenfeld, K. and Moss, F. (1995) Stochastic resonance and the benefit of noise: from ice ages to crayfish and SQUIDs. *Nature* **373**: 33–36.

Wuddel, I. and Apell, H. J. (1995) Electrogenicity of the sodium transport pathway in the Na,K-ATPase probed by charge-pulse experiments. *Biophys. J.* **69**: 909–921.

Xie, T. D., Chen, Y.-D., Marszalek, P. and Tsong, T. Y. (1997) Fluctuation-driven directional flow in biochemical cycle: further study of electric activation by Na,K-pumps. *Biophys. J.* **72**: 2496–2502.

Xie, T. D., Marszalek, P., Chen, Y.-D. and Tsong, T. Y. (1994) Recognition and processing of randomly fluctuating electric signals by Na,K-ATPase. *Biophys. J.* **67**: 1247–1251.

Zhou, H. X. and Chen, Y.-D. (1996) Chemically driven motility of Brownian particles. *Phys. Rev. Lett.* **77**: 194–197.

Part IV

Nonlinear control of biological and other excitable systems

One of the most intriguing practical applications derived from nonlinear dynamics has been the development of nonlinear control techniques. The chapters in Part IV explain how they can be employed in the control of chemical dynamics and biological activity. Chapter 14 by Kenneth Showalter outlines the origins of the concept of chaos control and its applications in the control of *chemical chaos*. As a model system he describes the well-established nonlinear chemical oscillator, the Belousov–Zhabotinsky reaction. The chapter concludes with a discussion of prospects for controlling *spatiotemporal chaos*. In Chapter 15, William Ditto and Mark Spano present experiments on *biological chaos* and clinical applications that employ chaos control techniques, for example in the treatment of heart arrhythmias by feedback-controlled electrical stimulation. Finally, they review experiments demonstrating that electric fields can be used in the nonlinear control of neuronal activity. John Milton, in Chapter 16, discusses epilepsy as a dynamical disease within the framework of developing dynamical therapies. He describes theoretical work showing how time-delayed feedback generates multistable dynamical states. The chapter proposes that the therapeutic control of epilepsy might be achievable by exploiting the sensitivity of multistable states to weak stimuli. Chapter 17 by Oliver Steinbock and Stefan Müller focuses on the control of spatiotemporal dynamics, in particular of *spiral wave patterns*. First, the authors present an overview of spiral wave pattern formation in different chemical and biological systems. Then they review experiments that have achieved the control of wave patterns in the Belousov–Zhabotinsky reaction, for example with electric field and laser light perturbations. The book's final chapter reviews some of the major characteristics of biodynamics research. Chapter 18 explores the implications of the herein-discussed work for basic research, clinical applications and biological thinking. It emphasizes that modern medicine might greatly benefit from a better understanding of the self-organized biodynamical processes that appear to be involved at all levels of physiological organization.

14

Controlling chaos in dynamical systems[1]

KENNETH SHOWALTER

14.1 Introduction

A clown sits precariously atop his unicycle, stopping at the viewing stand to get a few more laughs out of the audience. The more discerning observers notice, however, that the clown never actually stops. Rather, he is in constant motion, moving slightly ahead and slightly back and slightly ahead again in a seemingly uncoordinated, jerky fashion. In fact, the clown's motions are precise corrections he makes to stabilize his otherwise unstable vertical perch. Many other examples of this type of action–reaction behavior come to mind in everyday experience, from the tightrope walker in the same circus to a waiter one-handedly whisking away a wobbling tray loaded with dishes.

The clown's balancing act provides a good analogy for recently developed methods for controlling a type of behavior called *deterministic chaos*, which was previously thought to be uncontrollable. Contrary to the traditional view of 'chaos', deterministic chaos is not totally random but has an underlying, intricate order. Despite this order, predicting the long-term behavior in such systems is essentially impossible, since chaotic systems exhibit an *extreme sensitivity to initial conditions*, where two systems that differ only infinitesimally evolve in time so as to diverge exponentially from one another. Chaotic behavior arises in systems as diverse as turbulent fluids, electronic circuits, lasers, and oscillatory chemical reactions.

The extreme sensitivity of chaotic systems suggests that they might be difficult if not impossible to control, since any perturbations used for control would grow exponentially in time. Indeed, this quite reasonable view was widely held until only a few years ago. Surprisingly, the basis for controlling

[1] Adapted from K. Showalter (1995) *Chemistry in Britain* **31**: 202–205 (with permission from the Royal Society of Chemistry), and from W. L. Ditto and K. Showalter (1997) *Chaos* **7**: 509–511 (with permission from the American Institute of Physics).

chaos is provided by just this property, which allows carefully chosen, tiny perturbations to be used for stabilizing virtually any of the unstable periodic states that are inherent in a chaotic system. Unstable periodic states of a chaotic chemical reaction, for example, can be stabilized by perturbing the operating conditions of the system according to a simple feedback algorithm. The result is a transformation of the irregular chaotic responses into regular periodic oscillations.

One of the most attractive features of feedback control is that it can be applied without knowing the mechanism of a system. Other control methods require an accurate model of the system so the governing equations can be appropriately modified to produce the desired stabilization. Feedback methods, however, require only a means to monitor the system and access to an operating condition that can be suitably perturbed. Thus, complex systems for which no accurate mathematical models are available can be readily controlled. In addition, the fact that no knowledge of the mechanism is necessary can be turned around: feedback control can be used as a powerful tool for investigating the underlying dynamical structure of a system, thereby gaining insights into the mechanism.

14.2 Control theory and experiments

Research on controlling chaotic systems has seen remarkable growth in a short time span, with the 'early' studies in the field appearing less than ten years ago. In the late 1980s, Alfred Hübler and coworkers (1988) carried out a series of studies on manipulating chaotic systems to achieve a desired 'goal dynamics', with forcing terms appropriately incorporated into the corresponding governing equations. In 1990, Edward Ott, Celso Grebogi and James A. Yorke (Ott *et al.*, 1990) showed how unstable periodic oscillations of a chaotic system can be stabilized with small, controlled perturbations by using what is now known as the *OGY method* (Shinbrot *et al.*, 1993).

A mere nine months after the OGY method appeared, William L. Ditto, Steven N. Rauseo and Mark L. Spano demonstrated the concept of controlling chaos in a laboratory experiment involving a chaotically oscillating magnetoelastic ribbon (Ditto *et al.*, 1990). Magnetoelastic materials stiffen in a magnetic field, and an oscillatory field causes a vertical magnetoelastic strip to oscillate, relaxing under the influence of gravity only to stiffen again. For some values of the field strength the strip oscillates chaotically, and when tiny perturbations to the field are imposed according to the OGY algorithm, various periodic oscillations can be stabilized. Several other applications quickly followed, including the stabilization of a chaotic diode resonator by

Earle R. Hunt (1991) and a chaotic laser by Rajarshi Roy and coworkers (1992).

Perhaps the greatest driving force behind the advances in chaos control has been the prospect of developing *practical* applications. It now seems obvious – especially with the benefit of hindsight – that the ability to transform chaotic behavior into periodic or steady-state behavior would be highly beneficial in many day-to-day circumstances. One vitally important application that immediately springs to mind is the use of control techniques to restore a regular heartbeat from the state of atrial or ventricular fibrillation, debilitating heart maladies that are often fatal. Alan Garfinkel and coworkers (1992) were the first to demonstrate the feasibility of using chaos control to *stabilize* periodic behavior from irregular heart muscle activity. Using rabbit heart tissue that had been induced to undergo irregular behavior related to the contractions observed in fibrillation, they identified the stable and unstable directions of an unstable periodic state and stabilized the 'regular heartbeat' by applying appropriately timed electrical perturbations.

Another application with enormous potential benefits involves the use of chaos control for *destabilizing* periodic behavior in the brain, where periodicity is abnormal and associated with epileptic seizure activity. Steven J. Schiff and coworkers (1994) were successful in doing just that in an *in vitro* preparation of hippocampal brain tissue in which the system was forced away from the periodic state by electrical perturbations. For details regarding the above applications, see Ditto and Spano, Chapter 15, this volume.

14.3 Chemical chaos

Dynamical studies of chemical reactions are often carried out with continuous-flow stirred tank reactors, much like those used in commercial chemical manufacturing, because particular states far from equilibrium can be maintained and carefully examined. Oscillatory chemical reactions in such reactors display myriad responses – from steady state to periodic to chaotic – as a *control parameter*, such as the concentration of a reactant, is varied. Many experimental and theoretical studies of oscillatory reactions have been carried out and a number of *bona fide* chaotic chemical systems are now known (Scott, 1993, 1994).

Figure 1 shows behavior typical of an oscillatory chemical reaction, where the concentration of an intermediate species C periodically rises and falls in a regular pattern. The response varies according to the reactant concentration, R. As R is swept through a set of values, qualitative changes in the oscillatory behavior arise at critical values called *bifurcation points*. At the first bifurcation

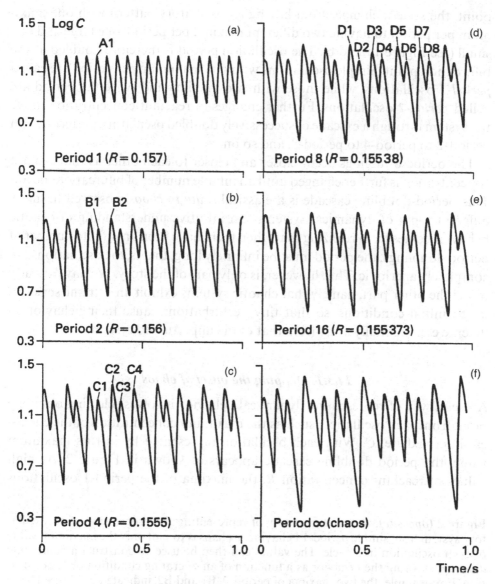

Figure 1. In an oscillatory chemical reaction, the concentration, C, of an intermediate species oscillates over time. Panels (a)–(e) show periodic oscillatory patterns for different values of the reactant concentration, R. An increasing period number not only corresponds to more oscillations per cycle, but also a longer time for each period. As R is changed further, chaos emerges when the number of oscillations per period, and hence the length of the period, becomes infinite; see panel (f). (Adapted from K. Showalter (1995) *Chemistry in Britain* **31**: 202–205, with permission from the Royal Society of Chemistry.)

point, the system changes from having an oscillatory pattern with one maximum per period to having two different maxima per period, one large and one small (see Figure 1a and b). The oscillation period is therefore doubled at the bifurcation point. Simple oscillations with one maximum per period are called *period-1* oscillations, while those with two maxima and twice the period are called *period-2* oscillations. Further changes in reactant concentration move the system through a cascade of successively doubled oscillatory patterns, from period-2 to period-4 to period-8, and so on.

The period doublings occur closer and closer to each other as the reactant concentration is further changed until an infinite number of bifurcations occur. This period-doubling cascade is a classical *route to chaos*, observed in many different types of dynamical systems. An intuitive understanding of chaotic behavior comes from realizing that when there have been an infinite number of period doublings, the period must be infinitely long; that is, the oscillations are completely aperiodic. This, however, is only part of the story. We have already noted the other part, namely that chaotic systems exhibit an extreme sensitivity to initial conditions, so that tiny perturbations cause their behavior to diverge exponentially away from that of the unperturbed system.

14.3.1 Mapping the onset of chaos

A useful way of representing dynamical behavior involves selecting only one point from each oscillation, such as the maximum concentration of the chemical intermediate, C. With each oscillation represented by just its maximum value, the period-doubling cascade appears as shown in Figure 2. At high values of reactant concentration R, the maxima of the period-1 oscillations

Figure 2 (*opposite*). A convenient way of representing the progression of an oscillatory system from simple period-1 behavior to chaos is to plot only the maximum value of each oscillation in a cycle. The values can then be used to construct a *bifurcation diagram*, showing the behavior as a function of an operating condition or concentration. For example, the two maxima of period-2, B1 and B2, indicated in Figure 1b are plotted in panel (a) at reactant concentration $R = 0.156$. As the control parameter is changed further, the system will display period-2 oscillations until R reaches about 0.15561, at which point the system exhibits period-4 oscillations. The points C1–C4 are the four maxima of period-4 at $R = 0.1555$ shown in Figure 1c, and the points D1–D8 correspond to the eight maxima of period-8 at $R = 0.15538$ shown in Figure 1d. The chaotic behavior shown in Figure 1f appears at $R = 0.1545$ in panel (a), where the points represent the maximum value of C in each oscillation. The solid curves represent stable states of the system at a given R-value, while the dashed curves represent unstable states. Panel (b) shows a blowup of a range of the behavior in panel (a). (Adapted from K. Showalter (1995) *Chemistry in Britain* 31: 202–205, with permission

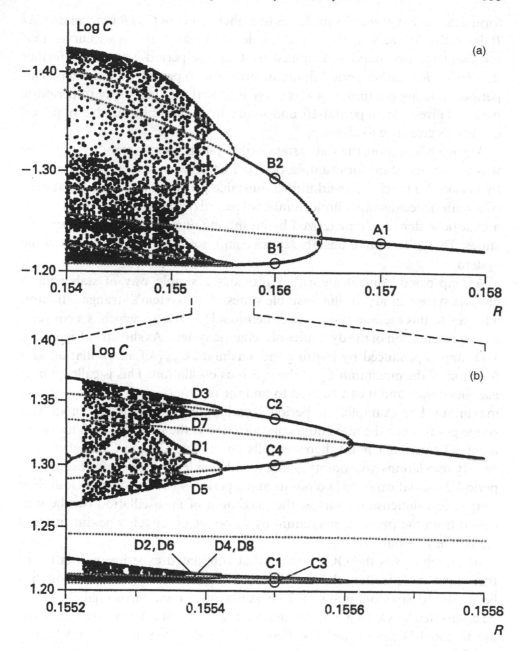

form a curve that gradually increases to higher values of C as R is decreased. As R is further decreased, the curve divides (bifurcates) into two curves that represent the two maxima assumed by C in the period-2 state. On further decreasing R, another period doubling gives rise to period-4 oscillations. This period doubling continues as R is decreased: period-4 gives way to period-8, period-8 gives way to period-16, and so on, until an infinite number of period doublings gives rise to chaos.

At each bifurcation, the stable state of the system loses its stability as it gives way to the next state; for example, period-1 becomes unstable as it is replaced by period-2. Therefore, an additional unstable periodic state is added at each bifurcation. Because an infinite number of period doublings gives rise to chaos, a chaotic system is characterized by an infinite number of unstable periodic states. This collection of unstable states comprises the *strange attractor* of the system.

A map-based control algorithm provides a simple way of stabilizing a chaotic system in any of the unstable states of the system's strange attractor. The key to this method is the one-dimensional (1-D) map, which is a conveni-ent representation of the dynamics of a chaotic system. As shown in Figure 3, a 1-D map is produced by plotting the maximum C_{n+1} of an oscillation as a function of the maximum C_n of the previous oscillation. This is called a *next maximum map*, and it can be used to find the next maximum from the current maximum. For example, the period-1 oscillation appears on the map as a single point, since the maximum concentration is the same for each measure-ment. The period-1 point therefore falls on the line $C_{n+1} = C_n$. Higher peri-odicity oscillations give points falling to either side of this line; for example, a period-2 oscillation gives two points and a period-4 oscillation gives four. For simple, low-dimensional chaos, the maximum of an oscillation can be pre-dicted from the previous maximum by using the effectively one-dimensional curve, the 1-D map.

Although it was the OGY theory that stimulated experimentalists to try their hand at controlling chaos in the laboratory, a reduction of the theory has been used to carry out many of the experiments. Low-dimensional chaos can be controlled by varying a parameter, such as reactant concentration, accord-ing to the 1-D map (Figure 4; Peng *et al.*, 1991; Petrov *et al.*, 1992). The stabilization procedure is greatly simplified with the map-based algorithm. This is especially important in controlling high-frequency chaos in diodes and lasers (Hunt, 1991; Roy *et al.*, 1992), since there is simply not enough time between oscillations to carry out extensive calculations. Even in relatively slow systems, like the chemical reaction considered below, the map-based method has an appealing intuitive basis.

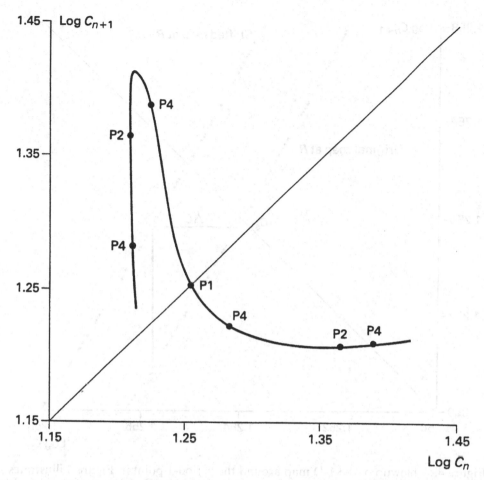

Figure 3. The 1-D map is a useful technique for representing a chaotic system, where each of the points on the curve represents two values: the maximum of the present oscillation and the maximum of the next oscillation. For period-1, the maximum of each oscillation is the same, so there is only one point. For period-2, one cycle includes two different maximum values, so the 1-D map has two points (labeled P2), which represent the changes from B1 to B2 and from B2 to B1 shown in Figure 1b. Similarly, four points appear for period-4. Plotting successive pairs of maxima for the chaotic behavior in Figure 1f yields the curve shown (schematically) by the solid line. The labeled points represent the unstable period-1, period-2, and period-4 orbits embedded in the strange attractor. (Adapted from K. Showalter (1995) *Chemistry in Britain* **31**: 202–205, with permission from the Royal Society of Chemistry.)

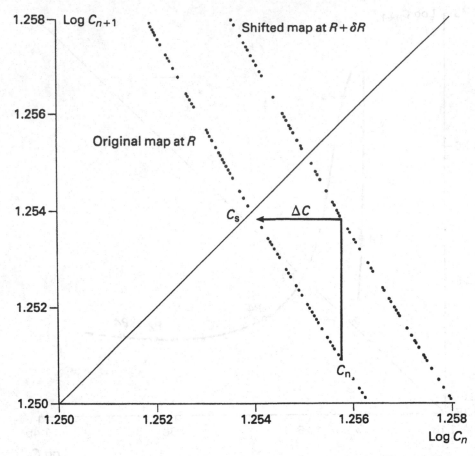

Figure 4. A blowup of the 1-D map around the period-1 point in Figure 3 illustrates the 1-D map method for controlling chaos. The unstable period-1 point, labeled C_s, is given by the intersection of the map and the diagonal line, where $C_n = C_{n+1}$. Also shown is a blowup of a second 1-D map produced from chaotic oscillations at a slightly different value of the reactant concentration, $R + \delta R$. The first step in controlling chaos with this method is to construct two such maps and measure the horizontal distance ΔC between them. This permits the determination of a proportionality constant, $g = \Delta C/\delta R$, which describes the shift of the map with the change in R. The second step is to use this proportionality constant to carry out control. To stabilize the unstable periodic state, it is necessary only to measure the current state C_n, and then change R by an amount δR such that the next maximum after C_n is the desired state C_s. The necessary change in R is calculated from the difference between the current state and the period-1 point, $C_n - C_s = \Delta C$, and the earlier determined proportionality constant, g, according to $\delta R = \Delta C/g$. This procedure is repeated for each cycle, or only as often as is necessary, to stabilize the periodic behavior. (Adapted from K. Showalter (1995) *Chemistry in Britain* **31**: 202–205, with permission from the Royal Society of Chemistry.)

14.3.2 Controlling chemical chaos

The famous oscillatory Belousov–Zhabotinsky (BZ) reaction, discovered and developed in the 1950s and 60s in the former Soviet Union (Zhabotinsky, 1985), is easily prepared from reagents found in most chemical laboratories. When the reagents are mixed together in a beaker, the reaction displays some 30 to 40 oscillations in the concentrations of a host of intermediate species. With cerium ion as the catalyst, the reaction mixture oscillates between pale yellow and colorless; however, with other metal ion catalysts, oscillations between red and blue or orange and green are displayed. In unstirred thin films of solution, the BZ reaction exhibits spectacular spiral waves and target patterns (Zaikin and Zhabotinsky, 1970; Winfree, 1972; see also Steinbock and Müller, Chapter 17, this volume).

The BZ reaction is often studied in a continuous-flow, stirred tank reactor, in which a particular oscillatory state can be maintained indefinitely for careful study. As a control parameter such as a reactant concentration or the rate of reactant inflow is systematically varied, the reaction undergoes a period-doubling cascade to chaos. Chaotic behavior in the BZ reaction has been extensively studied over the years, mainly in the laboratories of Harry L. Swinney at the University of Texas (Simoyi *et al.*, 1982) and John L. Hudson at the University of Virginia (Hudson *et al.*, 1979), as well as by Jean-Claude Roux and coworkers at the Centre de Recherche Paul Pascal in Bordeaux (Roux *et al.*, 1980).

The map-based control algorithm has been used to stabilize periodic oscillations in the period-doubling chaos of the BZ reaction (Petrov *et al.*, 1993). The period-1 and period-2 orbits were determined from appropriate 1-D maps constructed from electrode measurements. Once these were determined, the behavior could be readily switched between chaos and the two different periodic states at any time (see Figure 5). Additional periodic states could also be stabilized by determining the appropriate orbits and applying the algorithm.

An extension of the map-based algorithm allows the unstable states to be tracked as the operating conditions are varied (Petrov *et al.*, 1994a). This technique provides a means to experimentally characterize the dynamics of a system that is beyond the traditional methods of time series analysis. In addition, it allows a system to be stabilized even when the operating conditions uncontrollably drift, as often occurs in practical settings.

14.3.3 Controlling spatiotemporal chaos

Controlling chaos in homogeneous systems like the BZ reaction in a tank reactor can be understood in terms of descriptions based on ordinary differential equations, since the behavior varies in time only. The next level of complexity is found in *spatiotemporal* systems, where the behavior varies not only in time but also in space. Not only are spatiotemporal systems more complex, but they are also more likely to be encountered in everyday settings. While the homogeneous chemical reaction is typically relegated to the chemist's laboratory bench, spatiotemporal behavior is found throughout Nature – especially in living systems. The beating heart, for example, exhibits a wave of electrical activity that propagates through the excitable tissue to trigger a contraction of the muscle.

Applications of feedback methods have been developed to extend the range of stable burning in combustion systems by stabilizing oscillatory and chaotic flames (Petrov *et al.*, 1994b). Model flame systems exhibit period-doubling behavior and chaos as parameters such as the air–fuel ratio or reaction zone width are varied. As the flame chaotically flickers, the motion in time and space can be monitored just as the concentration of a chemical species in the BZ reaction was monitored. However, whereas varying a single parameter in time can control the BZ reaction, controlling the spatiotemporal variations of a flame requires varying parameters in both time and space. In this way it is possible to guide an otherwise chaotic flame into burning in a controlled, stable manner. It is not difficult to imagine how extending the range of stable burning could have a wide range of applications, from eking out higher efficiencies in power plants to providing more thrust in rocket engines.

14.4 Order out of chaos

Chaos is everywhere. Chaotic chemical reactions, lasers, electronic circuits, myocardial tissue, and flickering flames are just a few examples. Many more can be found by looking in the most commonplace settings. Just think of smoke rising from a cigarette, bubbling beer, or a dripping faucet. Less obvious is chaotic behavior in chemical manufacturing processes or combustion-based electrical power generation or irregular wave activity in living excitable tissue. Controlling chaos could be beneficial in many of these settings, and recently developed theoretical tools are now available for use in practical applications. Only a dozen years ago, researchers were just beginning to understand that chaos is as natural as order, and that the roots of order and chaos are intertwined. Now we are learning how to use chaos in real-world applications.

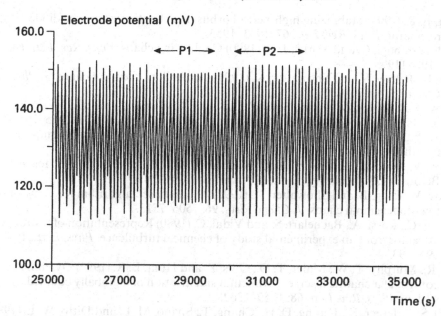

Figure 5. Stabilizing period-1 and period-2 in the Belousov–Zhabotinsky reaction. The plot shows electrode measurements of bromide ion concentration of the reaction in the chaotic state. Period-1 oscillations (P1) were stabilized from the chaotic behavior, which was followed by the stabilization of period-2 oscillations (P2), and then a return to chaotic behavior when the control algorithm was switched off. (Adapted from K. Showalter (1995) *Chemistry in Britain* **31**: 202–205, with permission from the Royal Society of Chemistry.)

Acknowledgments

I acknowledge the major contributions to the research described in this chapter by Valery Petrov and thank Stephen K. Scott for helpful discussions. I thank the National Science Foundation (CHE-9974336), the Office of Naval Research, and the Petroleum Research Fund for supporting this research.

References

Ditto, W. L., Rauseo, S. N. and Spano, M. L. (1990) Experimental control of chaos. *Phys. Rev. Lett.* **65**: 3211–3214.

Garfinkel, A., Spano, M. L., Ditto, W. L. and Weiss, J. N. (1992) Controlling cardiac chaos. *Science* **257**: 1230–1235.

Hübler, A., Georgii, R., Kuckler, M., Stelzl, W. and Lüscher, E. (1988) Resonant stimulation of nonlinear damped oscillators by Poincaré maps. *Helv. Phys. Acta* **61**: 897–900.

Hudson, J. L., Hart, M. and Marinko, D. (1979) An experimental study of multiple peak periodic and nonperiodic oscillations in the Belousov–Zhabotinskii reaction. *J. Chem. Phys.* **71**: 1601–1606.

Hunt, E. R. (1991) Stabilizing high-period orbits in a chaotic system: the diode resonator. *Phys. Rev. Lett.* **67**: 1953–1955.

Ott, E., Grebogi, C. and Yorke, J. A. (1990) Controlling chaos. *Phys. Rev. Lett.* **64**: 1196–1199.

Peng, B., Petrov, V. and Showalter, K. (1991) Controlling chemical chaos. *J. Phys. Chem.* **95**: 4957–4959.

Petrov, V., Crowley, M. F. and Showalter, K. (1994a) Tracking unstable periodic orbits in the Belousov–Zhabotinsky reaction. *Phys. Rev. Lett.* **72**: 2955–2958.

Petrov, V., Crowley, M. F. and Showalter, K. (1994b) Controlling spatiotemporal dynamics of flame fronts. *J. Chem. Phys.* **101**: 6606–6614.

Petrov, V., Gáspár, V., Masere, J. and Showalter, K. (1993) Controlling chaos in the Belousov–Zhabotinsky reaction. *Nature* **361**: 240–243.

Petrov, V., Peng, B. and Showalter, K. (1992) A map-based algorithm for controlling low-dimensional chaos. *J. Chem. Phys.* **96**: 7506–7513.

Roux, J. C., Rossi, A., Bachelart, S. and Vidal, C. (1980) Representation of a strange attractor from an experimental study of chemical turbulence. *Phys. Lett. A* **77**: 391–393.

Roy, R., Murphy, T. W., Maier, T. D., Gills, Z. and Hunt, E. R. (1992) Dynamical control of a chaotic laser: experimental stabilization of a globally coupled system. *Phys. Rev. Lett.* **68**: 1259–1262.

Schiff, S. J., Jerger, K., Duong, D. H., Chang, T., Spano, M. L. and Ditto, W. L. (1994) Controlling chaos in the brain. *Nature* **370**: 615–620.

Scott, S. K. (1993) *Chemical Chaos.* Oxford: Oxford University Press.

Scott, S. K. (1994) *Oscillations, Waves, and Chaos in Chemical Kinetics.* Oxford: Oxford University Press.

Shinbrot, T., Grebogi, C., Ott, E. and Yorke, J. A. (1993) Using small perturbations to control chaos. *Nature* **363**: 411–417.

Simoyi, R. H., Wolf, A. and Swinney, H. L. (1982) One-dimensional dynamics in a multicomponent chemical reaction. *Phys. Rev. Lett.* **49**: 245–248.

Winfree, A. T. (1972) Spiral waves of chemical activity. *Science* **175**: 634–636.

Zaikin, A. N. and Zhabotinsky, A. M. (1970) Concentration wave propagation in two-dimensional liquid-phase self-oscillating system. *Nature* **225**: 535–537.

Zhabotinsky, A. M. (1985) The early period of systematic studies of oscillations and waves in chemical systems. In *Oscillations and Traveling Waves in Chemical Systems* (ed. R. J. Field and M. Burger), pp. 1–6. New York: Wiley-Interscience.

15

Electromagnetic fields and biological tissues: from nonlinear response to chaos control

WILLIAM L. DITTO AND MARK L. SPANO

15.1 Introduction

Between regularity and randomness lies chaos. Our prevailing scientific paradigms interpret irregularity as randomness. When we see irregularity, we chauvinistically cling to randomness and disorder as explanations. Why should this be so? Why is it that, when the ubiquitous irregularity of biological systems is studied, instant conclusions are drawn about randomness and the whole vast machinery of probability and statistics is belligerently applied? Recently we have begun to realize that irregularity is much richer than mere randomness can encompass. Thus, we are brought to chaos.

Sustained irregularity has always upset our notions of how the world should behave. Yet it seems to be the canonical behavior of biological systems. One informal definition of chaos, *sustained irregular behavior*, although descriptive, is too vague to define the rich behavior of chaotic systems. A more precise defining feature of chaotic systems is *their sensitivity to initial conditions* (e.g., see Baker and Gollub, 1990). It is this definition that we will utilize for the characterization and control of chaotic systems.

Fleeting glimpses of order within disorder are quite common. We have all seen short stretches of almost periodic behavior in otherwise irregular systems. A tantalizing example lies in the stock market, where many hope to reap windfall fortunes from analyzing short-term order and predicting the volatile market. But short-term order is a profound, even defining, feature of chaotic systems. To be explicit, chaotic systems exhibit the following.

(1) *Sensitivity to initial conditions*, where the behavior of the system can change dramatically in response to small perturbations in the system's parameters and/or initial values. This makes long-term (but not short-term!) prediction impossible.
(2) *Complex geometric structure(s)* in the system's phase space (fractal objects whose composition includes an infinite variety of unstable periodic behaviors) to which the system's behavior is attracted.

341

The combination of sensitivity to initial conditions and complex geometry in state space can produce to the casual observer an *appearance of randomness*. However, upon closer inspection one glimpses a remarkable order. The flesh of chaotic systems adheres to a skeleton of infinite unstable periodic behaviors that seemingly come and go with no apparent pattern. That is to say, no apparent pattern until one looks with the tools of nonlinear dynamics. Armed with an understanding of unstable periodic motions, chaotic systems take on a more understandable, and consequently controllable, character. It is this order that allows the successful short-term prediction of a system's behavior and the subsequent control that would be impossible in a totally random system. In the following, we apply chaos theory to specific biological systems.

15.2 Cardiac dynamics

15.2.1 Introduction to ventricular fibrillation

Sudden cardiac death is a major health problem that claims one in six lives. Nearly all instances of sudden cardiac death do not occur in a hospital and the majority are due to an irregular and rapid heart rhythm termed *ventricular fibrillation* (VF). Heart attacks and primary heart muscle disease (cardiomyopathy) are the most common causes of VF. In VF, the main pumps of the heart, the ventricles, quiver in an irregular manner such that blood is not effectively pumped throughout the body. The only presently known treatment for VF is to pass an electrical signal with a large current through the heart muscle. This shock, if successful, effectively resets the heart back to a rhythm compatible with life. Without such treatment, sustained VF is always fatal.

15.2.2 Fibrillation as a dynamical state

Applications of nonlinear dynamical techniques to VF have, until recently, yielded contradictory results, primarily due to the inadequacies of current techniques to resolve determinism in short and noisy data sets. The mechanistic elucidation of VF has been hampered by its rapidly changing and markedly heterogeneous electrophysiological nature, rendering waveforms obtained during VF challenging to analyze quantitatively. In light of recent successes with the identification of unstable periodic motion embedded in chaotic systems and control of these motions in physical and biological systems (Ditto *et al.*, 1990; Ditto and Pecora, 1993; Ott and Spano, 1995; Pierson and Moss, 1995; So *et al.*, 1996, 1997), new techniques to decide whether experimentally obtained, irregular biological waveforms represent deterministic or stochastic (random) behavior abound. Previous quantitative measures for determinism, such as Fourier spectra, fractal dimension, and Lyapunov exponents, along with other statistical techniques, have proven uniformly inadequate for detecting determinism in these short, possibly nonstationary, time series. Elucidation of the presence of determinism during VF is important because it may make novel therapeutic and diagnostic strategies possible. However, whether VF represents a deterministic or stochastic process has been controversial (Guevara *et al.*, 1981; Goldberger *et al.*, 1986; Chialvo *et al.*, 1990; Kaplan and Cohen, 1990; Witkowski and Penkoske, 1990).

The presence of the infinite number of unstable periodic motions that comprise the skeleton of a chaotic system has recently been exploited experimentally. This corollary of a chaotic system's sensitivity to initial conditions is the key to the control of such systems (Ditto *et al.*, 1990; Ditto and Pecora, 1993; Ott and Spano, 1995). This form of stabilization is analogous to a baseball on a saddle. Control is achieved through movement of the saddle's position or adjustment of the ball's motion to keep the baseball constantly rolling back toward the unstable equilibrium point in the center of the saddle. This theoretical technique was originally pioneered by Ott, Grebogi and Yorke (Ott *et al.*, 1990). Their technique (and variations thereof) has been successfully applied to control chaos in the vibrations of a magnetoelastic ribbon (Ditto *et al.*, 1990), electrical circuits (Hunt, 1991), the output of a solid-state laser (Roy *et al.*, 1992), chemical oscillations (Parmananda *et al.*, 1993; Petrov *et al.*, 1993), drug-induced arrhythmias of *in vitro* rabbit ventricle (Garfinkel *et al.*, 1992), seizure-related population spiking of hippocampal slices (Schiff *et al.*, 1994) and many other systems. These successes all have in common a proportional feedback control around unstable periodic state space trajectories.

15.2.3 *Detection of deterministic dynamics in canine ventricular fibrillation*

Our experimental preparations consisted of open-chest, anesthetized dogs whose hearts were studied *in vivo* after VF was electrically induced (Witkowski *et al.*, 1995). Transmembrane cardiac current, I_m, was measured from the ventricular epicardium without cell disruption (Witkowski *et al.*, 1993). Minimal signal filtering was specifically employed to avoid the pitfalls associated with filtered noise (Rapp *et al.*, 1993). The I_m time series was then examined to detect activations (beats). The intervals between successive activations, $A(i)$, form a related time series that is most useful in diagnosing and controlling chaos.

We plotted the $(i + 1)$th interval, $A(i + 1)$, versus the previous interval, $A(i)$, in a *Poincaré map* (where deterministic points typically are attracted to a geometric structure known as an *attractor*). Such a Poincaré map provides a reduced view of the dynamics of the measured data. We searched for evidence of *unstable periodic orbits* (UPOs), which appear as *unstable fixed points* (UFPs) in a Poincaré map. These UFPs have associated directions along which the system state point approaches (stable manifold) and recedes from (unstable manifold) the fixed point. A typical example of such a sequence in the Poincaré map is demonstrated by following state points 23–29 in Figure 1a: from point 23 to point 25 the state of the system is drawn toward the UFP along the stable manifold; points 26 through 29 demonstrate exponential divergence away from the fixed point along the unstable manifold. This pattern is repeated time and again throughout the experimental run. The solid lines in the figure denote the positions of the stable and unstable manifolds as determined by fitting to a number of such sequences. Figure 1b–d displays similar behavior for other data sets. The resultant geometry is known as a *flip saddle* and is consistent with previous experimental results for *in vitro* rabbit hearts (Garfinkel *et al.*, 1992).

The detection of UFPs is only the beginning in the search to understand and control the physical mechanisms that underlie VF. While the canine model of VF is widely employed to test promising new defibrillation devices, the relevance of the findings in our study to the clinically important disturbance of VF in humans requires additional studies. VF can be induced in virtually all human hearts. It is fascinating to speculate that a greater understanding of the nonlinear dynamical behavior of VF might lead to the possibility that VF may be controlled through chaos control techniques. For this to become a reality we really need to know not merely the temporal dynamics but also the spatiotemporal dynamics that underlie VF. With that in mind we instituted a study to visualize the *spatiotemporal electrical patterns* long suspected of being the dynamical manifestation of fibrillation.

15.2.4 *Imaging of the spatiotemporal evolution of ventricular fibrillation*

Rotors, electrophysiological structures that emit rotating spiral waves, occur in a variety of systems that all share with the heart the functional properties of excitability and refractoriness. These reentrant waves, seen in numerical solutions of simplified models of cardiac tissue (Holden, 1997) are believed to occur during ventricular tachycardias (Winfree, 1994; Panfilov and Holden, 1997). The detection of such forms of reentry in fibrillating mammalian ventricles has been difficult (Gray *et al.*, 1995, 1998; Lee *et al.*, 1996). Here we show that in isolated perfused dog hearts, high spatial- and temporal-resolution optical transmembrane potential mapping can readily detect transiently

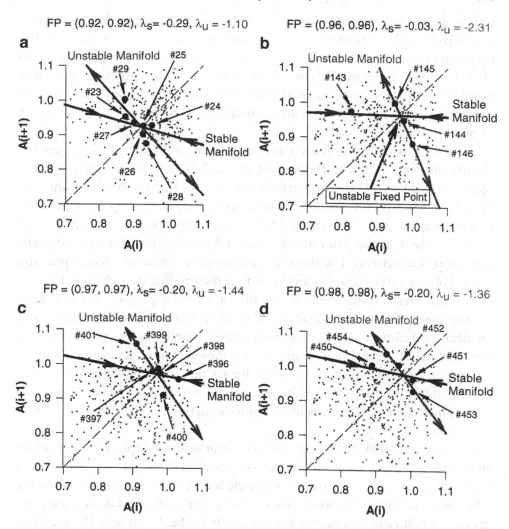

Figure 1. (a) Return plot of the time series $A(i)$ from a typical 1-min interval of *in vivo* ventricular fibrillation (VF) illustrating the local structure of the chaotic attractor. The range of data displayed was narrowed to more clearly demonstrate the points contributing to the local structure. Note the flip saddle structural appearance for points 23 through 29. Coordinates for the calculated unstable fixed point (UFP), the stable eigenvalue (λ_s) and the unstable eigenvalue (λ_u) for this visitation of the UFP are provided above the plot. The diagonal dashed line is the line of identity $(A(i + 1) = A(i))$. (b)–(d) Three separate flip saddle structures (points 143–146 in (b), 396–401 in (c), 450–454 in (d)) sequentially generated by a second representative VF time series with abbreviations as in (a).

erupting rotors during the early phase of VF. This activity is characterized by a relatively high spatiotemporal cross-correlation. During this early fibrillatory interval, frequent wavefront collisions and wavebreak generation are also dominant features (Pertsov et al., 1993). Interestingly, this spatiotemporal pattern undergoes an evolution to a less highly spatially correlated mechanism devoid of the epicardial manifestations of rotors, despite continued myocardial perfusion.

In 1930, Carl Wiggers used a movie camera operating at 32 frames/s to record the movements of the surface of the in situ heart in which VF was induced (Wiggers, 1930). He described four stages: (1) an initial stage consisting of 2 to 8 rapidly activated peristaltic waves; (2) a subsequent convulsive incoordination stage that lasted 14–40 s, so named because 'When the ventricles are held in the palm of the hand, a fluttering, undulatory, convulsive sensation is experienced' without the ability to generate any blood pressure; (3) and (4) are subsequent stages that reflect the progressive ischemia. In terms of clinical interventions the most significant of these stages is stage 2, when countermeasures can be instituted and sudden death aborted. Recently, using an electronic camera operating at 60 frames/s, together with voltage-sensitive dye staining of the heart, a single rapidly moving rotor, which produced an electrocardiographic pattern in rabbit hearts that resembled fibrillation, was described (Gray et al., 1995). However, similar rotating waves in larger mammalian hearts are described as uncommon occurrences in canine hearts (Lee et al., 1996).

We have recorded the electrical activity from a limited epicardial area of the right and left ventricles in isolated, blood-perfused canine hearts. The anterior right ventricle and part of the left ventricle were compressed under an optical window to minimize motion artifacts. (No pharmacological agents were employed to reduce mechanical motion artifacts in the dynamics.) The measurement technique employs an image-intensified charge-coupled device (CCD) optical recording system (Witkowski et al., 1998a) imaging an epicardial surface stained with the voltage-sensitive dye di-4-ANEPPS (Figure 2). An area of approximately 5.5 cm × 5.5 cm was imaged, which represents approximately 30% of the epicardial surface (Figure 3). At the levels of illumination utilized, continuous recordings lasting 10–15 min could be realized with no detectable phototoxic damage. A 128 × 128-pixel, frame-transfer CCD camera operating at 838 frames/s (1.19 ms/frame) was used. The analog video signal underwent 12-bit A/D conversion prior to transfer to a frame grabber.

Fluorescent images were obtained with a temporal resolution of 1.2 ms and a spatial resolution of approximately 0.5 mm (Witkowski et al., 1998a,b). VF was induced with a single, critically timed electrical pulse. Both the early onset

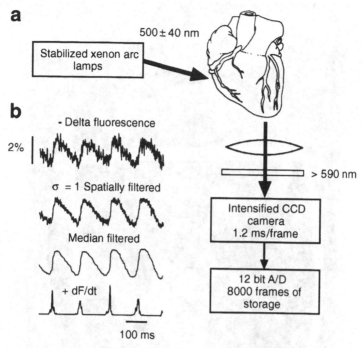

Figure 2. (a) Schematic representation of the experimental set-up. Two 1-kW stabilized xenon arc lamps were used to illuminate the epicardial cardiac surface under study with $\sim 100\,\mathrm{mW/cm^2}$ of quasi-monochromatic green light. This focused fluorescent source was imaged after barrier filter rejection of reflected light components and retention of fluorescent components with wavelengths $> 590\,\mathrm{nm}$, as illustrated. The cooled fiber optically coupled image-intensified CCD frame-transfer camera system was operated at 1.19 ms/frame with 12 bits of dynamic range. (b) Time series from the sequential processing steps for a representative single pixel from 500 frames during ventricular fibrillation (VF) are shown. The effects (on signal to noise) of subsequent image processing steps are depicted; those included 9×9 Gaussian spatial followed by 21-point median temporal filtering, and culminating with 5-point temporal derivative estimation with final clipping to maintain only the positive values (setting all negative values to 0). The optical calibration bar indicates a 2% change in fluorescence.

of VF (corresponding to Wiggers' stages 1 and 2) as well as sustained VF that lasted for more than 10 min in perfused hearts were imaged with this apparatus. The optical transmembrane potentials as well as their temporal derivatives were then viewed as videos. (Quicktime and AVI video sequences from this study are available at the web site: *http://www.physics.gatech.edu/chaos*)

The data from 15 frames of optical images clearly show the onset of spiral wave formation in our canine heart experiment (Figure 4a). The initial waves triggered by the induction of VF consistently produced a reentrant cycle with a 'figure-of-eight' morphology as in Chen *et al.* (1988). These are composed of two mirror-image rotors that share a common reentrant pathway. As an

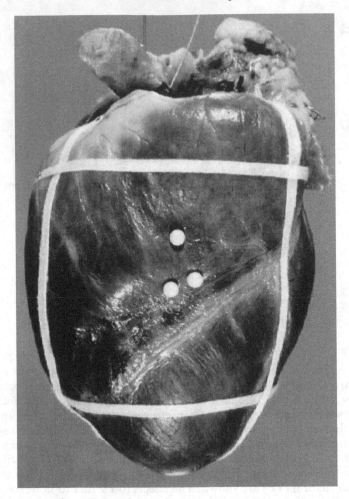

Figure 3. Photograph of formalin-fixed canine heart with the pacing electrodes marked by white-headed pins, and the approximate area of epicardium that was imaged is outlined with white tape.

example, this reentry might last for a total of eight cycles before being abolished by wavefronts that collided in the area of the initial reentry. These collisions often resulted in the subsequent emergence of two oppositely directed, spatially discrete wavefronts with observable dangling ends (Witkowski *et al.*, 1998a) as shown in Figure 4b. Each of the dangling ends of the emerging wavefronts is also called a *wavebreak* or *phase singularity* (Pertsov *et al.*, 1993). Thereafter, other rotors are formed. All episodes of induced VF were self-sustaining and terminated only when the heart was defibrillated.

A completely different electrophysiological pattern was also observed when perfused VF had persisted for 10 min. During this 'chronic' VF, no rotors were

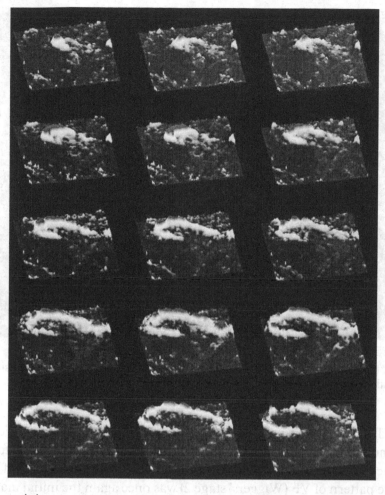

(a)

Figure 4. (a) Rotors, the source structures that immediately surround the core of rotating spiral waves, occur in a variety of systems that all share with the heart the functional properties of excitability and refractoriness. Here we show that in isolated, perfused dog hearts, high spatial and temporal resolution optical transmembrane potential mapping can readily detect transiently erupting rotors during the early phase of ventricular fibrillation (VF). This activity is characterized by a relatively high spatiotemporal cross-correlation.

Figure 4. (b) During this early fibrillatory interval frequent wavefront collisions and wavebreak generation are also dominant features.

observed. The source mechanism is still probably reentry, but we believe that its geometric aspect has become more three-dimensional. This pattern was reproducible in that, after defibrillation and a recuperation interval of 10 min, the acute pattern of VF (Wiggers' stage 2) was once again the initial manifestation after VF induction.

15.3 Control of chaos in cardiac systems

15.3.1 *Control of isolated cardiac tissue*

We have found that it is possible to control a chaotic cardiac arrhythmia using chaos control. Our cardiac preparation consisted of an isolated perfused portion of the interventricular septum from a rabbit heart as shown in Figure 5 (Garfinkel *et al.*, 1992). The heart was stimulated by passing a 3-ms constant voltage pulse, typically 10–30 V, at twice threshold between platinum electrodes embedded in the preparation. Electrical activity was monitored by recording monophasic action potentials with Ag-AgCl wires on the surface of the heart. Monophasic action potentials were digitized at 2 kHz and processed

Figure 5. Photograph of an isolated well-perfused portion of the interventricular septum from a rabbit heart, arterially perfused through the septal branch of the left coronary artery with a physiologically oxygenated Kreb's solution at 37 °C.

in real time by a computer to detect the activation time of each beat from the maximum of the first derivative of the voltage signal.

Arrhythmias were induced by adding 2 to 5 µM ouabain with or without 2 to 10 µM epinephrine to the arterial perfusate. The mechanism of ouabain/epinephrine-induced arrhythmias is probably a combination of triggered activity and nontriggered automaticity caused by progressive intracellular calcium (Ca^{2+}) overload from sodium (Na^+) pump inhibition and increased Ca^{2+} currents. Typically the ouabain/epinephrine combination induced spontaneous beating, initially at a constant interbeat interval and then progressing to period-2 and higher-order periodicity before developing a highly irregular aperiodic pattern of spontaneous activity. The duration of the aperiodic phase was variable, lasting up to several minutes before spontaneous electrical activity irreversibly ceased. The spontaneous activity induced by ouabain/epinephrine in this preparation showed a number of features symptomatic of chaos. Most importantly, in progressing from spontaneous beating at a fixed interbeat interval to highly aperiodic behavior, the arrhythmia passed through a series of transient stages that involved higher-order periodicities. These features are illustrated in Figure 6 in which the nth interbeat interval, I_n, has

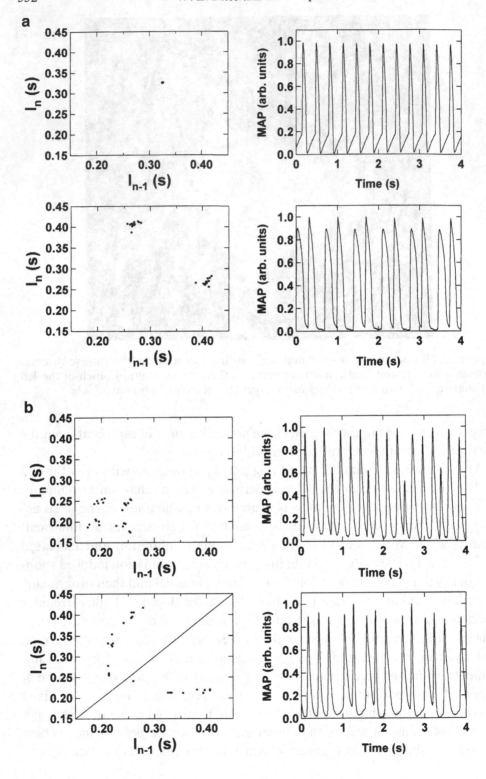

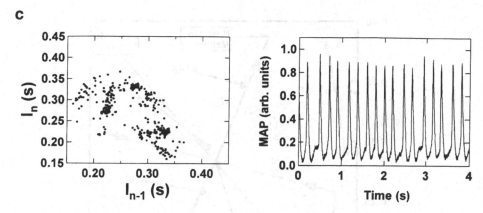

Figure 6. Time series of monophasic action potentials (right) and the Poincaré maps of interbeat intervals (left) at various stages during arrhythmias induced by ouabain/epinephrine in typical rabbit septa. Typically, the arrhythmia was initially characterized by spontaneous periodic beating at a constant interbeat interval (a) (*opposite*), proceeded through higher-order periodicities such as period-2, period-4, and period-5 (b) (*opposite*), and ending in a completely aperiodic pattern (c). Note that, in the Poincaré map of the final stage, the points form an extended structure that is neither point-like nor a set of points (i.e., is not periodic) and is not space-filling (i.e., is not random). This behavior is indicative of chaos.

been plotted against the previous interval, I_{n-1}, at various stages during ouabain/epinephrine-induced arrhythmias. As before, this Poincaré return map allows us to view the dynamics of the system as a sequence of pairs of points (I_n, I_{n-1}), thus converting the continuous dynamics of our system to a map. On such a map, chaotic data (Figure 6c) can easily be distinguished from periodic data (Figure 6a,b). Additionally, the sequence of the data points on such a plot reveals the stable and unstable directions; knowledge of those is required to implement control (Figure 7).

In this rabbit preparation, each control attempt consisted of a learning phase and a control phase. During the learning phase an UFP was identified and characterized (Garfinkel *et al.*, 1992). During the subsequent control phase, the computer waited until a close approach to the UFP was detected ($|I_n - I_{n-1}| < \varepsilon$ where ε defines the control region and is a fraction of the total attractor size). The algorithm then initiated a control stimulus that moved the next point on the Poincaré plot (as predicted by the local dynamics of the UFP) onto the stable manifold (i.e., the contracting direction) of the UFP, thereby allowing the natural dynamics of the system to subsequently draw the system state onto the UFP itself. Stimuli were administered on subsequent points to keep the system on the stable manifold. Thus, the system state was continually contracted towards the UFP, which is shown schematically in

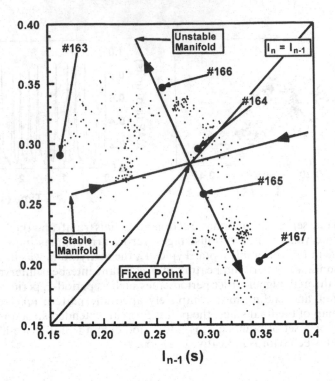

Figure 7. The Poincaré map of the aperiodic phase of a ouabain/epinephrine-induced arrhythmia in a typical rabbit heart preparation illustrating the structure of the chaotic attractor.

Figure 8. If the current point on the Poincaré plot strayed outside the control region, stimuli were then discontinued until the system state point reentered the control region. (It should be noted here that on-demand pacing is simply the special case of this algorithm where the stable manifold is horizontal.) An example of the application of this algorithm to the rabbit heart is shown in Figure 9.

Chaos control with this approach was complicated by the fact that in this experiment intervention was, of necessity, unidirectional. By delivering an electrical stimulus before the next spontaneous beat, the interbeat interval could be shortened, but it could not directly be lengthened. This is because a stimulus, which elicits a beat from the heart, shortens the interbeat interval between the previous spontaneous beat and the beat elicited by the stimulus. The effects of this limitation are apparent in Figure 9, where the best control we could achieve was a period-3.

Several observations should be made about the pattern of the stimuli delivered by the chaos control program. First, these stimuli did not simply overdrive the heart. Stimuli were delivered sporadically, not on every beat and

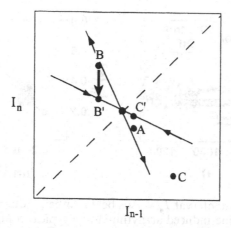

I_n

I_{n-1}

Figure 8. Schematic of the chaos control technique. I_n is the current and I_{n-1} the previous interval between activations. The central point at the intersection of the three lines represents an unstable fixed point (UFP). The stable (inward arrow) and unstable (outward arrow) manifolds are shown as calculated from multiple close returns to the UFP during the learning phase of the algorithm. Without intervention, the natural dynamics around the UFP carries the activation intervals from A to B to C, contracting inward along the stable manifold and expanding outward along the unstable manifold, as shown. Chaos control is implemented after a determination of these stable and unstable manifolds (or *eigenvectors*) and contraction and expansion rates along the manifolds (or *eigenvalues*), as determined by a data-derived least-squares linear fit or model of the dynamics in the vicinity of the UFP. A point occurring at A is predicted to move to B, as determined by this model of the local dynamics. A stimulus is introduced into the high right atrium to force a premature activation that directs A onto the stable manifold at location B' instead of allowing the uncontrolled dynamics to proceed to B. The contraction along the stable direction then pulls the next activation interval closer to C', which is closer to the UFP. Thus, the sequence of ABC is modified to AB'C', keeping the dynamics close to the UFP. This process is repeated in a feedback loop to stabilize the UFP.

never more than once in every three beats on average. In contrast, periodic pacing, in which stimuli were delivered at a fixed rate, was never effective at restoring a periodic rhythm and often made the original aperiodicity more marked. Nonchaos control, irregular pacing was similarly ineffective at converting chaotic to periodic behavior. Encouraged by both the evidence for UFPs in rabbit and canine hearts and by our success in controlling the rabbit tissue preparations, we decided upon the aggressive course of attempting to control chaos in fibrillating atria (upper chamber) of human hearts.

15.3.2 Control of atrial fibrillation in humans

Atrial fibrillation (AF) is the most common arrhythmia requiring treatment intervention (Prystowsky *et al.*, 1996). The occurrence of AF increases with

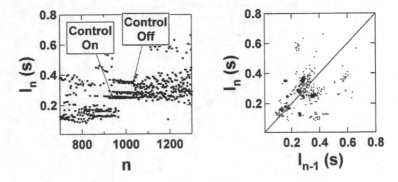

Figure 9. *Left:* Interbeat interval, I_n, versus beat number, n, during the chaotic phase of the ouabain/epinephrine-induced arrhythmia in a typical control run. The region of chaos control is indicated. *Right:* The corresponding Poincaré map for the time series on the left. The light points represent interbeat intervals during the uncontrolled arrhythmia while the dark points represent interbeat intervals during the chaos control. Note the transient period between the initiation of the chaos control and the establishment of a period-3 pattern that is apparent on both plots.

age, touching more than 5% of the population over the age of 65 (Kannel *et al.*, 1982). During AF the rapid and irregular ventricular rate as well as the loss of atrial mechanical function diminish overall cardiac performance and may cause palpitation, breathlessness, fatigue and lightheadedness. In addition to these disabilities, AF dramatically increases the risk of stroke and cardiovascular-related death (Kannel *et al.*, 1982).

Evidence has suggested that biological activity, including the beating of myocytes *in vitro* (Chialvo *et al.*, 1990; Chialvo, 1990), cardiac arrhythmias (Garfinkel *et al.*, 1992; Witkowski *et al.*, 1995; Hall *et al.*, 1997), and brain hippocampal electrical bursting (Schiff *et al.*, 1994; So *et al.*, 1997) exhibit deterministic dynamical behavior. Chaos is the deterministic collection of a large number of unstable periodic motions. Such unstable behavior, including its associated local dynamics, forms the basis for various chaos control techniques (for reviews, see Ditto and Pecora, 1993; Ott and Spano, 1995; Lindner and Ditto, 1995; Christini and Collins, 1995, 1996; Pei and Moss, 1996). Recent work on the control of chaos in low dimensions (Ott *et al.*, 1990), high dimensions (So and Ott, 1995; Ding *et al.*, 1996), and spatially extended systems (Petrov *et al.*, 1996; Petrov and Showalter, 1996) in physical and biological systems has enabled the application of chaos control to human AF.

The human AF study was performed on 25 patients undergoing clinically indicated electrophysiological testing. The study was conducted under a protocol approved by the Human Research Committee at Emory University

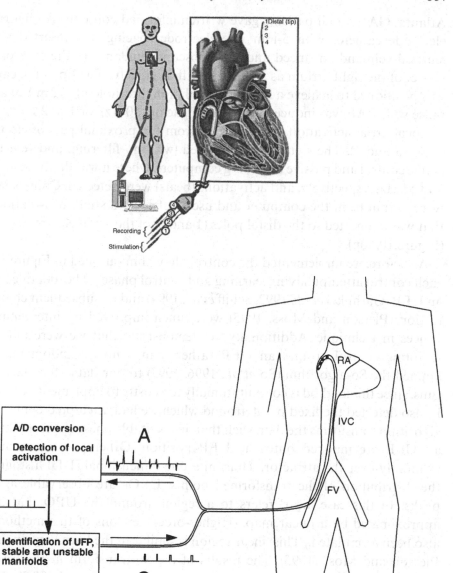

Figure 10. Summary of the human atrial fibrillation (AF) experiment. *Top:* A quadripolar electrode catheter was inserted in the femoral vein, advanced through the inferior vena cava, and positioned in the lateral right atrium. During AF, electrograms recorded from the proximal pair of electrodes were amplified, digitized, and local activations automatically detected. This timing information was used to characterize the chaotic dynamics of the system with identification of the UFP as well as the stable and unstable manifolds. The control algorithm then generated pacing pulses at times predicted to move the system towards a stable (periodic) state. *Bottom:* Details of the electrode and its placement.

Atlanta, GA, and all patients gave written informed consents. A quadripolar electrode catheter with 5-mm interelectrode spacing was inserted via the femoral vein and advanced under fluoroscopic guidance to the anterolateral aspect of the right atrium as is illustrated in Figure 10. The tip of the catheter was positioned to achieve a bipolar stimulation threshold of $\leq 2\,\text{mA}$ at a 2-ms pulse width. AF was induced using rapid pacing (50 Hz) for 1 to 2 s.

Local atrial activation was recorded from the proximal pair of electrodes (poles 3 and 4). The signal was amplified (with no filtering) and sent to the active control and passive recording computers, where it was digitized at 2 kHz and 5 kHz, respectively, and activations (beats) were detected. Control stimuli were output from the computer and used to trigger a stimulus isolation unit that was connected to the distal poles (1 and 2) of the atrial electrode catheter (Figure 10, top).

As before, we implemented the control algorithm outlined in Figure 8, with each control attempt having learning and control phases. The identification of an UFP (Garfinkel et al., 1992; Schiff et al., 1994) and its subsequent character-ization (Pierson and Moss, 1995) were much improved by intervening ad-vances in technique. Additionally, to demonstrate that we were indeed at-tempting control around an UFP rather than a noisy random point, we applied the So-algorithm (So et al., 1996, 1997) to our data after the control runs, since this method is computationally too costly to implement in real time. It also detected the fixed point around which we had attempted control. This algorithm transforms the data such that, in a suitable phase space, points near an UFP are mapped onto the UFP position. Other points are mapped randomly over the attractor. Thus, in a one-dimensional (1-D) histogram of the distribution of the transformed points, UPOs are observable as sharp peaks. In this case 'near' refers to a region around the UPO that can be approximated by a linear map. (Higher-order versions of this method have also been formulated.) This linear region is similar to the linear region used by Pierson and Moss (1995). The results of the So-transform method are dis-played in Figure 11. The large peak near an interbeat interval of 0.2 s denotes the period-1 UPO. The line where the error bars denote the standard deviation at each point represents the same transform applied to 100 surrogates. The peak exceeds the surrogate background by more than 40 standard deviations of the surrogate ensemble as shown in the inset in Figure 11. Also note that, since this method maps points near the UPO onto the UPO position, the fact that the transformed data on either side of the peak fall below the surrogate average provides additional confirmation that we have correctly detected a UPO in these data.

A typical outcome of human atrial chaos control is shown in Figure 12. The

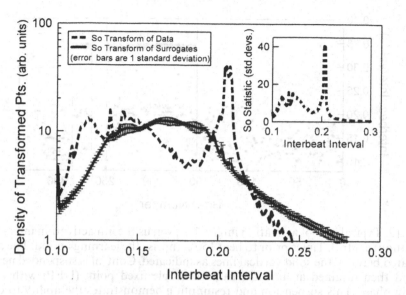

Figure 11. The So-transform of the human atrial fibrillation (AF) data. The large peak near an interbeat interval of 0.2 s denotes the period-1 unstable periodic orbit (UPO). The line with error bars denoting the standard deviation at each recorded point represents the average of 100 surrogates. (Inset) The So-statistic, which is the height of the So-transform minus the average of the surrogates and then normalized by the standard deviation of the surrogates for each value of the interbeat interval. Note that this is a slightly different definition from the one used by So *et al.* (1996).

outcomes of chaos control were categorized as follows. (1) *Excellent chaos control* was defined by successful capture, where a capture is an activation within 15 ms of the application of a control stimulus, for at least 25 sequential intervals around a UFP. The mean of the controlled intervals was equal to or longer than the mean of the activation intervals of spontaneous AF and the standard deviation from the mean was at least two times less than the standard deviation from the mean of the uncontrolled activation intervals. (2) *Partial chaos control* was defined as in (1) except with more frequent losses of control (10–50% of total intervals during chaos control were escapes from the control region) about the UFP. (3) *Unsuccessful chaos control* was defined as all other cases, including those with infrequent capture, lack of suitable UFPs, indiscernible dynamics about the UFP, and all other results. Out of the 25 patients in the study, excellent chaos control was achieved in 9/25 patients (36%), partial chaos control was achieved in 10/25 patients (40%) and unsuccessful chaos control was seen in the remaining 6/25 patients (24%). More than 80% of the interventions that exhibited partial control had evidence of incomplete activation *detection* as the cause for the frequent loss of control, rather than

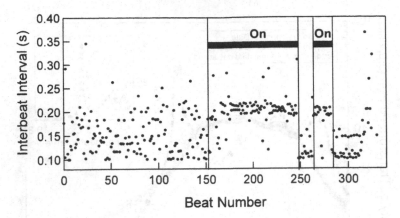

Figure 12. Typical plot of activation interval, I_n, versus the nth activation interval with and without control. Times prior to beat 150 comprise the learning period. The control is initiated between the solid vertical lines as indicated. Control is suspended near beat 250 and then resumed around the same unstable fixed point (UFP) with no new learning phase. This suspension and resumption demonstrates the ability to control about UFPs with activation intervals significantly longer than the mean uncontrolled intervals. The suspension and resumption of control confirms the validity of the UFP dynamics and the control using them.

any failure of the chaos control algorithm. Additional reasons for loss of control included poor characterization of UFPs and rapid changes in the (uncontrolled) dynamics. To diagnose poor characterization of the UFPs, control was turned off and then reinitiated with a new learning phase. After the second learning phase was completed, a dramatically different UFP was found, thus calling into question the accuracy of the original UFP characterization. In contrast, during excellent chaos control, we were able to discontinue chaos control and subsequently reacquire the same UFP, as shown in Figure 12. In these cases the UFP always had values for its position and its manifolds similar to those found previously. More significantly, the chaos control remained excellent. In the six unsuccessful control attempts, we were never able to both locate and control, for any significant length of time, a UFP with a mean cycle length at or above the uncontrolled mean.

Since this algorithm is implemented in a 2-D Poincaré section, it is useful to consider the results on that section which are displayed in Figure 13. The top panel shows the distribution of the data before control was implemented for a typical case, while the bottom panel gives the distribution as a result of the control algorithm. The symmetry is important in that it indicates that the deviations around the control point are truly random and not the result of poor control technique or bad control parameters.

It has been demonstrated that during AF conventional fixed-rate pacing

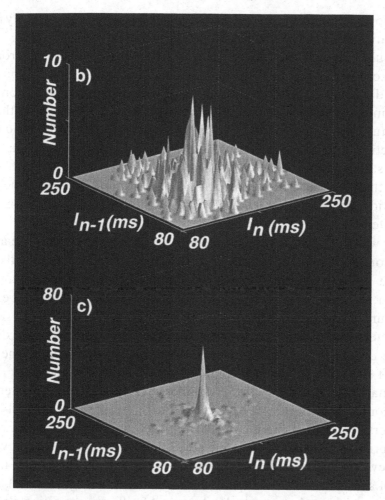

Figure 13. Histogram of the data on a Poincaré plot before (upper) and after (lower) implementation of chaos control of human atrial fibrillation (AF).

(stimulation at a constant cycle length) and on-demand pacing (a stimulation rate at which the intrinsic activation interval exceeds the programmed pacing interval) only local activations are entrained in a narrow window of cycle lengths around the mean activation interval (Allessie *et al.*, 1991; Kirchhof *et al.*, 1993). Both techniques suffer from inconsistent capture when pacing at shorter- and longer-cycle lengths (Allessie *et al.*, 1991; Kirchhof *et al.*, 1993). When pacing at intervals much shorter than the mean interval, either method only eliminates the long intervals, leaving the shorter ones unchanged. In contrast, our results have demonstrated the effectiveness of chaos control for entraining the atrium at intervals equal to and significantly longer than the mean spontaneous interval with the ability to eliminate both short and long fluctuations about the mean interval. Thus, chaos control dramatically reduces the variation from the mean cycle length.

It should be noted that chaos control operates quite differently from periodic or on-demand pacing by locating, characterizing and exploiting the natural dynamics around a UFP in the Poincaré plot. In addition, chaos control initiates a stimulus only when the system state comes near the UFP. The stimulus forces a predicted interval onto the stable manifold (the contracting direction) and allows the natural contraction along the stable manifold to pull the state point onto the UFP, thereby minimizing the number of control interventions needed. Chaos control uses control stimuli to exploit the natural contraction of the stable *direction* of a UFP rather than enforcing a rigid target *point* (interval).

Thus, we have shown that chaos control can be used to stabilize an UFP whose corresponding activation interval was equal to or significantly longer than the mean of the uncontrolled activation intervals during human AF. However, several unresolved questions remain. (1) It is unclear what extent of the atrium is captured during control. A previous study on the regional entrainment of AF in dogs has shown capture of a region 4 cm in diameter (Allessie *et al.*, 1991). We are currently working on determining the spatial extent of such chaos control in animal experiments. (2) While sinus rhythms occasionally follow chaos, further studies will be required to determine a causal connection. (3) It is an open question, in lieu of chaos control-induced cardioversion, whether chaos control can lower the energy threshold required for defibrillation of the atrium. The atrial defibrillation threshold, even with newly developed endocardial leads, remains sufficiently high to result in stimulation of skeletal muscles and patient discomfort (Wickelgren, 1996). Despite these uncertainties, chaos control in human AF offers a promising alternative for altering the dynamics of the arrhythmia. This alternative requires much less energy than the existing high-energy shock techniques, which are designed to overpower the dynamics of the atrium. Thus, a better understanding of the dynamics of human AF and its response to chaos control techniques presents us with an intriguing new direction for the study and treatment of human fibrillation (Langberg *et al.*, 1999).

15.4 Control of chaos in brain tissue

Our success in controlling chaotic cardiac systems led us to see whether a similar strategy could control chaotic behavior in brain tissue (Schiff *et al.*, 1994). One of the hallmarks of the human epileptic brain during periods between seizures is the presence of brief bursts of focal neuronal activity known

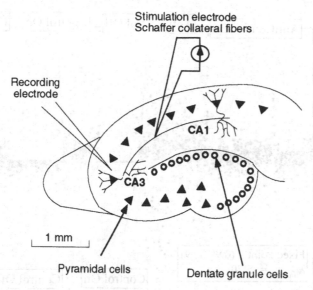

Figure 14. Schematic diagram of the transverse hippocampal slice and arrangement of recording electrodes.

as *interictal spikes*. Often such spikes emanate from the same region of the brain from which the seizures are generated. Several types of *in vitro* brain slice preparations, usually after exposure to convulsant drugs that reduce neuronal inhibition, exhibit population burst-firing activity similar to the interictal spike. One of these preparations is the high potassium (K^+) model, where slices from the hippocampus of the temporal lobe of the rat brain (a frequent site of epileptogenesis in human brain) are exposed to artificial cerebrospinal fluid containing a high K^+ concentration that causes spontaneous bursts of synchronized neuronal activity originating in a region known as the third part of the cornu ammonis or CA3, as shown in Figure 14.

If one observes the timing of these bursts, clear evidence for UFPs is seen in the return map. As reported, we were able to regularize the timing of such bursts through intervention with stimuli delivered by micropipette with timing as dictated by the chaos control algorithm to put the system onto the stable direction. As illustrated in Figure 15, not only were we able to regularize the intervals between spikes, but we were also able through a chaos '*anticontrol*' strategy to make the intervals more chaotic. It is the latter that might serve a useful purpose in breaking up seizure activity through the prevention or eradication of pathological order in the timing of the spikes. This original anticontrol or *chaos maintenance* strategy (Schiff *et al.*, 1994) has been further elaborated and expanded to high dimensions in the past few years (In *et al.*, 1995, 1998; Yang *et al.*, 1995).

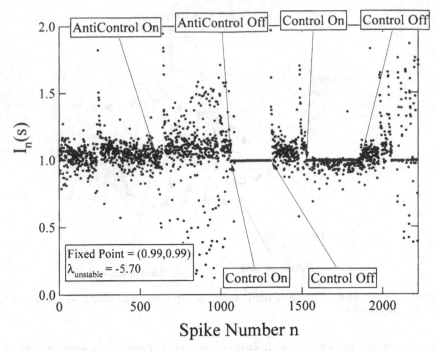

Figure 15. Demonstration of chaos anticontrol and chaos control in a hippocampal slice of a rat brain exposed to artificial cerebrospinal fluid containing a high concentration of K^+ and undergoing spontaneous chaotic population burst firing or spiking.

15.5 Electric field interactions with mammalian neuronal tissue

15.5.1 Stochastic resonance in rat hippocampal slices

Stochastic resonance (SR), a phenomenon in which the addition of noise enhances a nonlinear system's response to an otherwise subthreshold signal, has been suspected to play a role in the processing of information in neuronal systems (Benzi et al., 1981, 1982; Nicolis, 1982; Bezrukov and Vodyanoy, 1995; Moss and Wiesenfeld, 1995; Wiesenfeld and Moss, 1995; Bulsara and Gammaitoni, 1996). Theoretical models of single neurons (Bulsara et al., 1989, 1991; Chialvo and Apkarian, 1993) and simulations of neuronal networks (Riani and Simonotto, 1994; Collins et al., 1995) have all demonstrated SR. Although interspike-interval histograms (ISIHs) from the responses of periodically stimulated neuronal sensory systems have features consistent with SR, neuronal SR had been experimentally confirmed only in the sensory processes of invertebrate peripheral nervous systems (Douglass et al., 1993; Levin and Miller, 1996; see also Moss, Chapter 10, this volume). Recently, however, the observation of SR in a network of neurons from a mammalian brain has been reported (Gluckman et al., 1996b). A time-varying electric field with both periodic and stochastic components was used to deliver both signal and noise directly to the neuronal membranes in a hippocampal slice. As the magnitude of the stochastic component of the field was increased, *resonance* was observed in the response of the neuronal network to a weak periodic signal.

The brain is a noisy processor, and the idea that the brain might make use of such noise to enhance information processing is not new (Adey, 1972). In SR, the response of a nonlinear system to an otherwise subthreshold signal is optimized with the addition of noise. Since its proposal as a mechanism to explain how weak periodic variations in the Earth's orbit might be amplified by random meteorological fluctuations to produce ice ages, SR has been observed in a diverse range of physical systems. Despite a significant amount of theoretical work predicting that SR might be found in the brain, there had been no experimental demonstrations in the brain (Bulsara *et al.*, 1991; Longtin *et al.*, 1991; Chialvo and Apkarian, 1993; Riani and Simonotto, 1994; Collins *et al.*, 1995). Features suggestive of SR have been observed in ISIHs recorded from the auditory nerve (Longtin *et al.*, 1991), spinal cord (Chialvo and Apkarian, 1993), and visual (Longtin *et al.*, 1991) and somatosensory (Chialvo and Apkarian, 1993) cortex in response to periodic environmental stimuli. Direct observations of SR were made by detecting the activity of single peripheral sensory neurons from crayfish (Douglass *et al.*, 1993) and crickets (Levin and Miller, 1996). In these experiments, signal and noise were encoded into environmental fluid or gas motions, which were then detected by the neurons under study. An effort to increase internal neuronal noise in the crayfish experiment by raising temperature failed to demonstrate SR as a function of noise (Pantazelou *et al.*, 1995).

It was recently demonstrated that an electric field could be used to either suppress or enhance epileptiform activity in mammalian brain slices (Gluckman *et al.*, 1996a). The effect of an imposed electric field on neurons has been worked out in detail (Chan and Nicholson, 1986; Chan *et al.*, 1988), and can generally be understood as follows. The charged ions both inside and outside the neurons move under the influence of the imposed electric field. The cell membranes of the neurons act as containers, albeit leaky ones, opposing this motion, and the neurons are therefore polarized by the field. We aligned the field parallel to the axis between the apical dendrites, where signals come in from other neurons, and the soma, where these signals are integrated and translated into action potentials. The resulting polarization can be thought of as either a signal similar to neuronal synaptic activation or an offset in the somatic transmembrane potential. The overall effect of the electric field on the network is somewhat analogous to the injection of current into each of the neurons (Chan *et al.*, 1988). Because the field can interact with neurons at magnitudes

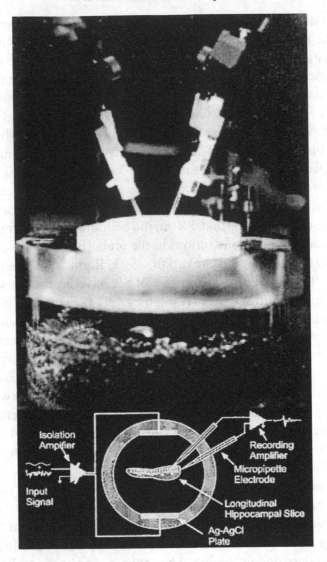

Figure 16. Photograph (as viewed from the side) and schematic of perfusion chamber (as viewed from above) for the hippocampal slice preparation. The hippocampal slice rests just below the upper surface of the bath. An electric field is imposed by a potential between parallel Ag-AgCl plates submerged in the bath.

insufficient to trigger action potentials, it is an ideal means to introduce a subthreshold signal into an entire network to probe for the existence of SR.

A picture and a schematic of our set-up for studying SR in mammalian brain tissue are shown in Figure 16 (Gluckman *et al.*, 1996b). Longitudinally or transversely cut hippocampal slices were placed in the center of a field produced by parallel Ag-AgCl electrode plates submerged in the perfusate. The neuronal layers of the slice can be visually identified and therefore easily oriented with respect to the field. The potential between the plates, and therefore the field amplitude, was set by a computer-generated signal applied through an isolation amplifier.

Synchronous activity of the network was monitored from the extracellular potential in the cell body layer of the slices. The recordings were made with respect to a point in the bath on nearly the same isopotential of the imposed field as the measurement electrode. This configuration minimized the amount of artifact in the measurement from the stimulus field. Because some remnant of the input signal leaked into the recording, care was taken in the choice of input signals to ensure that synchronous burst-firing neuronal activity could be differentiated from stimulus artifacts in the recordings. The bursts typically last 10 to 30 ms, occur as frequently as a few hertz, and can be identified from characteristic oscillations near 250 Hz. We therefore chose an input signal composed of a sinusoid with amplitude, A_{sin}, and frequency, $f_0 < 4$ Hz, and a noise signal with a high frequency cutoff, f_{nmax}, with $f_0 \ll f_{nmax} \ll 250$ Hz. The noise was Gaussian distributed in amplitude.

No neuronal population events were observed for this network in response to a sinusoidal signal with $A_{sin} = 3.75$ mV/mm, $f_0 = 3.3$ Hz; a sinusoidal field with approximately twice that amplitude was required to excite bursts. With a pure noise input ($A_{noise} = 10$ mV/mm, $f_{nmax} = 26$ Hz), randomly occurring bursts were observed. With the combination of both of these signals, bursts occurred preferentially near the maxima of the sinusoidal component of the signal. This is the essence of SR – the behavior of a noise-driven system being coherently modulated by the introduction of an otherwise subthreshold signal.

The signal-to-noise ratio (SNR) as a function of A_{noise}, with constant A_{sin}, is shown in Figure 17a. A series of these optimization curves, corresponding to different values of A_{sin}, is shown in Figure 17b. In each case, a maximum is observed in the SNR at intermediate noise levels. Also, as would be expected, as A_{sin} is increased, the maximum of the SNR curve increases in amplitude and occurs at lower noise levels.

In contrast to previous biological experiments that have demonstrated SR in the sensory processes of peripheral sensory neurons, this experiment shows SR in the behavior of a network of neurons taken from the brain. Although SR for individual nonlinear elements is fairly well understood, the effects of different types of noise and coupling in arrays or networks of devices are still being worked out. Noise in an array of elements can be either local, where the noise sources for each element are independent and uncorrelated (Pantazelou *et al.*, 1993; Collins *et al.*, 1995; Lindner *et al.*, 1995), or global, where the noise is uniform across the array (Bezrukov and Vodyanoy, 1995; Inchiosa and

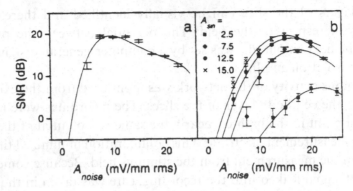

Figure 17. Signal-to-noise ratio (SNR) in decibels (dB) as a function of Gaussian noise amplitude, A_{noise}, for various values of noise and input sinusoidal signal amplitude, A_{sin}, in the hippocampal slice experiment as shown in Figure 16.

Bulsara, 1995). The mammalian brain experiments correspond to global noise, where the random fluctuations in the external electric field produce correlated noise at each element in the neuronal array. Although one might anticipate that global noise would make the detection of SR more difficult in an array, this did not prove to be an impediment to the identification of SR in these experiments.

Much interest has been generated in recent years concerning the so-called 40-Hz endogenous oscillations in the brain. Neurons have been shown to fire action potentials in phase with these oscillations, which are thought to be of importance in the binding of neuronal events across disparate regions of the brain (Jefferys *et al.*, 1996). SR, as illustrated by our experiments, provides a means to enhance this phase locking in the presence of noise. Also of great interest are the possible effects of 60-Hz environmental electromagnetic fields on the brain (Wiesenfeld and Moss, 1995; see also Gailey, Chapter 6, this volume). Here too, SR provides the capability of enhancing the impact of such weak signals on neurons, especially when SR in arrays of elements is considered.

All nervous systems, from those of invertebrates to those of humans, are noisy – membrane potentials fluctuate, membrane channels open and close, quantal release at synapses is probabilistic. Two hypotheses suggest themselves: either nervous systems evolved to include noise within their circuits as an advantage to processing, or, perhaps more palatably, the components that all nervous systems had to use in their evolution were inherently noisy and brains had to make the best of it. Regardless, the findings presented here show that random noise can enhance the response to a signal within a mammalian neuronal network.

15.6 Conclusions

In less than a decade since the original chaos control experiments, the experimental applications of chaos control have exploded in number. Advances in the lore of chaos control, especially extensions to higher-dimensional systems, the development of the maintenance of chaos, and the beginnings of techniques for controlling spatially extended systems, have enabled chaos control to be applied to reasonably complex systems, including many biomedical systems of clinical interest. Applications range from AF, for which devices incorporating chaos control appear to be quite viable, through VF, which awaits practical extensions of chaos control theory to spatiotemporal systems, all the way to epilepsy, which may benefit from techniques for maintaining chaos.

SR also comprises a new and exciting subfield of nonlinear dynamics, having been studied for less than two decades. We know that SR is definitely employed in several sensory systems and now understand that it can occur even in mammalian brain tissue. One hope is that these results might shed some light on the means by which the brain communicates both externally and internally.

We have found that while the techniques of nonlinear dynamics can be applied with success to many different biological systems, it is also true that these selfsame biological systems can teach us much about the nonlinear dynamics of complex systems in general. Many of the techniques discussed here were either developed for (maintenance of chaos) or discovered (SR) in biological systems and later applied to physical systems. Thus, there is a synergy between our efforts to apply the techniques of nonlinear dynamics to biology and our understanding of nonlinear dynamics itself.

Acknowledgments

We acknowledge support from ONR Physical Sciences Division and Control Dynamics Inc. M.L.S. acknowledges support from the NSWC ILIR Program. We also acknowledge the leadership and hard work of Jonathon Langberg and Francis Witkowski. Finally, W.L.D. acknowledges Robin Ditto, without whose patience and love none of this work could have come about.

References

Adey, W. R. (1972) Organization of brain tissue: is the brain a noisy processor? *Int. J. Neurosci.* **3**: 271–284.

Allessie, M., Kirchhof, C., Scheffer, G. J., Chorro, F. and Brugada, J. (1991) Regional control of atrial fibrillation by rapid pacing in conscious dogs. *Circulation* **84**: 1689–1697.

Baker, G. L. and Gollub, J. P. (1990) *Chaotic Dynamics: An Introduction.* Cambridge: Cambridge University Press.

Benzi, R., Parisi, G., Sutera, A. and Vulpiani, A. (1982) Stochastic resonance in climatic change. *Tellus* **34**: 10–16.

Benzi, R., Sutera, A. and Vulpiani, A. (1981) The mechanism of stochastic resonance. *J. Phys. A: Math. Gen.* **14**: L453–L457.

Bezrukov, S. M. and Vodyanoy, I. (1995) Noise-induced enhancement of signal transduction across voltage-dependent ion channels. *Nature* **378**: 362–364.

Bulsara, A., Boss, R. D. and Jacobs, E. W. (1989) Noise effects in an electronic model of a single neuron. *Biol. Cybern.* **61**: 211–222.

Bulsara, A. R. and Gammaitoni, L. (1996) Tuning in to noise. *Phys. Today* **49**: 39–45.

Bulsara, A., Jacobs, E. W., Zhou, T., Moss, F. and Kiss, L. (1991) Stochastic resonance in a single neuron model: theory and analog simulation. *J. Theor. Biol.* **152**: 531–555.

Chan, C. Y., Houndsgaard, J. and Nicholson, C. (1988) Effects of electric fields on transmembrane potential and excitability of turtle cerebellar Purkinje cells *in vitro*. *J. Physiol. (Lond.)* **402**: 751–771.

Chan, C. Y. and Nicholson, C. (1986) Modulation by applied electric fields of Purkinje and stellate cell activity in the isolated turtle cerebellum. *J. Physiol. (Lond.)* **371**: 89–114.

Chen, P.-S., Wolf, P. D., Dixon, E. G., Danieley, N. D., Frazier, D. W., Smith, W. M. and Ideker, R. E. (1988) Mechanism of ventricular vulnerability to single premature stimuli in open-chest dogs. *Circ. Res.* **62**: 1191–1209.

Chialvo, D. R. (1990) Non-linear dynamics of cardiac excitation and impulse propagation. *Nature* **330**: 749–752.

Chialvo, D. R. and Apkarian, A. V. (1993) Modulated noisy biological dynamics: three examples. *J. Stat. Phys.* **70**: 375–391.

Chialvo, D. R., Gilmour, R. F. and Jalife, J. (1990) Low dimensional chaos in cardiac tissue. *Nature* **343**: 653–657.

Christini, D. J. and Collins, J. J. (1995) Using noise and chaos control to control nonchaotic systems. *Phys. Rev. E* **52**: 5806–5809.

Christini, D. J. and Collins, J. J. (1996) Using chaos control and tracking to suppress a pathological nonchaotic rhythm in a cardiac model. *Phys. Rev. E* **53**: R49–R52.

Collins, J. J., Chow, C. C. and Imhoff, T. T. (1995) Stochastic resonance without tuning. *Nature* **376**: 236–238.

Ding, M., Yang, Y., In, V., Ditto, W. L., Spano, M. L. and Gluckman, B. (1996) Controlling chaos in high dimensions: theory and experiment. *Phys. Rev. E* **53**: 4334–4344.

Ditto, W. L. and Pecora, L. M. (1993) Mastering chaos. *Sci. Am.* **269**: 78–84.

Ditto, W. L., Rauseo, S. N. and Spano, M. L. (1990) Experimental control of chaos. *Phys. Rev. Lett.* **257**: 3211–3214.

Douglass, J., Wilkens, L., Pantazelou, E. and Moss, F. (1993) Noise enhancement of information transfer in crayfish mechanoreceptors by stochastic resonance. *Nature* **365**: 337–340.

Garfinkel, A., Spano, M. L., Ditto, W. L. and Weiss, J. N. (1992) Controlling cardiac chaos. *Science* **257**: 1230–1235.

Gluckman, B. J., Neel, E. J., Netoff, T. I., Ditto, W. L., Spano, M. L. and Schiff, S. J. (1996a) Electric field suppression of seizure activity in brain tissue. *J. Neurophysiol.* **96**: 4202–4205.

Gluckman, B. J., Netoff, T. I., Neel, E. J., Ditto, W. L., Spano, M. L. and Schiff, S. J. (1996b) Stochastic resonance in a neuronal network from mammalian brain.

Phys. Rev. Lett. **77**: 4098–4101.

Goldberger, A. L., Bhargava, V., West, B. J. and Mandell, A. J. (1986) Some observations on the question: is ventricular fibrillation 'chaos'? *Physica D* **19**: 282–289.

Gray, R. A., Jalife, J., Panfilov, A. V., Baxter, W. T., Cabo, C., Davidenko, J. M. and Pertsov, A. M. (1995) Mechanisms of cardiac fibrillation. *Science* **270**: 1222–1223.

Gray, R. A., Pertsov, A. M. and Jalife, J. (1998) Spatial and temporal organization during cardiac fibrillation. *Nature* **392**: 75–78.

Guevara, M. R., Glass, L. and Schrier, A. (1981) Phase locking, period-doubling bifurcations, and irregular dynamics in periodically stimulated cardiac cells. *Science* **214**: 1350–1353.

Hall, K., Christini, D. J., Tremblay, M., Collins, J. J., Glass, L. and Billete, J. (1997) Dynamic control of cardiac alternans. *Phys. Rev. Lett.* **78**: 4518–4521.

Holden, A. V. (1997) The restless heart of a spiral. *Nature* **387**: 655–656.

Hunt, E. R. (1991) Stabilizing high-period orbits in a chaotic system: the diode resonator. *Phys. Rev. Lett.* **67**: 1953–1956.

In, V., Mahan, S. E., Ditto, W. L. and Spano, M. L. (1995) Experimental maintenance of chaos. *Phys. Rev. Lett.* **74**: 4420–4423.

In, V., Spano, M. L. and Ding, M. (1998) Maintaining chaos in high dimensions. *Phys. Rev. Lett.* **80**: 700–703.

Inchiosa, M. E. and Bulsara, A. R. (1995) Nonlinear dynamic elements with noisy sinusoidal forcing: enhancing response via nonlinear coupling. *Phys. Rev. E* **52**: 327–339.

Jefferys, J. G. R., Traub, R. D. and Whittington, M. A. (1996) Neuronal networks for induced '40 Hz' rhythms. *Trends Neurosci.* **19**: 202–208.

Kannel, W. B., Abbott, R. D., Savage, D. D. and McNamara, P. M. (1982) Epidemiologic features of chronic atrial fibrillation. *New Engl. J. Med.* **306**: 1018–1022.

Kaplan, D. T. and Cohen, R. J. (1990) Is fibrillation chaos? *Circ. Res.* **67**: 886–892.

Kirchhof, C., Chorro, F., Scheffer, G. J., Brugada, J., Konings, K., Zetelaki, Z. and Allessie, M. (1993) Regional entrainment of atrial fibrillation studied by high-resolution mapping in open-chest dogs. *Circulation* **88**: 736–749.

Langberg, J. J., Bolmann, A., McTeague, K., Spano, M. L., In, V., Neff, J., Meadows, B. and Ditto, W. L. (1999) Control of human atrial fibrillation. *Int. J. Bifurc. Chaos*, in press.

Lee, J. J., Kamjoo, K., Hough, D., Hwang, C., Fan, W., Fishbein, M. C., Bonometti, C., Ikeda, T., Karagueuzian, H. S. and Chen, P. S. (1996) Reentrant wave fronts in Wiggers' stage II ventricular fibrillation: characteristics and mechanisms of termination and spontaneous regeneration. *Circ. Res.* **78**: 660–675.

Levin, J. E. and Miller, J. P. (1996) Broadband neural encoding in the cricket cercal sensory system enhanced by stochastic resonance. *Nature* **380**: 165–168.

Lindner, F. J. and Ditto, W. L. (1995) Removal, suppression and control of chaos by nonlinear design. *Appl. Mech. Rev.* **48**: 795–808.

Lindner, J. F., Meadows, B. K., Ditto, W. L., Inchiosa, M. E. and Bulsara, A. R. (1995) Array enhanced stochastic resonance and spatiotemporal synchronization. *Phys. Rev. Lett.* **75**: 3–6.

Longtin, A., Bulsara, A. and Moss, F. (1991) Time-interval sequences in bistable systems and the noise-induced transmission of information by sensory neurons. *Phys. Rev. Lett.* **67**: 656–659.

Moss, F. and Wiesenfeld, K. (1995) The benefits of background noise. *Sci. Am.* **273**:

66–68.

Nicolis, C. (1982) Stochastic aspects of climatic transitions-response to a periodic forcing. *Tellus* **34**: 1–16.

Ott, E., Grebogi, C. and Yorke, J. A. (1990) Controlling chaos. *Phys. Rev. Lett.* **64**: 1196–1199.

Ott, E. and Spano, M. L. (1995) Controlling chaos. *Phys. Today* **48**: 34–40.

Panfilov, A. V. and Holden, A. V. (eds.) (1997) *Computational Biology of the Heart*. Chichester: Wiley.

Pantazelou, E., Dames, C., Moss, F., Douglass, J. and Wilkens, L. (1995) Temperature dependence and the role of internal noise in signal transduction efficiency of crayfish mechanoreceptors. *Int. J. Bifurc. Chaos* **5**: 101–108.

Pantazelou, E., Moss, F. and Chialvo, D. (1993) *AIP Conference Proceedings on Noise in Physical Systems and 1/f Fluctuations* (ed. P. Handel and A. Chung), No. 285, pp. 549–552.

Parmananda, P., Sherard, P., Rollins, R. W. and Dewald, H. D. (1993) Control of chaos in an electrochemical cell. *Phys. Rev. E* **47**: R3003–R3006.

Pei, X. and Moss, F. (1996) Characterization of low-dimensional dynamics in the crayfish caudal photoreceptor. *Nature* **379**: 618–621.

Pertsov, A. M., Davidenko, J. M., Salomonsz, R., Baxter, W. T. and Jalife, J. (1993) Spiral waves of excitation underlie reentrant activity in isolated cardiac muscle. *Circ. Res.* **72**: 631–650.

Petrov, V., Gaspar, V., Masere, J. and Showalter, K. (1993) Controlling chaos in the Belousov–Zhabotinsky reaction. *Nature* **361**: 240–243.

Petrov, V., Schatz, M. F., Muehlner, K. A., VanHook, S. J., McCormick, W. D., Swift, J. B. and Swinney, H. L. (1996) Nonlinear control of remote unstable state in a liquid bridge convection experiment. *Phys. Rev. Lett.* **77**: 3779–3782.

Petrov, P. and Showalter, K. (1996) Nonlinear control of dynamical systems from time series. *Phys. Rev. Lett.* **76**: 3312–3315.

Pierson, D. and Moss, F. (1995) Detecting periodic unstable points in noisy chaotic and limit cycle attractors with applications to biology. *Phys. Rev. Lett.* **75**: 731–734.

Prystowsky, E. N., Benson, W. J., Fuster, V., Hart, R. G., Kay, G. N., Myerburg, R. J., Naccarelli, G. V. and Wyse, G. (1996) Management of patients with atrial fibrillation. *Circulation* **93**: 1262–1277.

Rapp, P. E., Albano, A. M., Schah, T. I. and Farwell, L. A. (1993) Filtered noise can mimic low-dimensional chaotic attractors. *Phys. Rev. E* **47**: 2289–2297.

Riani, M. and Simonotto, E. (1994) Stochastic resonance in the perceptual interpretation of ambiguous figures: a neural network model. *Phys. Rev. Lett.* **72**: 3120–3123.

Roy, R., Murphy, T. W., Maier, T. D., Gills, Z. and Hunt, E. R. (1992) Dynamical control of a chaotic laser: experimental stabilization of a globally coupled system. *Phys. Rev. Lett.* **68**: 1259–1262.

Schiff, S. J., Jerger, K., Duong, D. H., Taeun, C., Spano, M. L. and Ditto, W. L. (1994) Controlling chaos in the brain. *Nature* **370**: 615–620.

So, P. and Ott, E. (1995) Controlling chaos using time delay coordinates via stabilization of periodic orbits. *Phys. Rev. E* **51**: 2955–2966.

So, P., Ott, E., Sauer, T., Gluckman, B. J., Grebogi, C. and Schiff, S. J. (1997) Extracting unstable periodic orbits from chaotic time series data. *Phys. Rev. E* **55**: 5398–5417.

So, P., Ott, E., Schiff, S. J., Kaplan, D. T., Sauer, T. and Grebogi, C. (1996) Detecting unstable periodic orbits in chaotic experimental data. *Phys. Rev. Lett.* **76**: 4705–4708.

Wickelgren, I. (1996) New devices are helping transform coronary care. *Science* **272**: 668–670.

Wiesenfeld, K. and Moss, F. (1995) Stochastic resonance and the benefits of noise: from ice ages to crayfish and SQUIDs. *Nature* **373**: 33–36.

Wiggers, C. J. (1930) Studies of ventricular fibrillation produced by electric shock. II. Cinematographic and electrocardiographic observations of the natural process in the dog's heart. Its inhibition by potassium and the revival of coordinated beats by calcium. *Am. Heart J.* **5**: 351–365.

Winfree, A. T. (1994) Electrical turbulence in three-dimensional heart muscle. *Science* **266**: 1003–1006.

Witkowski, F. X., Kavanagh, K. M., Penkoske, P. A. and Plonsey, R. (1993) *In vivo* estimation of cardiac transmembrane current. *Circ. Res.* **72**: 424–439.

Witkowski, F. X., Kavanagh, K. M., Penkoske, P. A., Plonsey, R., Spano, M. L., Ditto, W. L. and Kaplan, D. T. (1995) Evidence for determinism in ventricular fibrillation. *Phys. Rev. Lett.* **75**: 1230–1233.

Witkowski, F. X., Leon, L. J., Penkoske, P. A., Clark, R. B., Spano, M. L., Ditto, W. L. and Giles, W. R. (1998a) A method for visualization of ventricular fibrillation: design of a cooled fiberoptically coupled image intensified CCD data acquisition system incorporating wavelet shrinkage based adaptive filtering. *Chaos* **8**: 94–102.

Witkowski, F. X., Leon, L. J., Penkoske, P. A., Giles, W. R., Spano, M. L., Ditto, W. L. and Winfree, A. T. (1998b) Spatiotemporal evolution of ventricular fibrillation. *Nature* **392**: 78–82.

Witkowski, F. X. and Penkoske, P. A. (1990) Activation patterns during ventricular fibrillation. In *Mathematical Approaches to Cardiac Arrhythmias* (ed. J. Jalife), pp. 219–231. New York: New York Academy of Sciences.

Yang, W., Ding, M., Mandell, A. and Ott, E. (1995) Preserving chaos: control strategies to preserve complex dynamics with potential relevance to biological disorders. *Phys. Rev. E* **51**: 102–110.

16

Epilepsy: multistability in a dynamic disease

JOHN G. MILTON

16.1 Introduction

Suddenly the person sitting beside you falls to the ground in the grip of a generalized convulsion. Within seconds to minutes, the cataclysm ends as abruptly as it began. Why did the seizure occur when it did, and why, once started, did it stop? Even more puzzling is the centuries-old observation that a brief sensory stimulus, such as noise, given close to the onset of a seizure might have stopped it (Figure 1). The clinical challenge is to prevent seizures from occurring and, thus, to restore a normal life to the sufferer. The scientific challenge is to understand how sudden qualitative changes in brain dynamics, reflected by changes in the electroencephalogram (EEG), occur. The hope is that insights into mechanism translate into the development of effective therapeutic strategies.

A sudden change in qualitative dynamics in response to a clinical maneuver, or to a change in an endogenous factor, is the hallmark of a *dynamic disease*. Dynamic diseases can arise because of alterations in underlying physiological control mechanisms (Mackey and Glass, 1977; Glass and Mackey, 1979; Mackey and an der Heiden, 1982; Mackey and Milton, 1987; Milton and Mackey, 1989; Glass, 1991; Bélair *et al.*, 1995; Milton and Black, 1995). By analogy with mathematical models, changes of certain physiologically important parameters into critical ranges result in the sudden appearance of qualitatively different dynamical behaviors. In the mathematical models these changes in qualitative dynamics correspond to *bifurcations*.

There are over 30 diseases of the nervous system in which recurrence of symptoms or the appearance of oscillatory signs are a defining feature (Milton and Black, 1995). The significance of identifying which of these disorders is a dynamic disease is that it raises the possibility of devising strategies to treat these diseases based on the manipulation of the underlying control mechan-

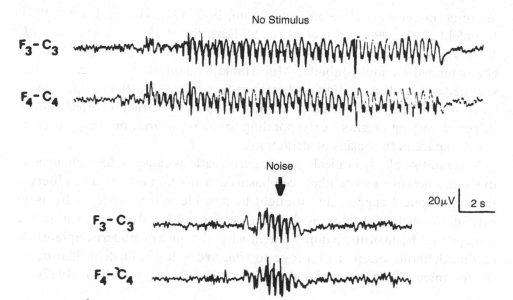

Figure 1. Effect of an auditory stimulus on length of absence seizures in an adolescent with primary generalized epilepsy. The top traces show two channels of the electroencephalogram (EEG) during a typical seizure: length corresponds to the average length of 71 spontaneously occurring seizures. The bottom traces show the EEG when a loud noise (↓) is given near seizure onset: length corresponds to an average of 69 noise-shortened seizures.

ism. The precise nature of the control strategies that are possible depends on the type of bifurcation that has occurred and the nature of the abnormal dynamics that arose from the bifurcation. The simplest therapeutic strategy is to manipulate the altered *control parameter* back into a range associated with healthy dynamics. In most cases this is not possible, since the identity of the altered parameter is not known. However, it may be possible to devise control strategies that exploit the properties of the abnormal dynamics that arise; for example, chaotic dynamics can be controlled using *control-of-chaos techniques* (Schiff *et al.*, 1994) and multistable dynamics can be controlled using brief perturbations (Guttman *et al.*, 1980).

The fact that seizures can be aborted using brief stimuli is very suggestive of an underlying *multistable* dynamical system. Multistability refers to the co-existence of multiple attractors. Multistability arises because of a bifurcation that results in the appearance of more than one possible dynamical behavior, e.g., there may be a *subcritical Hopf bifurcation*. More than 25 years of experimental and theoretical work indicates that the onset of oscillations in neurons (FitzHugh, 1969; Best, 1979; Guttman *et al.*, 1980; Hounsgaard *et al.*, 1988; Canavier *et al.*, 1993, 1994; Booth and Rinzel, 1995; Lechner *et al.*, 1996) and in neuronal populations (Wilson and Cowan, 1972, 1973; Hopfield, 1982, 1984; Kleinfeld *et al.*, 1990; Kelso *et al.*, 1992; Destexhe *et al.*, 1993; Milton *et al.*, 1993; Zipser *et al.*, 1993; Stadler and Kruse, 1995; Barnes *et al.*, 1997) is characterized by multistability. Multistable dynamical systems can be envisioned as a landscape, shown schematically in Figure 2, with multiple valleys (corresponding to the basins of attraction for each attractor) separated by ridges of varying heights (corresponding to the *separatrix*, or energy barrier, which separates the basins of attraction).

In a multistable dynamical system, perturbations cause sudden changes in dynamics because a switch between basins of attraction occurs. The observations in Figure 1 suggest that it might be possible to treat epilepsy by using carefully honed perturbations designed to confine the dynamics within the nonepileptic basin of attraction. Thus, much in the spirit of modern implantable cardiac defibrillators, it might prove possible to develop brain defibrillators for the treatment of patients with medically intractable epilepsy (Glanz, 1997).

16.2 Time-delayed feedback

It has long been felt that seizures arise because the delicate mechanisms of integrative control break down, resulting in the mass discharge of thousands or millions of neurons (Jasper, 1969). These control mechanisms operate through the neuronal circuitry between neurons in local aggregates, interactions between distant ganglionic centers, and changes in the neurochemical

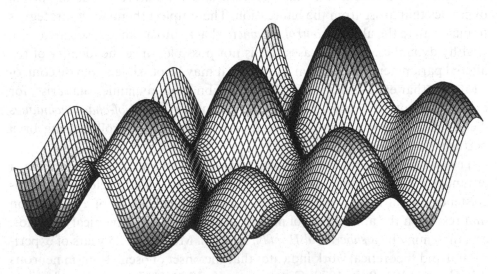

Figure 2. Schematic representation of the energy landscape for a hypothetical multistable dynamical system. For details, see text in Section 16.1.

environment of neurons regulated by neuroglia. Since finite distances separate neurons and ganglionic centers, and conduction velocities are finite as well, *time delays* are necessarily present. These time delays range from 10s to 100s of milliseconds (Gotman, 1983; Miller, 1994; Eurich and Milton, 1996). Even an arbitrarily small delay can have dramatic effects on the behavior of feedback control mechanisms (Hale, 1977; Kolmanovskii and Nosov, 1991). Time delays play a crucial role in the in-phase synchronization of neuronal populations (Ernst *et al.*, 1995; Gerstner *et al.*, 1996).

As a consequence of the above observations, mathematical models of neuronal feedback mechanisms take the form of delay differential equations (DDEs) (Glass and Mackey, 1988; Milton *et al.*, 1989; Milton *et al.*, 1990; Milton, 1996); an example is the first-order DDE

$$\dot{V}(t) + \alpha V(t) = f(V(t - \tau)), \tag{1}$$

where $V(t)$, $V(t - \tau)$ are, respectively, the values of the state variable (such as membrane potential) at times t, $t - \tau$. The time delay is τ, α is a rate constant, and f describes the feedback. $\dot{V}$ is the first differential of V. In order to obtain the solution of Equation (1) it is necessary to specify an initial function, ϕ, on the interval $[-\tau, 0]$.

An important, but only recently emphasized, property of delayed feedback control mechanisms, is the occurrence of multistability (an der Heiden and Mackey, 1982; Ikeda and Matsumoto, 1987; Aida and Davis, 1992; Losson *et al.*, 1993). Indeed, multistability arises even in a damped harmonic oscillator with a delayed monotonic restoring force (Campbell *et al.*, 1995), i.e.,

$$\ddot{x} + b\dot{x} + ax = F[x(t - \tau)],$$

where a, b are constants and F is negative feedback. $\ddot{x}$ is the second differential of x. Equations of this form arise from considerations of mechanical or neuromechanical systems, operating under the influence of a delayed restoring force. The necessary conditions for these complex dynamics to occur, i.e., an underdamped control system ($b < \sqrt{4a}$) with low-gain feedback, are easily satisfied by many mechanical systems.

16.2.1 Multistability in delayed recurrent loops

Recurrent inhibitory loops play a role in epileptic seizures arising from the amygdala–hippocampal complex (Figure 3a): an excitatory neuron, E, gives off collateral branches that excite an inhibitory interneuron, I, which, in turn, inhibits the firing of E (Mackey and an der Heiden, 1984; Mackey and Milton, 1987; Milton *et al.*, 1990). To illustrate how multistability arises in a delayed

a)   b)

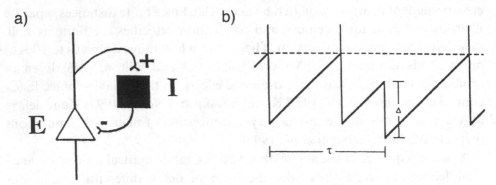

Figure 3. (a) Schematic representation of a neuronal recurrent inhibitory feedback loop. (b) The time course of the membrane potential (vertical axis) for an integrate-and-fire approximation of the recurrent inhibitory loop shown in (a). The dashed line is the firing threshold and τ indicates the time for neuronal activity to traverse the recurrent loop to deliver an inhibitory pulse of magnitude Δ. For details, see text.

recurrent loop, consider the integrate-and-fire approximation shown in Figure 3b. Here the membrane potential of the neuron increases linearly until it reaches the firing threshold at which point the neuron fires, and the membrane potential is reset to its resting value. The firing of the neuron excites the inhibitory interneuron, I, which, in turn, at a time τ later, delivers an inhibitory pulse to the excitatory neuron, E. The advantage of this simple model is that considerable analytical insight can be obtained into its dynamics (Foss *et al.*, 1996, 1997; Milton and Foss, 1997).

In dimensionless form, the dynamics of this model depend on only two parameters: the magnitude of the inhibitory pulse, Δ, and the time delay, τ (Foss *et al.*, 1996, 1997). Multistability arises when τ is longer than the firing period of the integrate-and-fire neuron (equal to 1 in the dimensionless model). This complex behavior becomes possible because the inhibitory pulses are not necessarily the result of the immediately preceding excitatory spike. It can be shown that the solutions that arise can be constructed from segments of length τ (Foss *et al.*, 1997): each segment must satisfy an equation of the form

$$\tau = x + m + n\Delta, \tag{2}$$

where n, m are positive integers and $0 < x < 1$. For τ, Δ fixed, the total number of (m,n) pairs that satisfy Equation (2) is $\lceil \tau/\Delta \rceil$, where the notation $\lceil \ \rceil$ denotes the smallest integer greater than τ/Δ. Since the number of (m,n) segments is finite for fixed τ, Δ, it follows that all solutions are periodic. Moreover, since there is a one-to-one relationship between excitatory spikes and inhibitory pulses, and since each inhibitory pulse prolongs the period by Δ, the period of these solutions is $S(1 + \Delta)$, where S is a positive integer equal to the number of excitatory spikes per period. The mean interspike interval is therefore $(1 + \Delta)$.

Once the (m,n) pairs have been determined from Equation (2), it is possible to construct the periodic spike trains that arise (Foss *et al.*, 1997). Figure 4 shows the four qualitatively different periodic spike trains that can be obtained when $\tau = 4.1$ and $\Delta = 0.8$: regular spiking (Figure 4a), bursting (Figure 4b) and two more irregular spiking patterns (Figure 4c and d). The regions in τ–Δ space for which these four spiking patterns occur are shown, respectively in Figure 4e–g. Figure 4h shows the intersection, $\cap$, of these regions. The fact that $\cap$ has nonzero measure lies at the basis of the multistability in this model. For choices of (τ,Δ) in $\cap$, all four of the above solutions coexist (see Figure 4h).

Multistable DDEs evolve in a functional space in which functions of length τ are mapped onto functions of length τ, and the basins of attraction correspond to sets of functions. Typically, the space of these functions is very complex (Losson *et al.*, 1993; Foss *et al.*, 1996); however, in the simple example discussed here it has been possible to construct this space since the functions, $\phi(s)$, consist of spiking patterns of length τ (e.g., see Milton and Foss, 1997). The complexity of these initial function spaces emphasize that perturbations must be very carefully timed, and be of the right magnitude, to cause a switch between basins of attraction.

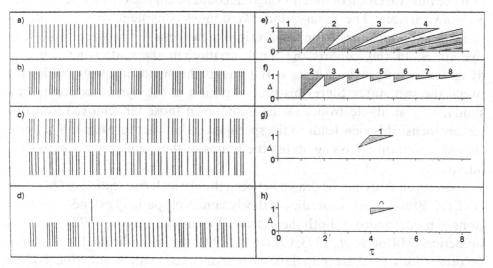

Figure 4. ((a)–(d)) Periodic solutions coexisting in the integrate-and-fire model for a recurrent inhibitory loop when $\tau = 4.1$ and $\Delta = 0.8$. Only the times of spiking of the excitatory neuron are shown. (e)–(f) Values of τ and Δ for which the solutions shown, respectively, in (a), (b) and (c),(d) exist. (h) The values of (τ,Δ) for which all four solutions coexist.

16.3 Noise-induced switching

The nervous system is a very noisy environment (Fatt and Katz, 1950; Stark *et al.*, 1958; Calvin and Stevens, 1968; Moss *et al.*, 1994; Areili *et al.*, 1996). The fact that relatively large regions of the initial function space can give rise to the same solution implies that multistable dynamical systems have a certain robustness to noise. However, noise will eventually cause a switch to occur between basins of attraction (Kapral *et al.*, 1986; Foss *et al.*, 1997). Indeed, the conclusion that *noise-induced switching* between basins of attraction plays a major role in shaping the dynamics of the nervous system seems both inescapable and self-evident (Longtin *et al.*, 1991; Moss *et al.*, 1994; Eurich and Milton, 1996). One of the earliest demonstrations for noise-induced switching between attractors in the human brain is the observation that the distribution of switching times for visually ambiguous figures is given by a γ-distribution (Borsellino *et al.*, 1972).

The possibility that noise-induced switches occur between attractors would be a simple explanation for the clinical observation that the timing of seizure recurrence in many patients with medically refractory epilepsy is random (Milton *et al.*, 1987). Recently, direct evidence for switching between attractors in the brains of epileptic patients has been obtained from intracranial EEG recordings (Manuca *et al.*, 1998). In these studies, time series analysis suggested that the time variation of the EEG signals could be characterized by changes in a *single* variable. The observations were most consistent with a model of *bistability* in which mesoscopic collections of neurons flip between two collective states. In this context the time variation in the underlying neuronal dynamics corresponds to time variations in the switching probabilities between the two states. Since this time variation in the dynamics occurs in a similar way at all electrodes in the brain (even those far removed from the seizure focus), this view leads to the speculation that there are *excitation waves* that are constantly passing across the brains of these epilepsy patients, even *interictally*.

A strong analogy can be drawn between the clinical observations of Manuca *et al.* (1998) and previous studies of the dynamics of spatially extended populations of model neurons. Both theoretical (Wilson and Cowan, 1972, 1973) and numerical (Milton *et al.*, 1993; Chu *et al.*, 1994) studies indicate that the onset of oscillations in these populations is associated with a subcritical Hopf bifurcation. The presence of such a bifurcation implies the existence of multistability. These oscillations lead to the propagation of activity through the population in the form of waves of neuronal activity (an der Heiden, 1991; Milton *et al.*, 1995; Milton, 1996).

Although the above observations are all consistent with the notion that noise-induced switching between attractors occurs in the human nervous system, perhaps the most elegant demonstration was recently obtained through a study of human postural sway (Eurich and Milton, 1996). The experiment requires that a subject stand quietly on a pressure platform so that fluctuations in the center of pressure can be measured (Collins and De Luca, 1993, 1994). The movement of an inverted pendulum subjected to both noisy perturbations and a time-delayed restoring force can model human postural sway. By taking into account the overdamped nature of the postural sway control mechanisms and the threshold-type response of the neurons that detect joint position, the model for postural sway in the front to back direction, x, becomes (in dimensionless form)

$$\dot{x} = \begin{cases} x + \sqrt{2D}\xi(t) + C & \text{if } x(t - \tau) < -1 \\ x + \sqrt{2D}\xi(t) & \text{if } -1 \leq x(t - \tau) \leq 1 \\ x + \sqrt{2D}\xi(t) - C & \text{otherwise} \end{cases} \quad (3)$$

where $\sqrt{2D}\xi(t)$ is the δ-correlated Gaussian noise of intensity $\sqrt{2D}$. The model implies that postural sway feedback operates by allowing the system to drift for small displacements (open-loop control) with stabilizing feedback (closed-loop control) becoming operational for larger displacements.

The dynamics of Equation (3) depends on three parameters only: the noise intensity, D, the time delay, τ, and the magnitude of the restoring force, C. In the absence of noise, i.e., $D = 0$, the solution of Equation (3) is

$$x(t) = \begin{cases} -C + [x(t_0) + C]\exp(t - t_0) & \text{if } x(t - \tau) < -1 \\ x(t_0)\exp(t - t_0) & \text{if } -1 \leq x(t - \tau) \leq 1 \\ C + [x(t_0) - C]\exp(t - t_0) & \text{otherwise.} \end{cases} \quad (4)$$

For a relatively large range of C and τ, two limit attractors coexist (Figure 5a): one corresponds to the subject swaying off center to the front, the other to swaying off center to the back.

Figure 5b shows the two-point correlation function, $K(s)$, where $s = \Delta t$, as a function of the noise intensity, D. For intermediate noise levels, $K(s)$ contains three scaling regions and is quantitatively identical with that observed experimentally ($\diamond$ in Figure 5b). From the model it is estimated that $\tau \approx 230$ ms and the threshold for the joint position neurons is ~ 6 mm, which agree remarkably well with the experimental observation, respectively, 200–300 ms and 5–6 mm. The mechanism for the three scaling regions in the two-point correlation function is intuitively clear. For s shorter than one period of the limit cycle (vertical line in Figure 5b), transitions occur only in one direction between

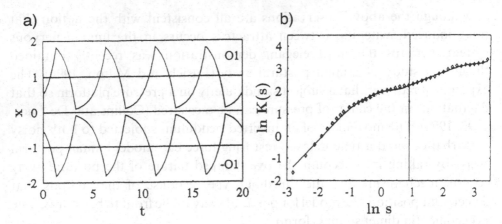

Figure 5. (a) Two coexistent limit-cycle attractors arise in Equation (3) in the absence of noise when τ, C satisfies $\tau < \ln C$ and $0 < \tau < \ln (C/(C-1))$. (b) Two-point correlation function predicted by Equation (3) (vertical line) compared with that observed experimentally ($\diamond$). In order to be able to directly compare experimental observation with the prediction the unscaled version of Equation (3) is used (for details, see Eurich and Milton, 1996).

basins of attraction, for example $1 \to 2$. These transitions are reflected by an increase in $K(s)$. For s longer than the period of the limit cycle, transitions of the form $1 \to 2 \to 1$ begin to occur and hence $K(s)$ increases less rapidly. Finally, for long s it becomes equally probable that the swayer is in either basin of attraction, and $K(s)$ reflects the mean displacement.

16.4 Brain defibrillators

I believe that if you asked patients with epilepsy what would be the seizure frequency that would make the biggest difference to their lives, they would emphatically answer zero! Benchtop research involving single neurons and single ion channels naturally leads to the development of new anticonvulsant medications. Unfortunately, for many, medications alone do not free them from the burden of unpredictable seizure recurrences. Moreover, the very fact that seizures occur paroxysmally and typically with low frequency questions the logic behind therapeutic modalities administered on a *daily* basis. After all, why should someone take medications every day to reduce the risk of developing an event that occurs once every few weeks or months?

The most important implication of the hypothesis that epileptic seizures arise in a multistable nervous system is that it directly implies the possibility of developing a brain defibrillator for the treatment of epilepsy. If multistability lies at the heart of epilepsy, then it should be possible to construct a brain defibrillator capable of first detecting seizure onset and then of delivering an appropriate stimulus (electrical or chemical) to a localized area of the brain to abort the seizure. A major advantage of this therapeutic approach is that it is called upon only when needed. Indeed the possibility of using brief electrical pulses to control the dynamics of an epileptic hippocampal slice has already been demonstrated (Schiff *et al.*, 1994; for details, see Ditto and Spano, Chapter 15, this volume). Moreover, many patients with implanted vagal nerve stimulators have already noted that many of their seizures can be aborted if the stimulator is turned on during their epileptic aura.

The role of pencil and paper in the development of therapeutic strategies for application at the bedside receives little attention in modern day clinical research. Indeed, medical students are not even required to know the simplest concepts of control. In contrast, I strongly believe that the performance of critically important experiments requires a theoretical knowledge of the response of the underlying control mechanisms to perturbations. Only through the efforts of dedicated teams involving theorists, basic scientists, and bedside clinicians, will the scourge called epilepsy be conquered.

Acknowledgments

The author thanks Christian Eurich, Jennifer Foss, John Hunter and Robert Savit for useful comments. This research was supported by grants from the Brain Research Foundation and the National Institutes of Mental Health.

References

Aida, T. and Davis P. (1992) Oscillation modes of laser diode pumped hybrid bistable systems with large delay and application to dynamic memory. *IEEE J. Quantum Elect.* **28**: 686–699.

an der Heiden, U. (1991) Neural networks: flexible modelling, mathematical analysis, and applications. In *Neurodynamics: Proceedings of the Summer Workshop on Mathematical Physics* (ed. F. Paseman and H. D. Donner), pp. 49–95. Singapore: World Scientific.

an der Heiden, U. and Mackey, M. C. (1982) The dynamics of production and destruction: analytic insights into complex behavior. *J. Math. Biol.* **16**: 75–101.

Areili, A., Sterkin, A., Grinvald, A. and Aertson, A. (1996) Dynamics of ongoing activity: explanation of the large variability in evoked cortical responses. *Science* **273**: 1868–1871.

Barnes, C. A., Suster, M. S., Shen, J. and McNaughton, B. L. (1997) Multistability of cognitive maps in the hippocampus of old rats. *Nature* **388**: 272–275.

Bélair, J., an der Heiden, U., Glass, L. and Milton, J. (eds.) (1995) *Dynamical Diseases: Mathematical Analysis of Human Illness*. Woodbury, NY: American Institute of Physics.

Best, E. N. (1979) Null space in the Hodgkin–Huxley equations: a critical test. *Biophys. J.* **27**: 105–116.

Booth, V. and Rinzel, J. (1995) A minimal compartmental model for a dendritic origin of bistability of motoneuron firing patterns. *J. Comput. Neurosci.* **2**: 299–312.

Borsellino, A., De Marco, A., Allazetta, A., Rinesi, S. and Bartolini, B. (1972) Reversal time distribution of visual ambiguous stimuli. *Kybernetik* **10**: 139–144.

Calvin, W. H. and Stevens, C. F. (1968) Synaptic noise and other sources of randomness in motoneuron interspike intervals. *J. Neurophysiol.* **31**: 574–587.

Campbell, S. A., Bélair, J., Ohira, T. and Milton, J. G. (1995) Complex dynamics and multi-stability in a damped harmonic oscillator. *Chaos* **5**: 640–645.

Canavier, C. C., Baxter, D. A., Clark, J. W. and Byrne, J. H. (1993) Nonlinear dynamics in a model neuron provide a novel mechanism for transient synaptic inputs to produce long-term alterations of postsynaptic activity. *J. Neurophysiol.* **69**: 2252–2257.

Canavier, C. C., Baxter, D. A., Clark, J. W. and Byrne, J. H. (1994) Multiple modes of activity in a model neuron suggest a novel mechanism for the effects of neuromodulators. *J. Neurophysiol.* **72**: 872–881.

Chu, P. H., Milton, J. G. and Cowan, J. D. (1994) Connectivity and the dynamics of integrate-and-fire neural networks. *Int. J. Bifurc. Chaos* **4**: 237–243.

Collins, J. J. and De Luca, C. J. (1993) Open-loop and closed-loop control of posture: a random-walk analysis of center-of-pressure trajectories. *Exp. Brain Res.* **95**: 308–318.

Collins, J. J. and De Luca, C. J. (1994) Random walking during quiet standing. *Phys. Rev. Lett.* **73**: 764–767.

Destexhe, A., Babloyantz, A. and Sejnowski, T. J. (1993) Ionic mechanisms for intrinsic slow oscillations in thalamic relay neurons. *Biophys. J.* **65**: 1538–1552.

Ernst, U., Pawelzik, K. and Geisel, T. (1995) Synchronization induced by temporal delays in pulse-coupled oscillators. *Phys. Rev. Lett.* **74**: 1570–1573.

Eurich, C. W. and Milton, J. G. (1996) Noise-induced transitions in human postural sway. *Phys. Rev. E* **54**: 6681–6684.

Fatt, P. and Katz, B. (1950) Some observations on biological noise. *Nature* **166**: 597–598.

FitzHugh, R. (1969) Mathematical models of excitation and propagation in nerve. In *Biological Engineering* (ed. H. P. Schwan), pp. 1–85. New York: McGraw Hill.

Foss, J., Longtin, A., Mensour, B. and Milton, J. (1996) Multistability and delayed recurrent loops. *Phys. Rev. Lett.* **76**: 708–711.

Foss, J., Moss, F. and Milton, J. (1997) Noise, multistability and delayed recurrent loops. *Phys. Rev. E* **55**: 4536–4543.

Gerstner, W., Kempter, R., van Hemmen, J. L. and Wagner, H. (1996) A neuronal learning rule for sub-millisecond temporal coding. *Nature* **383**: 76–78.

Glanz, J. (1997) Mastering the nonlinear brain. *Science* **277**: 1758–1760.

Glass, L. (ed.) (1991) Chaos focus issue on nonlinear dynamics of physiological function and control. *Chaos* **1**: 247–334.

Glass, L. and Mackey, M. C. (1979) Pathological conditions resulting from instabilities in physiological control systems. *Ann. NY Acad. Sci.* **620**: 22–44.

Glass, L. and Mackey, M. C. (1988) *From Clocks to Chaos: The Rhythms of Life.* Princeton, NJ: Princeton University Press.

Gotman, J. (1983) Measurement of small time differences between EEG channels: method and application to epileptic seizure propagation. *Electroencephalogr. Clin. Neurophysiol.* **79**: 403–412.

Guttman, R., Lewis, S. and Rinzel, J. (1980) Control of repetitive firing in squid axon membrane as a model for a neuron oscillator. *J. Physiol.* **305**: 377–395.

Hale, J. K. (1977) *Theory of Functional Differential Equations.* New York: Springer-Verlag.

Hopfield, J. J. (1982) Neural networks and physical systems with emergent collective computational properties. *Proc. Natl. Acad. Sci. USA* **79**: 2554–2558.

Hopfield, J. J. (1984) Neural networks with graded responses have collective computational properties like those of two-state neurons. *Proc. Natl. Acad. Sci. USA* **81**: 3088–3092.

Hounsgaard, J., Hultborn, H., Jesperson, B. and Kiehn, O. (1988) Bistability of α-motoneurons in the decerebrate cat and in the acute spinal cat after intravenous 5-hydroxytryptophan. *J. Physiol.* **405**: 345–367.

Ikeda, I. and Matsumoto, K. (1987) High dimensional chaotic behavior in systems with time-delayed feedback. *Physica D* **29**: 223–235.

Kapral, R., Celarier, E., Mandel, P. and Nardone, P. (1986) Noisy delay-differential equations in optical bistability. *SPIE Optical Chaos* **667**: 175–182.

Kelso, J. A. S., Bressler, S. L., Buchanan, S., DeGuzman, G. C., Ding, M., Fuchs, A. and Holroyd, T. (1992) A phase transition in human brain and behavior. *Phys. Lett. A* **169**: 134–144.

Kleinfeld, D., Raccuia-Behling, F. and Chiel, H. J. (1990) Circuits constructed from identified *Aplysia* neurons exhibit multiple patterns of persistent activity. *Biophys. J.* **57**: 697–715.

Lechner, H. A., Baxter, D. A., Clark, J. W. and Byrne, J. H. (1996) Bistability and its regulation by serotonin in the endogenously bursting neuron R15 in *Aplysia. J. Neurophysiol.* **75**: 957–962.

Kolmanovskii, V. B. and Nosov, V. R. (1986) *Stability of Functional Differential Equations.* New York: Academic Press.

Longtin, A., Bulsara, A. and Moss, F. (1991) Time interval sequences in bistable systems and the noise induced transmission of information by sensory neurons. *Phys. Rev. Lett.* **67**: 656–659.

Losson, J., Mackey, M. C. and Longtin, A. (1993) Solution multistability in first order nonlinear delay differential equations. *Chaos* **3**: 167–176.

Mackey, M. C. and an der Heiden, U. (1982) Dynamic diseases and bifurcations in physiological control systems. *Funkt. Biol. Med.* **1**: 156–164.

Mackey, M. C. and an der Heiden, U. (1984). The dynamics of recurrent inhibition. *J. Math. Biol.* **19**: 211–225.

Mackey, M. C. and Glass, L. (1977) Oscillation and chaos in physiological control systems. *Science* **197**: 287–289.

Mackey, M. C. and Milton, J. (1987) Dynamical diseases. *Ann. NY Acad. Sci.* **504**: 16–32.

Manuca, R., Casdagli, M. and Savit, R. (1998) Nonstationarity in epileptic EEG and implications for neural dynamics. *Math. Biosci.* **147**: 1–22.

Miller, R. (1994) What is the contribution of axonal conduction delay to temporal structure in brain dynamics? In *Oscillatory Event-Related Brain Dynamics* (ed. C. Pantev), pp. 53–57. New York: Plenum Press.

Milton, J. G. (1996) *Dynamics of Small Neural Populations. CRM Monographs in Applied Mathematics.* Providence, RI: American Mathematical Institute.

Milton, J. G., an der Heiden, U., Longtin, A. and Mackey, M. C. (1990) Complex dynamics and noise in simple neural networks with delayed mixed feedback. *Biomed. Biochim. Acta* **49**: 697–707.

Milton, J. and Black, D. (1995) Dynamic diseases in neurology and psychiatry. *Chaos* **5**: 8–13.

Milton, J. G., Chu, P. H. and Cowan, J. D. (1993) Spiral waves in integrate-and-fire neural networks. In *Advances in Neural Information Processing Systems*, Vol. 5 (ed. S. J. Hanson, J. D. Cowan and C. L. Giles), pp. 1001–1007. San Mateo, CA: Kaufmann.

Milton, J. and Foss, J. (1997) Oscillations and multistability in delayed feedback control. In *Case Studies in Mathematical Modeling: Ecology, Physiology, and Cell Biology* (ed. H. G. Othmer, F. R. Adler, M. A. Lewis and J. C. Dallon), pp. 179–198. Upper Saddle River, NJ: Prentice-Hall.

Milton, J. G., Gotman, J., Remillard, G. M. and Andermann, F. (1987) Timing of seizure recurrence in adult epileptic patients: a statistical analysis. *Epilepsia* **28**: 471–478.

Milton, J. G., Longtin, A., Beuter, A., Mackey, M. C. and Glass, L. (1989) Complex dynamics and bifurcations in neurology. *J. Theor. Biol.* **138**: 129–147.

Milton, J. G. and Mackey, M. C. (1989) Periodic haematological diseases: mystical entities or dynamical disorders? *J. R. Coll. Phys. Lond.* **23**: 236–241.

Milton, J. G., Mundel, T., an der Heiden, U., Spire, J.-P. and Cowan, J. D. (1995) Activity waves in neural networks. In *The Handbook of Brain Theory* (ed. M. A. Arbib), pp. 994–997. Boston, MA: MIT Press.

Moss, F., Pierson, D. and O'Gorman, D. (1994) Stochastic resonance: tutorial and update. *Int. J. Bifurc. Chaos* **4**: 1383–1397.

Schiff, S. J., Jerger, K., Duong, D. H., Chay, T., Spano, M. L. and Ditto, W. L. (1994) Controlling chaos in the brain. *Nature* **370**: 615–620.

Stadler, M. and Kruse, P. (eds.) (1995) *Ambiguity in Mind and Nature: Multistable Cognitive Phenomena*. New York: Springer-Verlag.

Stark, L., Campbell, F. W. and Atwood, J. (1958) Pupillary unrest: an example of noise in a biological servo-mechanism. *Nature* **182**: 857–858.

Wilson, H. R. and Cowan, J. D. (1972) Excitatory and inhibitory interactions in localized populations. *Biophys. J.* **12**: 1–23.

Wilson, H. R. and Cowan, J. D. (1973) A mathematical theory of the functional dynamics of cortical and thalamic nervous tissue. *Kybernetik* **13**: 55–80.

Zipser, D., Kehoe, B., Littlewort, G. and Fuster, J. (1993) A spiking network model of short-term memory. *J. Neurosci.* **13**: 3406–3420.

17

Control and perturbation of wave propagation in excitable systems

OLIVER STEINBOCK AND STEFAN C. MÜLLER

17.1 Introduction

A physicist who wants to get a rough estimate of the acoustics of a concert hall just claps his hands and listens to the echoes. With this simple experiment he can obtain valuable information from the acoustic answer of the hall (at least if his ears are well trained). In more technical terms, one perturbs an unknown system with a very short pulse that in the ideal case would be a delta-function describing a perturbation of infintesimal short duration. Regardless of its specific nature, *external perturbations* belong to the most important methods for analyzing unknown systems. In this context, the perturbation by a short pulse is only one example for enforcing a characteristic answer. Other approaches include *periodic* perturbations where systematic variations of *amplitude* and *frequency* open ample possibilities to interrogate the system.

Although external perturbations have long proven to be a valuable tool for science and engineering, many experiments on novel systems studied in young emerging fields of science are dedicated to pure observation as a starting point. Investigations on *spatial pattern formation* in chemical and biological systems are an excellent example for this characteristic development of a young branch of experimental science (Field and Burger, 1985). Most of the current research activities in this field were triggered by the early work of Zhabotinsky and Winfree, who reported the observation of *chemical waves* in a reaction system that today is known as the *Belousov–Zhabotinsky* (BZ) reaction (Zaikin and Zhabotinsky, 1970; Winfree, 1972). A main effort in the subsequent research activities was then the systematic variation of reactant concentrations and the exchange of certain compounds. Most of these important studies focused on the accurate observation of *chemical patterns* and *self-organization* as complex functions of the latter parameters. Today, however, a significant portion of experimental research in this field is benefiting from the possibilities of external control and perturbation (see, e.g., Kapral and Showalter, 1995).

In this chapter, examples for the control and perturbation of chemical wave patterns will be reviewed. The Belousov–Zhabotinsky reaction (Tyson, 1976) serves as an experimental model system, where several features of this system, such as the presence of ionic species or its photosensitivity are used for the purpose of applying external perturbations. The focus of our interest are chemical wave patterns that have the geometry of *spirals* (Winfree, 1974). Both, spiral waves and the BZ reaction are briefly described in the following.

17.2 Spatial patterns in excitable systems

Propagating waves of excitation are known to occur in a variety of chemical or biological systems (Murray, 1993; Kapral and Showalter, 1995). The classical example is the propagation of action potentials along nerve fibers, without which the reader would not be capable of reading this text or turning the pages. The key idea of an excitable medium is that an initially localized impulse spreads through the system by stimulating neighboring areas to a fast response, thus keeping the initial impulse alive. In the wake of this travelling front the system is refractory and cannot be excited again. This is the phase when the system is recovering. In chemical systems the fast response could be caused by autocatalysis, producing a certain chemical compound in a self-accelerating fashion. In excitable reaction–diffusion systems, this autocatalytic activity can propagate due to diffusive coupling with adjacent areas. It usually has a constant wave velocity that is determined by the rate of the autocatalytic reaction and the diffusion coefficient of the autocatalytic compound.

The most striking wave patterns known from (quasi) two-dimensional media are structures consisting of expanding *concentric circles* (target patterns) and *rotating spiral waves* (Tyson and Keener, 1988). Figure 1 shows six examples of spiral waves in quite different systems ranging from heterogeneous catalysis on platinum to slime mold colonies and the mammalian heart (Winfree, 1987; Murray, 1993; Scott, 1995). The occurrence of wave patterns that are not only similar in their geometry but also in their dynamic evolution is not a coincidence. The mathematical description of the underlying local processes and transport phenomena (e.g., chemical reactions and short-range diffusion) gives rise to partial differential equations that have the same nonlinear structure, thus explaining the intriguing similarities exemplified in Figure 1.

Of particular interest is the dynamics of the spiral tip, which is the open wave end in the center of the pattern (Winfree, 1991). The tip location is defined by the point of highest curvature of the isoconcentration line at the average level of concentration. One might think that the rotation of the spiral occurs around a well-defined point that does not change in time. In that case, the reader could easily simulate the dynamics of the spiral by locating the tip position in one of the pictures of Figure 1 and rotating the book around this fixed point. Unfortunately, this simple scheme is wrong. The tip is actually describing a trajectory that is either a circle with a fixed center (rigid rotation) or a curve of higher complexity. Rigid rotation leads to a circular area that is known as the *spiral core* (Müller *et al.*, 1985). This core shows only small concentration changes in time and essentially remains in an unexcited state. Note, that trajectories of higher complexity are not necessarily closed, even if one could wait for an infinite time.

For the Belousov–Zhabotinsky reaction, where spiral waves have wavelengths of the order of 0.1 to 10 mm, various scenarios have been observed (e.g., Nagy-Ungvarai *et al.*, 1993). Under certain experimental conditions, rigid rotation occurs, where the trajectory diameters are typically 0.5 mm or less. In reaction solutions with a very low activity, however, circular trajectories with diameters of 5 mm or larger have been observed. Nevertheless, these observations demarcate extreme cases of BZ wave dynamics. Another mode of spiral wave rotation has been named *meandering*. This slightly misleading term summarizes a class of tip trajectories that are either hypocycloidal or epicycloidal (Winfree, 1991). An example for meandering motion of spirals is shown in Figure 2, where the trajectory (white curve) is superimposed on a single snapshot of the wave pattern. The trajectory has little loops pointing outwards, indicating the involvement of at least two major frequencies. The underlying motion is similar to the motion of our moon that rotates around the Earth, which is again orbiting around the sun. The question of whether the loops are pointing outward or inward is determined by the relative sense of rotation (opposite rotation, for example, gives rise to hypocycles as shown in Figure 2). There is also evidence for various other modes of spiral tip dynamics, including loopy lines or irregular trajectories that might indicate *deterministic chaos* (Winfree, 1991; Nagy-Ungvarai *et al.*, 1993).

17.2.1 The Belousov–Zhabotinsky reaction

One of most intensively studied model systems for the experimental investigation of chemical wave propagation is the BZ reaction. It consists of the bromination of certain organic compounds, such as malonic acid ($CH_2(COOH)_2$) in a sulfuric acid solution employing appropriate redox catalysts. A frequently used catalyst is the redox couple ferroin/ferriin which leads to striking color differences (red/blue) along the profile of chemical waves.

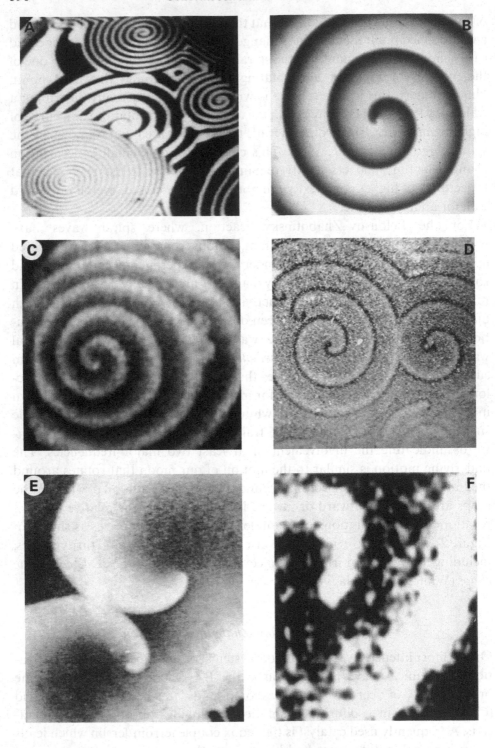

Figure 1 (*opposite*). Six examples of spiral waves in excitable media. (A) Population of spirals with different rotation periods and wavelengths in the catalytic carbon monoxide oxidation on a platinum surface, visualized with photoemission electron microscopy (PEEM). Image area: 450 µm × 400 µm; rotation period of spirals with intermediate wavelength ≈ 20 s (from Nettesheim *et al.*, 1993; reprinted with permission). (B) Spiral wave in the Belousov–Zhabotinsky reaction. The distribution of light intensity reveals the front of oxidized cerium catalyst at $\lambda = 344$ nm as a black band moving through a reduced solution layer of 0.7-mm thickness (bright background). Wavelength: 2.1 mm; rotation period ≈ 40 s. (C) Spiral Ca^{2+}-wave pattern observed in *Xenopus laevis* oocytes (wavelength ≈ 60 µm, period ≈ 3 s). IP_3-mediated Ca^{2+}-release is detected by confocal laser scanning microscopy (from Lechleiter and Clapham, 1992; reprinted with permission). (D) Aggregation of social amoebae in the cellular slime mold *Dictyostelium discoideum* observed with darkfield optics. In bright areas cells move chemotactically toward the spiral core, while in dark bands no directed cell migration is found. Wavelength ≈ 2.5 mm; rotation period ≈ 7 min (from Foerster *et al.*, 1990; reprinted with permission). (E) Colliding spiral-shaped fronts in neuronal tissue: these 'spreading depression' waves on chicken retina are visualized by white light scattered in zones of increased turbidity. The waves are moving through the otherwise transparent medium at a speed of 2.2 mm/min. Image area: 9 mm × 10 mm. (F) Clockwise rotating wave in a slice (20 mm × 20 mm × 0.5 mm) of isolated canine cardiac muscle, visualized by use of a potentiometric dye (with fluorescence excited at 490 nm; measured at 645 nm). Rotation period: 180 ms (from Davidenko *et al.*, 1992; reprinted with permission).

These differences can be readily detected by two-dimensional spectrophotometry (Müller *et al.*, 1985). In this technique the absorption at a given wavelength (usually 490 nm for maximal contrast) is detected as a function of space and time by using video cameras. Sequences of video images are digitized and then available for computer analysis.

A catalyst that has proven to enable a valuable modification of the BZ system is the metalloorganic ruthenium complex $[Ru(bpy)_3]^{2+/3+}$ (Gaspar *et al.*, 1983). Using this catalyst the reaction system becomes *photosensitive*, thus opening fascinating possibilities for external control and perturbation by light (Kuhnert *et al.*, 1989).

The chemical mechanism of the BZ reaction has been discussed in great detail elsewhere (Field and Burger, 1985; Scott, 1995). Therefore, we simply want to mention the most important chemical species involved in wave propagation. The unstable compound $HBrO_2$ acts as the autocatalytic propagator species transmitting local excitations by diffusion and is unfortunately not easily detectable. Bromide ions play the role of the inhibitor, and, last but not least, the concentration of the oxidized catalyst defines a control variable that regulates the recovery process of the system.

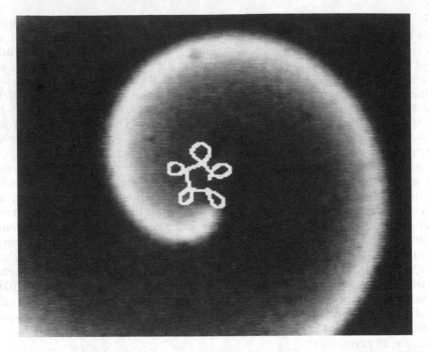

Figure 2. Snapshot of a rotating spiral wave in the Belousov–Zhabotinsky (BZ) reaction. The bright and dark regions indicate high and low concentrations of the oxidized catalyst, respectively. The trajectory of the spiral tip (overlaid white curve) is similar to a hypocycloid. Such noncircular trajectories are typical for spiral meandering. In this particular experiment a photosensitive ruthenium complex catalyzes the reaction. Image area: 3.8 mm × 3.0 mm.

17.3 External control of excitation waves

In the following, three major approaches for the external control and perturbation of excitation waves are presented. The presentation begins with a rather modest example of external control that is established by variations of reactant concentrations. The subsequent section describes recent research activities aimed at control with electric fields. Finally, local and global perturbations of photosensitive BZ systems are presented.

17.3.1 Tuning chemical parameters

Chemical oscillations in well-stirred BZ solutions reveal pronounced dependencies on the initial concentrations of sulfuric acid, bromate, bromide, and malonic acid (Field and Burger, 1985). An increase in sulfuric acid concentration, for example, usually speeds up the oscillations, and the oscillation frequency might increase by a factor of 10 or more. Molecular oxygen is a chemical parameter that is often ignored in the discussion of experiments on spatiotemporal pattern formation in BZ systems. Dissolved oxygen reacts with organic radicals that are formed during the oxidation of malonic acid and bromomalonic acid by the oxidized catalyst (Neumann *et al.*, 1995). While the detailed chemistry of these reactions is not yet understood, there are several studies reporting an overall trend of oxygen-induced inhibition of oscillations and wave patterns. For spatially homogeneous systems this means that oscillation periods are usually increased or oscillations are suppressed. In spatially extended media, such as thin layers of the reaction solution or thin BZ gel systems, wave propagation is either suppressed or propagation velocities are decreased due to the presence of oxygen. More interestingly, recent experiments have shown that thin layers of the aerobic BZ medium (that usually behave as a quasi two-dimensional system) can undergo *stratification* (typical thickness: 1 mm; Zhabotinsky *et al.*, 1991). This stratification gives rise to two, even thinner, sublayers of excitable medium that develop independent wave patterns with only weak interactions. A typical example for the resulting patterns is shown in Figure 3. It should be emphasized again that the visual impression of crossing waves is misleading, since the patterns evolve in two separate layers. Interference of waves, common in acoustic or electromagnetic systems, does not occur in excitable BZ media, because excitation waves annihilate upon mutual collision. The stratification of thin BZ gel systems is strongly related to an oxygen concentration gradient that is formed by the interplay of diffusive inflow from the atmosphere into the gel system and the consumption of oxygen within the BZ system. Furthermore, one of the BZ intermediates (bromine) is leaving the gel by diffusion into the atmosphere. Both compounds, oxygen and bromine, act as inhibitors of wave propagation. We are just beginning to understand the details of this interesting interaction, but the described phenomena seem to lead to a nonmonotonic profile of excitability that induces the observed stratification (Zhabotinsky *et al.*, 1994).

We believe that the role of molecular oxygen in the context of the BZ reaction opens a variety of quite promising research activities, in particular for control experiments. The involved free-radical chemistry is far from being understood and implies challenges for future kinetic and mechanistic investigations. Furthermore, oxygen is an intriguing chemical parameter that gives rise to unexpected modes of self-organization in the BZ reaction such as the described stratification of thin BZ gel systems. As we will discuss in the following, this behavior may also assist other experimental approaches on external control and perturbation exploiting photosensitivity.

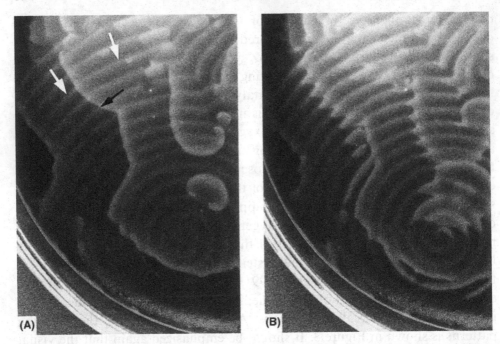

Figure 3. Under certain experimental conditions and in the presence of oxygen, thin-layered BZ gel systems can undergo a vertical stratification. The top view (A) shows the formation of two pattern-forming layers, where wave propagation proceeds without strong interaction. White (bottom layer) and black arrow (top layer) indicate an example of these 'crossing' waves. In the course of the reaction the chemical patterns begin to couple, giving rise to sawtooth-shaped fronts (B). Interval between photographs: 40 s (Zhabotinsky *et al.*, 1991).

17.3.2 Control of excitation waves by electric fields

Numerous chemical species in the complex BZ reaction mechanism are of an ionic nature. The central species are the bulky iron-complex ferroin that carries a positive charge of two or three depending on its oxidation state and the small negatively charged bromide ion. The autocatalytic species $HBrO_2$ is another important actor taking part in the chemical events that lead to wave propagation. This species, however, is electrically neutral. The question we discuss in this section is: what happens to propagating waves and rotating spirals if an *external electric field* is applied?

In 1981 Feeney *et al.* performed experiments in which they applied parallel electric fields ($E \approx 10$ to $50\,V/cm$) to spatially extended BZ media. They observed an increased velocity of waves propagating toward the positive anode, while waves propagating toward the cathode were decelerated. Sevcikova and Marek (1984) continued this work and found more recently that, at higher field strength, waves can reverse their propagation direction or split in a fairly complex fashion (Sevcikova *et al.*, 1992, 1996).

Figure 4 illustrates the effect of an externally applied electric field on spiral waves in the ferroin-catalyzed BZ reaction. In these experiments a constant electric field ($E = 0$ to $6\,V/cm$) was applied to the BZ gel system via two parallel electrodes that were realized as simple salt bridges to avoid the contamination of the BZ medium by products of undesired electrochemical reactions. One central finding from this work demonstrated that spiral waves are drifting toward the anode (Steinbock *et al.*, 1992). Figure 4A shows a typical snapshot of a pair of drifting spiral waves (anode oriented parallel to the bottom side of the figure). The overall drift toward the anode has an additional component perpendicular to the field. The direction of this perpendicular drift depends on the chirality of the spirals, as shown in Figure 4B. Although the trajectory of both spirals points in the x-direction, one finds that the clockwise rotating pattern is also pulled to the left, while the counterclockwise spinning wave is pulled to the right. Switching off the electric field stops the drift immediately. Changing the polarity of the field causes a drift back toward the initial locations. For typical experimental conditions, such as those of the experiments described in Figure 4, the drift velocity of spiral tips is found to be in the range 0 to $0.3\,mm/min$. Notice, that the drift does not occur along a straight line, but is rather characterized by a continuous trajectory with successive loops. Although present, these loops are not fully resolved in the experimental data displayed in Figure 4B.

Another interesting phenomenon that has been observed in experiments on electric field-induced spiral drift is the deformation of spiral geometry (Steinbock *et al.*, 1992). While spiral waves in unperturbed systems have usually an Archimedian shape (i.e., constant pitch), the drift of its tip is generating variations of the wavelength. The deviations from the unperturbed wavelength reach a maximum in the back of the drift direction (compare Figure 4A). Apparently, the increased wavelength is due to a Doppler effect. Furthermore, the shape of the drifting spiral tip can vary significantly during different phases of its rotation. Depending on the direction of its relative orientation to the electric field vector, the curved tip is either elongated or strongly curled.

The perturbation of chemical wave patterns by electric fields not only opens an interesting field of research but can also be used for the external control of wave propagation. An intriguing example for exploiting spiral drift is discussed in the following. Since, it is possible to induce spiral drift, one should be able to send a pair of spirals on a collision course. Figure 5 gives a sequence of four snapshots illustrating the outcome of a *spiral wave collision* (Schütze *et al.*, 1992). Snapshot (A) shows the initial wave pattern consisting of a pair of

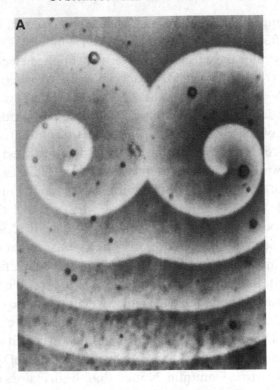

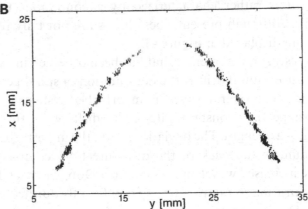

Figure 4. A pair of spiral waves in the ferroin-catalyzed BZ system is perturbed by a constant electric field. Field lines are parallel and oriented vertically with the anode located at the bottom side of the figure. The electric field is inducing a spiral drift toward the anode and a strong deformation of the Archimedian spiral geometry. (A) Snapshot of the pair of drifting spirals. (B) Trajectory of the corresponding spiral tips. Notice, that the drift direction is also influenced by the sense of spiral rotation (Steinbock *et al.*, 1992).

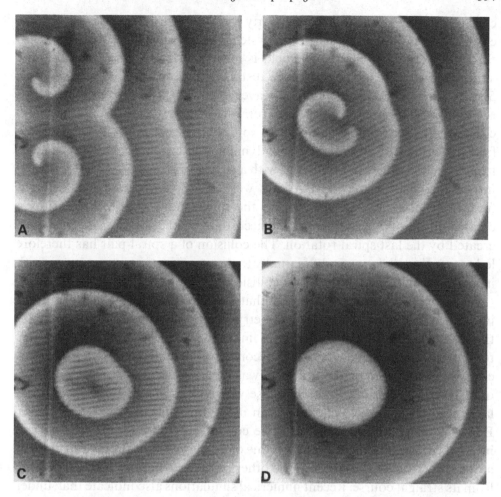

Figure 5. Evolution of spiral wave annihilation due to electric field-induced drift. The electric field is oriented with the anode to the right and the cathode to the left. Opposite perpendicular motions of spiral tips in the initial structure (A) reduce their relative distance (B). Annihilation occurs when the separation distance of the tips is below a certain critical size (C). In a truly excitable system, the remaining circular waves would continue to propagate outward leaving behind a 'quiet' region without chemical activity. Since this particular BZ medium can show autonomous oscillations, we were able to observe the birth of a nonrotating trigger wave (D) in the central region of spiral wave annihilation. The location of its pacemaker is determined by the local phase information created by the last spiral rotation (Schütze *et al.*, 1992).

counterrotating spirals that have nearly identical size and phase. A constant electric field is applied to the BZ gel system with the anode located parallel to the right side of the figure. The field is now pulling the spirals towards the anode and is decreasing the relative distance between the spiral tips (Figure 5B). At a certain critical distance ($\ll$ wavelength) the spiral tips annihilate, thus removing the spinning pacemakers from the system (C). The resulting unexcited area in the former region of tip rotation triggers a new pattern of low frequency (Figure 5D), since the bulk dynamics of this particular BZ system is, in fact, not excitable but oscillatory with a rather long oscillation period. These intrinsic oscillations of the BZ bulk were earlier suppressed by the high-frequency spiral waves. Now, where the spirals have vanished, the system creates an autonomous pacemaker according to the local phase gradients created by the last spiral rotation. The collision of a spiral pair has therefore led to *spiral annihilation* and eventually to the creation of a target pattern, with its pacemaker located in the spiral collision region.

Additional experiments revealed that the electric field strength and the initial geometry (i.e., phase and symmetry) determine whether spiral annihilation can occur. If the initial symmetry line between both spirals is not parallel to the electric field, annihilation becomes more unlikely. Furthermore, a critical field strength of about 2 V/cm was found below which no annihilation occurred. Disregarding the trivial case where the spirals pass each other at large distance, weak electric fields can induce another unexpected response. Experiments showed that under these conditions colliding spiral waves can interact without annihilation. While one of the spiral tips is proceeding its drift along the expected straight line, the other tip is forced to deviate significantly from its straight course. Recent numerical simulations also indicate that, under ideal conditions, two spirals can form a bound state that keeps the tips at a constant average distance (of the order of one wavelength) and moves them toward the anode along a trajectory that is parallel to the electric field vector (Schmidt and Müller, 1997).

We want to point out some striking similarities between spiral waves and particles. A single, one-armed spiral has a topological charge (either $+1$ or -1) that is determined by its sense of rotation. It is a well-known fact that the total sum of topological charges in an excitable medium (without boundaries) is conserved. Therefore, only spirals having a different sense of rotation can annihilate. Depending on the sign of its topological charge, the drifting spiral tip is experiencing different contributions to the vector of drift velocity. From a phenomenological point of view this behavior is quite similar to moving electrically charged particles that are subjected to a magnetic field. Furthermore, interacting spirals can form a bound state that has a topological charge of $1 - 1 = 0$. This spiral pair is drifting with a velocity that has no component perpendicular to the field, similar to an uncharged atom that is experiencing no magnetic forces that would alter its velocity. It will be interesting to see how far the concept of particles can be used fruitfully for understanding the dynamics of spiral waves in excitable systems.

The microscopic driving force of the observed phenomena is *electromigration* of ionic species – an effect that is exploited by certain analytical techniques such as electrophoresis. Ions are pulled towards the electrode along the field lines of the electric field. This force can result in a local change of the concentration c that is given by the equation

$$\partial c/\partial t = \mu \mathbf{E} \mathbf{V}_c, \qquad (1)$$

where $\mathbf{E}$ is the electric field vector and $\mathbf{V}_c$ the gradient vector of the local ion concentration oriented parallel to the line of highest concentration increase. The parameter μ is the ionic mobility of the charged species defined as the proportionality factor between the velocity of ions and the applied electric field (Atkins, 1994). Hence, the value of $\mu \mathbf{E}$ corresponds to the migration velocity of the particular ion in an electric field, E. Note that the electric field causes concentration changes only in the presence of a local concentration gradient. By adding the right-hand term of Equation (1) to the set of reaction–diffusion equations describing the chemical system, one obtains a mathematical model that is suitable for the numerical simulation of experimental data. On this basis, numerous numerical studies have been carried out to reproduce and understand the effects of electric fields on pattern formation in chemical systems. Generally, these investigations are in good agreement with experiments, although certain quantitative features of numerical studies are not quite satisfactory yet.

Bromide ions seem to be the key species of electric field effects in the BZ reaction. Electric fields are forcing the inhibitor bromide to migrate toward the anode following the mathematical description discussed above. A BZ wavefront traveling toward the cathode is now experiencing an electric field-induced flow of bromide that is slowing down its propagation. On the other hand, a front traveling toward the anode is always expanding into regions of lowered inhibitor concentration and, therefore, propagating with an increased velocity. The presented drift of spiral waves is directly related to this effect.

17.3.3 Exploiting photosensitivity

While the perturbation of spiral rotation by electric fields is an example for perturbations by a *vectorial* quantity, the following section describes a powerful approach for the control by a *scalar* quantity, which is the intensity of an external light field. As mentioned before, it has been found that the $Ru(bpy)_3^{2+}$-catalyzed BZ reaction is photosensitive (Gaspar *et al.*, 1983). By illuminating the system with visible light one can stimulate the production of additional bromide, which is inhibiting the excitable behavior of the system. The maximum effect is established at a wavelength of about 454 nm via the formation of a photochemically excited ruthenium complex (Jinguji *et al.*, 1992). The detailed chemistry of the light-induced inhibition is not understood yet. Nevertheless, working models exist that were derived from the Oregonator model of the BZ reaction by adding an additional source of bromide, where the rate of bromide production is assumed to be simply proportional to the light intensity (Krug *et al.*, 1990).

17.3.3.1 Control with a laser spot

The initial goal of our experiments with the light-sensitive BZ reaction was to develop tools for the controlled creation of spiral waves. An argon-ion laser of relatively high intensity (0.8 W attenuated by a neutral density filter, OD 2.0) is used to irradiate small circular areas (typical diameter 0.1–2 mm) of the BZ medium. These areas are now inhibited in the sense that, locally, wave propagation becomes impossible. The time required for reaching efficient inhibition is short (about 2 s) and the effect is reversible. If a solitary chemical wave is propagating across the inhibited spot, it breaks apart and two open wave ends are generated. After the laser is switched off, these wave ends develop into a pair of rotating spirals (Steinbock and Müller, 1992). Hence, the argon laser can be used as a highly controllable tool for generating spirals. Furthermore, it is possible to pin spiral tips to laser-inhibited spots. In this context the laser spot acts as an artificial spiral core that forces a circular wave rotation around the boundary of the inhibited spot. By these means, the spiral rotation period and the wavelength of the chemical pattern can be adjusted to desired values. Figure 6 shows a pair of spiral waves in the photosensitive BZ reaction where this laser control is carried out. The unperturbed spiral (left) rotates with a characteristic period of 26 s and has a wavelength (pitch) of 1.3 mm. The perturbed spiral (right) rotates around an unexcitable laser spot having a diameter of 1.2 mm. The perturbation leads to an increase in period to 49 s as well as in wavelength to 3.4 mm. In the subsequent stages of the experiment the area covered by the perturbed spiral is continuously decreased. Eventually, the spiral on the left conquers the whole observation area, leaving a defect at the position of the laser spot. A series of experiments confirmed that the period of spiral rotation (and also the wave velocity) increases monotonically, the diameter of the laser spot constituting the artificial rotation center (artificial core).

Figure 6. Argon-ion lasers allow an efficient manipulation of spiral waves in the photosensitive BZ reaction. This example shows a pair of spiral waves, where the left spiral is unperturbed and rotates freely. The right spiral spins around a small laser spot (diameter 1.2 mm) that creates an unexcitable disk (artificial core). This constant perturbation induces an increase in rotation period and wavelength (Steinbock and Müller, 1992).

An additional function of the argon laser allows shifting of the center of the spirals through the system. If a spiral tip is pinned by the laser spot, one can slowly move the BZ probe and prevent further rotation (Steinbock and Müller, 1993). It is crucial that the probe is translated with a speed that is roughly identical with the propagation velocity of planar waves. Under this condition the tip is experiencing a permanent obstacle and follows the relative motion of the laser spot. We found this technique to be a useful tool for removing undesired spirals from the observation area and thus preventing the interaction of spirals that might otherwise influence the measurement. In addition, we found that a spiral that is moved to the physical boundary of the BZ system (e.g., the boundary of a Petri dish) is transformed into a defect rotating around the entire circumference of the probe. Using this procedure one can accumulate numerous spirals of identical topological charge in the central region of the system and further use the laser beam for creating multi-armed spirals (Steinbock and Müller, 1993).

17.3.3.2 Control with periodic light stimuli

A quite different approach for exploiting the photosensitivity of the ruthenium-catalyzed BZ reaction focuses on spatially homogeneous perturbations that are modulated in time (Steinbock *et al.*, 1993). We investigated a system in which spiral tips are describing a roughly five-lobed trajectory if subject to an average (comparatively low) light intensity (compare Figure 2). For the minimum and maximum intensities applied in the following example, we observed four- and six-lobed trajectories, respectively. Periodic modulations of the light intensity within these bounds cause dramatic changes in the dynamics of spiral rotation. The small variations in the trajectory shape that occur during each period of modulation accumulate with time. As a result, the modulated perturbation forces the tip to follow trajectories that differ significantly from those observed at constant intensities. The shape depends strongly on the modulation period, T_m. The underlying effects are strongly related to the *entrainment phenomena* discussed by Kaiser (Chapter 1, this volume). In contrast to the nonlinear oscillators discussed by Kaiser, however, we now have to understand the response of a *spatiotemporal* system, having an infinite number of degrees of freedom, to periodic perturbations.

Figure 7 shows five examples of tip trajectories of spiral waves traced in a BZ system under periodic modulation of light intensity. The trajectories of Figure 7B–D are members of an entrainment band of hypocyloids with one lobe corresponding to one external period. The number of lobes continuously increases from three to more than 12 with increasing values of T_m. The exact one-to-one agreement between the number of lobes and the number of modulation cycles has been observed only in a well-defined interval of modulation periods ($T_m = 20$ to 35 s). For smaller periods a quite different behavior occurs, as shown in Figure 7A. This trajectory is a deformed five-lobed curve with exactly one lobe described during two modulation periods. Figure 7E illustrates one possible mode of spiral tip dynamics at slow modulation, showing a surprising trajectory with alternating distances between neighboring lobes. In this small frequency range the spiral tip describes a pair of lobes during one external modulation. For periods between those shown in Figure 7D and E, we observed irregular motion with epicyclic segments of the trajectories. In all examples given in Figure 7, however, the tip motion is synchronized (or entrained) by the external rhythm.

Numerical simulations of reaction–diffusion systems are useful supplements for experimental investigations, since they allow a thorough scan through a broad parameter region in a tolerable time. Based on a modified Oregonator model describing the photosensitive BZ medium, V. Zykov has carried out

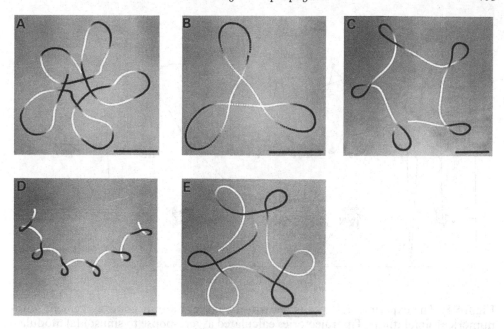

Figure 7. Sequence of spiral tip trajectories measured under a sinusoidal modulation of light intensity with a period T_m of: 17.0 s (A), 26.2 s (B), 30.4 s (C), 34.5 s (D), and 52.2 s (E). The shading of the trajectories indicates the current intensity of illumination and clearly shows the one-to-one phase locking of the lobes to the external rhythm ((B)–(D)). The traces in (A) and (E) obey a 1:2 and a 2:1 phase locking, respectively. Period of unperturbed spiral (0.93 mW cm^{-2}) as measured in its symmetry center: $T_0 = 24.5$ s. Scale bar: 0.2 mm (Steinbock *et al.*, 1993).

extensive computer simulations in order to achieve a more complete picture of the entrainment of meandering spiral waves (Zykov *et al.*, 1994). Figure 8 shows some tip trajectories obtained numerically by changing the modulation period, T_m, as well as the modulation amplitude, A. The value T_0 denotes the wave period at the center of the unperturbed spiral. Notice that, owing to the complex motion of a meandering spiral wave, different excitation periods are detected at the center and at infinite distance (T_∞). For a n-lobed hypocycloidal trajectory the periods obey the equation: $T_0 = T_\infty(n-1)/n$. The reader can easily perform 'experiments' to check this dependence. Take a pen (or better a slightly curved object) that will represent the spiral tip. Then move the pen along the four-lobed tip trajectory (i.e., $n = 5$) shown in Figure 2 until you are back to the starting point. The pen should be always perpendicular to the trajectory in the plane. While carrying out this little test count how often the pen crosses the symmetry center of the trajectory and how often it crosses the upper right corner of the figure which is representing a point far from the center. The reader will find these numbers to be five and four corresponding to

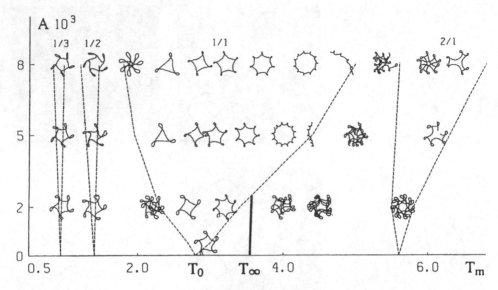

Figure 8. The experimental results shown in Figure 7 are in good agreement with numerical simulations. Tip trajectories calculated as a response to sinusoidal modulation of bromide production are shown in the plane of modulation period, T_m, and modulation amplitude, A. The center of each tip trace is arranged at the actual parameter pair (T_m, A) used for calculations. Their size was varied to allow a better resolution of details. Dashed lines indicate boundaries of entrainment bands related to different ratios $n:m$, where n is the number of lobes per m periods of modulation. T_0 and T_∞ are the excitation periods of the unperturbed spiral as measured in its symmetry center and at a point far away from the center, respectively. The thick black line indicates parameters for which resonance drift was found (for further details, see text in Section 17.3.3.2; Zykov *et al.*, 1994).

the periods $T_0 = \tau/5$ and $T_\infty = \tau/4$ that actually fulfill the above equation (τ represents the total time it takes to move the pen).

Coming back to the numerical results shown in Figure 8 one finds that the entrainment band of trajectories that obey a one-to-one phase locking is located around the period $T_m = T_0$. The trajectories of this band correspond to the experimental observations shown in Figure 7B–D. Figure 8 depicts three additional entrainment bands where the number of lobes and the number of modulation cycles have the constant ratios $1:3$, $1:2$ and $2:1$. Also notice the good agreement between the simulated curves in the $1:2$ and $2:1$ band and the experimental data shown in Figure 7A and B, respectively. Similar to the experimental observations the simulations also reveal irregular trajectories in between the entrainment bands. The corresponding motion in these parameter regions shows no phase-locking with respect to the rhythm of external modulation.

The intrinsic period T_0 is of central importance for the structure of entrainment bands. An obvious question to ask now is whether additional phenomena occur at a modulation period of $T_m = T_\infty$. Figure 9A shows a typical example of a spiral tip trajectory observed for this particular forcing period (Grill *et al.*, 1996). The experimental data reveal a complex drift of the tip. Despite the complex local structure of the trajectory one finds an overall translation of the tip along a straight line. This effect has been named *resonance drift* and should be clearly distinguished from entrainment phenomena. The parameter region, where resonance drift is found, has been indicated in Figure 8 as a thick black line. Resonance drift can be readily understood for the example of rigid rotation (i.e., circular tip trajectories). If the external modulation is limited to situations in which the spiral is showing rigid rotation, one observes a periodic increase and decrease of the radius of the trajectory. If the modulation period is furthermore identical with the rotation period of the

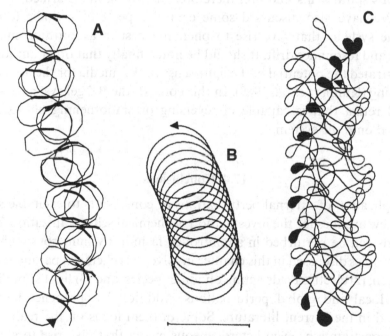

Figure 9. Resonance drift of spiral waves differs significantly from the dynamics of entrainment patterns shown in Figures 7 and 8. Parts (A) and (C) give a typical example of resonance drift of a meandering spiral as obtained from experiments (A) and numerical calculations (C). The simpler resonance behavior of a spiral wave with rigid rotation is shown in (B). While entrainment is mainly determined by the period T_0 of the unperturbed pattern, resonance drift occurs for modulation periods that are approximately identical with the excitation period T_∞. Scale bar: 0.8 mm (Zykov *et al.*, 1994; Grill *et al.*, 1996).

unperturbed spiral, a drift along a straight line is generated. A typical example for this simple resonance drift is shown in Figure 9B.

For a further comparison of experimental and numerical data, computer simulations were performed (Zykov *et al.*, 1994) that yielded the precise trajectories of spiral tips under resonance conditions (e.g., compare the black line at $T_m = T_\infty$ in Figure 8). Figure 9C shows the resonance trajectory found for $A = 0.0001$, which is in good qualitative agreement with our experimental observations.

The presented experimental results demonstrate that the photosensitive BZ reaction is an excellent model system for the investigation of excitable reaction–diffusions systems by means of external control and perturbation. First of all, spiral waves can be generated in a reproducible fashion and their location in the system can be chosen at will. Secondly, artificial cores, created by unexcitable laser-irradiated spots, can be exploited for the control of the rotation period and the wavelength of spiral waves. These cores can also host numerous spiral arms and are, therefore, stabilizing multi-armed wave patterns. We have also discussed some central aspects of periodic forcing of excitable systems that give rise to phenomena such as entrainment, phase locking and resonance drift. It should be noted finally that other authors have demonstrated the potential of the light-sensitive BZ media for the use of image processing (Kuhnert *et al.*, 1989). In this context, the BZ gel system acts as an artifical retina that is capable of reversing or smoothening photographies projected onto the system.

17.4 Conclusions

The application of external perturbations or constraints to excitable systems opens new avenues for the investigation of chemical self-organization. The BZ reactions can be perturbed in a controlled fashion by numerous techniques. The examples discussed in this chapter involved: (1) chemical parameters, such as oxygen; (2) electric fields acting on ionic species; and (3) light stimuli giving rise to local and global perturbations. Additional approaches have been discussed in the current literature. Some of them focus on the realization of spatial constraints or even heterogeneous media that give rise to surprising wave patterns (Steinbock *et al.*, 1995).

All of these research activities have the potential to yield important insights into the rules that govern *biological self-organization*, since living systems are strongly influenced by periodic rhythms and constant gradients in their natural environment (such as circadian rhythms). Other perturbations might occur from artificial sources. Our knowledge, however, of the response of nonlinear and, in particular, living systems to external forcing is still relatively poor. A major complication is that the response of nonlinear systems is often hard to anticipate by human intuition, since our intuition relies strongly on linear extrapolation. On the basis of the intriguing similarities between certain biological and chemical systems (compare Figure 1) it seems reasonable to broaden our understanding of self-organization by highly reproducible experiments with systems such as the Belousov–Zhabotinsky reaction.

Acknowledgment

We thank V. S. Zykov and H. Sevcikova for useful discussions.

References

Atkins, P. (1994) *Physical Chemistry*. New York: Freeman.

Davidenko, J. M., Pertsov, A. V., Salomonsz, R., Baxter, W. and Jalife, J. (1992) Stationary and drifting spiral waves of excitation in isolated cardiac muscle. *Nature* 355: 349–351.

Feeney, R., Schmidt, S. and Ortoleva, P. (1981) Experiments on electric field-BZ chemical wave interaction: annihilation and the crescent wave. *Physica D* 2: 536–544.

Field, R. J. and Burger, M. (eds.) (1985) *Oscillations and Traveling Waves in Chemical Systems*. New York: Wiley.

Foerster, P., Müller, S. C. and Hess, B. (1990) Curvature and spiral geometry in aggregation patterns of *Dictyostelium discoideum*. *Development* 109: 11–16.

Gaspar, V., Bazsa, G. and Beck, M. (1983) The influence of visible light on the Belousov–Zhabotinskii oscillating reaction applying different catalysts. *Z. Phys. Chem.* 264: 43–48.

Grill, S., Zykov, V. and Müller, S. C. (1996) Spiral wave dynamics under pulsatory modulation of excitability. *J. Phys. Chem.* 100: 19082–19088.

Jinguji, M., Ishihara, M. and Nakazawa, T. (1992) Primary process of illumination effect on the $Ru(bpy)_3^{2+}$-catalyzed Belousov–Zhabotinskii reaction. *J. Phys. Chem.* 96: 4279–4281.

Kapral, R. and Showalter, K. (eds.) (1995) *Chemical Waves and Patterns*. Dordrecht: Kluwer.

Krug, H.-J., Pohlmann, L. and Kuhnert, L. (1990) Analysis of the modified complete Oregonator accounting for oxygen sensitivity and photosensitivity of the Belousov–Zhabotinsky systems. *J. Phys. Chem.* 94: 4862–4866.

Kuhnert, L., Agladze, K. I. and Krinsky, V. I. (1989) Image processing using light-sensitive chemical waves. *Nature* 337: 244–247.

Lechleiter, J. D. and Clapham, D. E. (1992) Molecular mechanisms of intracellular calcium excitability in *Xenopus laevis* oocytes. *Cell* 69: 283–294.

Müller, S. C., Plesser, T. and Hess, B. (1985) The structure of the core of the spiral wave in the Belousov–Zhabotinskii reaction. *Science* 230: 661–663.

Murray, J. D. (1993) *Mathematical Biology*. Berlin: Springer-Verlag.

Nagy-Ungvarai, Zs., Ungvarai, J. and Müller, S. C. (1993) Complexity in spiral wave dynamics. *Chaos* 3: 15–19.

Nettesheim, S., von Oertzen, A., Rotermund, H. H. and Ertl, G. (1993) Reaction diffusion patterns in the catalytic CO-oxidation on Pt(110): front propagation and spiral waves. *J. Chem. Phys.* **98**: 9977–9982.

Neumann, B., Hauser, M., Steinbock, O., Simoyi, R., Dalal, N. and Müller, S. C. (1995) Identification and kinetic study of the peroxymalonyl radical in the aerobic oxidation of malonic acid by Ce(IV). *J. Am. Chem. Soc.* **117**: 6372–6373.

Schmidt, B. and Müller, S. C. (1997) Forced parallel drift of spiral waves in the Belousov–Zhabotinsky reaction. *Phys. Rev. E* **55**: 4390–4393.

Schütze, J., Steinbock, O. and Müller, S. C. (1992) Forced vortex interaction and annihilation in an active medium. *Nature* **356**: 45–47.

Scott, S. K. (1995) *Oscillations, Waves, and Chaos in Chemical Kinetics.* Oxford: Oxford University Press.

Sevcikova, H. and Marek, M. (1984) Chemical front waves in an electric field. *Physica D* **13**: 379–386.

Sevcikova, H., Marek, M. and Müller, S. C. (1992) The reversal and splitting of waves in an excitable medium caused by an electrical field. *Science* **257**: 951–954.

Sevcikova, H., Schreiber, I. and Marek, M. (1996) Dynamics of oxidation Belousov–Zhabotinsky waves in an electric field. *J. Phys. Chem.* **100**: 19153–19164.

Steinbock, O., Kettunen, P. and Showalter, K. (1995) Anisotropy and spiral organizing centers in patterned excitable media. *Science* **269**: 1857–1860.

Steinbock, O. and Müller, S. C. (1992) Chemical spiral rotation is controlled by light-induced artificial cores. *Physica A* **188**: 61–67.

Steinbock, O. and Müller, S. C. (1993) Multi-armed spirals in a light-controlled excitable reaction. *Int. J. Bifurc. Chaos* **3**: 437–443.

Steinbock, O., Schütze, J. and Müller, S. C. (1992) Electric-field-induced drift and deformation of spiral waves in an excitable medium. *Phys. Rev. Lett.* **68**: 248–251.

Steinbock, O., Zykov, V. and Müller, S. C. (1993) Control of spiral-wave dynamics in active media by periodic modulation of excitability. *Nature* **366**: 322–324.

Tyson, J. J. (1976) *The Belousov–Zhabotinskii Reaction. Lecture Notes in Biomathematics,* Vol. 10. Berlin: Springer-Verlag.

Tyson, J. J. and Keener, J. P. (1988) Singular perturbation theory of travelling waves in excitable media (a review). *Physica D* **32**: 327–361.

Winfree, A. T. (1972) Spiral waves of chemical activity. *Science* **175**: 634–636.

Winfree, A. T. (1974) Rotating chemical reactions. *Sci. Am.* **230**: 82–96.

Winfree, A. T. (1987) *When Time Breaks Down.* Princeton, NJ: Princeton University Press.

Winfree, A. T. (1991) Varieties of spiral wave behavior: an experimentalist's approach to the theory of excitable media. *Chaos* **1**: 303–308.

Zaikin, A. N. and Zhabotinsky, A. M. (1970) Concentration wave propagation in two-dimensional liquid-phase self-oscillating system. *Nature* **225**: 535–537.

Zhabotinsky, A. M., Györgyi, L., Dolnik, M. and Epstein, L. (1994) Stratification in a thin-layered excitable reaction–diffusion system with transverse concentration gradient. *J. Phys. Chem.* **98**: 7981–7990.

Zhabotinsky, A. M., Müller, S. C. and Hess, B. (1991) Pattern formation in a two-dimensional reaction–diffusion system with a transversal chemical gradient. *Physica* **49**: 47–51.

Zykov, V., Steinbock, O. and Müller, S. C. (1994) External forcing of spiral waves. *Chaos* **4**: 509–518.

18

Changing paradigms in biomedicine: implications for future research and clinical applications

JAN WALLECZEK

18.1 Introduction

A revolution is underway in the physical sciences, based on insights from nonlinear dynamics, which includes the areas popularly known as chaos and complexity studies. As described in the previous chapters, this revolution is beginning to affect greatly the biological and medical sciences as well. Prominent examples are the discovery of deterministic chaos in physiological time series data, of fractal properties of living processes, and of dynamical information processing in single cells and coupled cell signaling networks. The key feature of this work is the treatment of a living system as a dynamical system of nonlinearly interacting elements. As I have proposed in the Introduction to this book, the field of *biodynamics* might therefore be defined as the study of the complex web of nonlinear dynamical interactions between and among molecules, cells and tissues, which give rise to the emergent functions of a biological system as a whole.

This concluding chapter reviews major characteristics of this quickly developing research area and explores implications for basic research, clinical applications and biological thinking. Before highlighting selected findings that emphasize the nonlinear dynamical view, I first draw attention to the nonequilibrium foundations of living processes.

18.2 Life as a dynamical, nonequilibrium process

Many standard textbooks of cell biology or biochemistry still stress ideas that are based upon biochemical reaction–diffusion processes for equilibrium conditions in closed systems. Yet, we now know that the decay to biochemical equilibrium is a poor representation of living systems. As discussed in this book, the complex, dynamical organization of matter that we recognize as life, which does not exist as units smaller than the cell, requires a continuous inflow of energy and matter; hence living systems are thermodynamically open at all levels in the hierarchy of biological organization. In any open system, whether it is physical, chemical or biological in nature, mechanisms such as positive or negative feedback, autocatalysis and time delays may generate self-organized dynamical behaviors. Furthermore, energy is continuously dissipated during the generation of such self-stabilizing states – also known as dissipative structures – and in cells they prevent biochemical reactions from reaching thermodynamic equilibrium (see, e.g., Prigogine and Nicolis, 1971). Thus, as Schrödinger (1944) put it in his famous essay *What is Life?* over 50 years ago, 'living matter evades the decay to equilibrium'. Indeed, the only time when a cell ever reaches equilibrium is when it has died.

The view put forward in this book suggests that it is time for mainstream biochemistry and biology to embrace the fact that the nonequilibrium, nonlinear character of intra- and extracellular control processes is crucial to biological function and regulation. Specifically, any comprehensive model of regulatory biochemical dynamics, the complex responsiveness to stimuli, and the information-processing capacity of living systems must be able to account for the out-of-equilibrium, dissipative nature of biological processes. It is therefore of interest to discuss the impact of this understanding on the classical notion of homeostatic, quasi-equilibrium regulation from a historical perspective.

18.3 From homeostasis to homeodynamics?

The traditional view of physiological control systems in organisms and single cells holds that they are governed by *homeostasis*. The concept that physiological parameters are feedback regulated so that they remain close to a *constant* value has a long history. Seventy years ago, Cannon (1929) introduced the term 'homeostasis' to describe the 'coordinated physiological arrangements for attaining constancy' in a biological organism. Cannon sought to coin a term that could appropriately define the general condition of the living state as described, for example, by Richet (1900):

The living being is stable. It must be in order not to be destroyed, dissolved or disintegrated by the colossal forces, often adverse, which surround it. By an apparent contradiction it maintains its stability only if it is excitable and capable of modifying itself according to external stimuli and adjusting its response to the stimulation. In a sense it is stable because it is modifiable – the slight instability is the necessary condition for the true stability of the organism.

The true scope and power of *instabilities* in biological function and regulation, however, could only be recognized several decades later as a result of the application to biological studies of concepts from nonequilibrium thermodynamics (for an early perspective, see Prigogine and Nicolis, 1971). Importantly, *novel* biochemical patterns and dynamical control processes may arise at a point of instability; that is, at a *bifurcation* point. The recognition of bifurcations, at which point the evolution of a biodynamical system may take different paths, has thus revealed that instabilities are an indispensable *source* of biological function and order (e.g., see Kaiser, Chapter 1, this volume). During the past decade, researchers have become increasingly aware of the fact that the dissipative formation of periodic or complex oscillations is at the root of signal detection and processing events in many types of biochemical and coupled cell signaling networks. Some examples are discussed in Sections 18.4 to 18.6, including coordinated interactions across the *extracellular* space that lead to the long-range correlated dynamics of heart and brain function.

We can now recognize that, frequently, biological systems do not function by simply dampening out or counteracting oscillations in biochemical and physiological parameters through homeostatic feedback. On the contrary, *sustained* oscillatory dynamics and complex pattern formation resulting from instabilities under nonequilibrium constraints may be involved in biological control at all levels of physiological organization. While it is certainly the case that cells and organisms control biological parameters within only a limited range of values, often the immediate or long-term goal is *not* the establishment of constancy, or even regularity, in intra- or extracellular parameters. Therefore, instead of 'homeostasis', the term 'homeodynamics' may be a more accurate definition that captures the nonlinear regulatory principles governing the dynamical stability of a living system (see, e.g., Bassingthwaighte *et al.*, 1994). Some striking examples from the previous chapters and from the related scientific literature that emphasize this viewpoint are highlighted in subsequent sections.

18.4 Order and chaos in health and disease

We normally associate healthy physiological function with order, regularity and stability. Many instances are now known, however, where *regular, periodic dynamics* and *stability*, i.e., resistance to change, has been linked with certain disease states. For example, there exists good evidence that normal, healthy function can be accompanied by irregular, chaotic dynamics and that, in turn, loss of irregularity and complex behavior may be associated with pathological function (for an overview, see Goldberger, 1997). One specific observation comes from a physiological system whose seemingly clocklike, regular motion is familiar to us all: the human heartbeat. Unexpectedly, time series recorded from *healthy* subjects, however, revealed nonstationary, irregular fluctuations in heart rate. In contrast, regular periodic dynamics was detected in similar time series obtained from subjects with congestive heart failure. Furthermore, the studies by Goldberger and coworkers reviewed in Chapter 3 and by others (e.g., Poon and Merril, 1997) found that the fluctuating behavior in normal heart rate exhibited fractal characteristics, including deterministic chaotic dynamics, and that loss of chaotic behavior or of fractal complexity was indicative of heart failure.

An analogous observation comes from neurology: the onset of an epileptic seizure is characterized by a transition from irregular fluctuations to periodic, synchronous oscillations in brain electrical activity as described in Chapters 2 and 16 by R. Larter *et al.* and J. Milton, respectively. Milton suggests that this rapid transition could be symptomatic of a multistable, dynamical state of an underlying neuronal control system, where a minute deviation in a control parameter might lead to the dramatic change in global brain activity. As in the case of heart dynamics, a loss of complex brain dynamics could therefore again be related to dysfunction. Additionally, a decrease in the complexity of brain electrical activity, as established by nonlinear analysis of electroencephalo-gram time series recordings, was recently proposed to allow the prediction of the onset of seizure activity (Lehnertz and Elger, 1998; Hively *et al.*, 1999).

Together the above examples serve to show that the emergence of periodic-ity, or a loss of biodynamical complexity, can be symptomatic of severe pathological conditions. Importantly, the above-mentioned cases indicate that nonlinear time series analysis, including dynamical complexity measures, can be employed as a form of 'dynamical diagnosis'. Clearly, more research is needed to establish whether this concept might have broad applications in the identification and characterization of disease states in other systems, from the level of the single cell to the whole organism. The examples available so far are encouraging and they propose a rewarding research direction for biomedicine.

18.4.1 Dynamical diseases and nonlinear control

The ability to define certain disease conditions in terms of pathological dy-namical states has led to the concept of 'dynamical diseases' (Glass and Mackey, 1979; Mackey and Glass, 1979). Over 30 clinical disorders have been classified as dynamical diseases in neurology and psychiatry alone (Milton and Black, 1995). One particularly exciting prospect concerns the application of nonlinear control techniques in the treatment of pathophysiogical dynamics.

A few such approaches are already under serious consideration. For example, the synchronous electrical activity associated with an epileptic seizure might be influenced by dynamical interventions with the goal of returning brain electrical activity to a more complex state. In Chapter 16, J. Milton discussed the possibility of a brain 'defibrillator' for the treatment of epilepsy based on this idea (see also Glanz, 1997). One type of intervention, which eventually might be developed into such a clinical tool, was already demonstrated in an *in vitro* model, a seizing hippocampal brain slice: Schiff and coworkers (1994) successfully applied a chaos 'anticontrol' strategy, and returned the regular spiking activity of epileptic brain tissue into a more chaotic regime (see also Ditto and Spano, Chapter 15, this volume).

On the other hand, disease states can also be characterized by decoherence, for example by a loss of periodicity. In the case of atrial heart fibrillation, a common form of arrhythmia, an unstable, chaotic motion defines the dynamics of the arrhythmic heart. The finding that deterministic chaos may underlie an arrhythmia has opened the door to the development of chaos control techniques for regularizing the irregular dynamics of fibrillating heart tissue (Garfinkel *et al.*, 1992). As described in Chapter 15, the principal feasibility of this approach in the treatment of heart arrhythmias has already been confirmed in a first clinical trial.

In summary, a loss of either irregular, chaotic dynamics or regular, periodic behavior can result in a breakdown in the control of healthy function. Therefore, the question of whether order or chaos are markers of healthy or pathological dynamics depends on the specific situation. Importantly, the new observations demonstrate that the tools and ideas from nonlinear dynamics, including those from chaos studies, have opened up a new line of clinical investigations that may fundamentally change how we view and treat certain life-threatening illnesses.

18.5 Random fluctuations as a contributor to optimal biological function

In traditional models of homeostatic regulation, the effective maintenance of an equilibrium-like, orderly state is seen as a sign of optimal, healthy function. In this view, random fluctuations, due to either environmental or internal factors, are interpreted as unwanted but unavoidable disturbances of biological processes, which adversely affect their function. Several examples are discussed in this book, suggesting that, under certain conditions, biological function can instead be enhanced or optimized by stochastic perturbations.

The findings in Part III describe that the exposure to noisy fluctuation of signal detection processes in biological or biochemical systems can increase the efficiency of signal transmission. The electromyography experiments reviewed in Chapter 10 by F. Moss indicate an improvement in the signal-to-noise ratio in the detection of electric signals by human sensory neurons at an optimal, intermediate noise level – the signature of stochastic resonance (Chiou-Tan et al., 1996). In addition, recent psychophysical experiments demonstrated an enhancement in human tactile sensation by imposition of an optimal dose of noise (Collins et al., 1996). The first demonstration of stochastic resonance in a brain tissue preparation was reviewed in Chapter 15 by W. Ditto and M. Spano. The experiment established that a certain amount of electrical noise enhanced the signal-to-noise ratio in the detection by tissue of an applied stimulus, as measured by the noise-enhanced increase in synchronous burst-firing of neuronal activity (Gluckman et al., 1996).

Evidence for stochastic resonance at the *molecular level* was reviewed in Chapter 11 by S. Bezrukov and I. Vodyanoy. Their experiments showed that electrical signal detection by voltage-gated ion channels could also be improved by electrical noise addition (Bezrukov and Vodyanoy, 1995, 1997). The theoretical analysis by D. Astumian in Chapter 12 presented molecular ratchet mechanisms by which random, nonequilibrium fluctuations might be harnessed by molecules such as ion pumps to perform work such as unidirectional transport against a concentration gradient (e.g., Astumian, 1997). An experimental demonstration of fluctuation-driven transport by a membrane ion pump, a Na^+,K^+-ATPase, was described in Chapter 13 by T. Tsong, who observed that the exposure of the ATPase to stochastically fluctuating electric fields significantly enhanced the pump's transport activity (Xie et al., 1994, 1997).

A. Arkin (Chapter 5) predicted a potentially constructive role of stochastic processes in gene expression as a consequence of the inherent uncertainties involved in the reaction probabilities of the small numbers of interacting molecules during RNA transcriptional activity. Specifically, a recent theoretical analysis by Arkin and coworkers (1998) showed 'how molecular level thermal fluctuations can be exploited by the regulatory circuit designs of developmental switches to produce different phenotypic outcomes'. Thus, the indeterminism intrinsic to gene activation events might have a positive function in producing phenotypic diversity.

Many cases of noise-facilitated processes in chemical, biochemical and biological systems have been identified in recent years. For an overview of experimental results and of theoretical advances see a recent issue of *Chaos* edited by D. Astumian and F. Moss (1998). In summary, the evidence for a constructive role of random fluctuations at the molecular, cellular and tissue level suggests that a more sophisticated view of physiological control mechanisms than is commonly believed may be required to account for the highly nonlinear, noise-sensitive processes that underlie critical biological operations.

18.6 Nonlinear frequency detection and processing in excitable, nonequilibrium systems

The formation of excitable and oscillatory states in nonequilibrium systems allows these systems to interact with external perturbations in nonlinear dependence on frequency information. This principle holds true for chemical, biochemical as well as biological systems. The Chapter 17 by O. Steinbock and S. Müller demonstrated this phenomenon in a nonlinear chemical reaction, the Belousov–Zhabotinsky system, where pulsed laser light triggered transitions in spatiotemporal pattern formation which strictly depended on pulse frequency (Steinbock *et al.*, 1993). At the level of isolated enzyme activity, Chapter 8 discussed the frequency-dependent control of oscillatory enzyme dynamics by physical perturbations such as oscillating magnetic stimuli (Eichwald and Walleczek, 1998). While this prediction has not yet been tested for magnetic stimuli, the nonlinear frequency sensitivity of collective enzyme dynamics was confirmed in experiments with electrical and chemical oscillating stimuli (e.g., Förster *et al.*, 1994, 1995). At a higher level of organization, Chapter 7 by H. Petty reviewed work showing that coherent metabolic oscillations in migrating neutrophils could be enhanced or diminished by an oscillating electrical stimulus. He also found that applied electrical signal oscillations, which were in phase with internal cellular oscillations, stimulated cellular migration, whereas anti-phasic signals did not (see also Adachi *et al.*, 1999).

For electrically excitable biological systems such as muscle and neuronal cells the sensitivity to the periodicity of an electric field perturbation, as discussed by P. Gailey (Chapter 6), has been known for some time. A particularly interesting example from the recent literature has dealt with frequency-specific cellular amplification: oscillating electric fields centered at 60 Hz, but not at much higher or lower frequencies, were found to enhance neuronal amplification processes (Haag and Borst, 1996). With respect to the sensitivity to oscillating magnetic fields of neuronal processes, Chapter 9 by Engström and coworkers reported that hippocampal brain tissue was capable of discriminating between 1-Hz and 60-Hz oscillating magnetic fields. Many more examples can be found in the literature, reporting frequency sensitivities of diverse biological processes to oscillating inputs. With respect to practical applications, nonlinear control techniques could be developed for influencing molecular, cellular or tissue dynamics, which exploit the interaction specificity on coherent frequency information (see Chapter 8).

18.6.1 Biological frequency encoding and decoding

The possibility that cellular signaling processes may rely on frequency encoding of information in the control of fundamental cellular processes is another important aspect of biological frequency sensitivities (e.g., Arkin, Chapter 5, this volume). The nonlinear control of gene expression by the ubiquitous second messenger calcium (Ca^{2+}) is an important recent example. For many cell types, biologists discovered that cellular Ca^{2+} 'homeostasis' in response to external stimuli is regulated *dynamically*. For example, rather than responding by a simple change in intracellular Ca^{2+} concentration, $[Ca^{2+}]_i$, repetitive spiking or oscillations in $[Ca^{2+}]_i$, which typically cover a range of frequencies from 0.01 to 3 Hz, were observed in single, stimulated cells. Since the Ca^{2+} oscillation period was observed to be a function of stimulus strength, it was suspected that these oscillations might control basic cell functions by frequency-encoding and decoding processes (see, e.g., Berridge and Gallione, 1988; Goldbeter *et al.*, 1990; Meyer and Stryer, 1991). Evidence in support of this hypothesis was recently provided by a series of remarkable experiments. Two groups of investigators independently observed that periodically generated $[Ca^{2+}]_i$ perturbations were capable of controlling gene activation events in nonlinear dependence on perturbation frequency (Dolmetsch *et al.*, 1998; Li *et al.*, 1998).

The results with the cellular Ca^{2+} oscillator and more complex systems such as heart and neuronal tissues suggest that the traditional view of homeostatic regulation for maintaining equilibrium states is, at a minimum, incomplete. Specifically, there is mounting evidence that many tissue and cell types have evolved to actively sustain nonequilibrium dynamical states such as the oscillations in $[Ca^{2+}]_i$ as part of complex biological regulatory networks. Importantly, such states are capable of detecting, amplifying and processing dynamical information that is contained in many types of intra- or extracellular signals. It may be difficult, if not impossible, for traditional, quasi-equilibrium biological models to account for the nonlinear cell and tissue dynamics that is at the core of important physiological control systems. The work discussed in this book offers general principles that can be employed in the development of more appropriate biological models.

18.7 Toward a new epistemology for biology

In the Introduction to this book, I offered several reasons why exclusively reductionistic strategies and explanatory models were unlikely to deliver complete answers to fundamental biological questions such as (1) how living systems function *as a whole*, (2) how they transduce and process *dynamical information*, and (3) how they respond to *external perturbations*. To some, this statement could pose a difficult conceptual challenge because many scientists have made a habit of equating rigorous scientific thinking with only reduction-istic thinking. Modern biology, in particular, is skeptical of any explanation that cannot be derived, at least in principle, from the sum of the microscopic actions of individual molecules alone.

The present dominance of reductionist approaches is not surprising for two additional reasons. The most obvious one is the remarkable success of modern cell and molecular biology. As a result, most scientists are convinced that current concepts and models are sufficient to deal with any research problem that biology may encounter. The other reason is a subtler one and is related to the apprehension of researchers to embrace any kind of 'whole-systems' or holistic thinking in biology, probably for historical reasons. During the first half of the twentieth century, scientists and philosophers contemplated the view that new physical laws would be required to explain the seemingly miraculous functionality and efficiency of living organisms. The fact that the highly organized, self-stabilizing temporal and spatiotemporal patterns, which are the hallmark of any living system, could spontaneously emerge from random microscopic interactions of organic chemicals was simply inconceiv-able then. Instead, they proposed a role for an unspecified unifying force or organizing field, which would be *unique* to living matter and could account for the properties of the whole that could not be derived from the parts. As a consequence, present-day biologists still have a tendency to reject out of hand even scientifically sound notions of holism, possibly because holistic ideas still remind them of the mysticism often associated with the unsubstantiated neovitalistic proposals of the past (for reviews, see, e.g., Mayr, 1982, 1988).

Today, of course, we know that the emergent formation of macroscopic, self-stabilizing patterns in biochemical and biological systems can theoreti-cally be accounted for by the science of nonequilibrium thermodynamics in complex, nonlinear systems; no unknown forces or fields from outside the realm of modern physics have to be postulated to explain many of the aston-ishing and, frequently, counter-intuitive features of living organisms such as nonlinear sensitivities to weak stimuli and spontaneous pattern formation. The nonlinear dynamical systems view, which I also referred to in the Intro-duction to this book as the paradigm of self-organization, thus provides biology with a theoretically sound approach toward a 'holistic biology' for the first time in the history of science. This new holism is firmly grounded in and consistent with the physical laws that also govern the material processes of the nonliving world.

From an epistemological perspective, there is the prospect that the century-old philosophical conflicts between holistic and reductionistic ways of knowing in biology could finally be resolved. Ironically, the solution might satisfy both holists and reductionists, although concessions have to be made on either side. Holistically inclined thinkers will have to accept the fact that all currently known properties of living matter are fully consistent with the microscopic properties of nonliving matter and the laws of modern chemistry and physics. Strict reductionists, on the other hand, will have to concede that qualitatively novel, macroscopic features may arise spontaneously in complexly organized, nonlinear systems such as living organisms, which cannot be reduced in explanation to the mere sum of the individual actions of the system's microscopic elements.

Finally, the paradigm of self-organization offers a powerful set of tools for thinking about the irreducible, whole-systems properties of living processes. By definition, the systematic development of this paradigm and its application to biomedicine is crucial if we wish to understand (1) the power and role of self-organization in health and disease, (2) how clinical diagnosis can benefit from a better understanding of the nonlinear dynamics of disease processes, and (3) how external stimuli may be best used to influence or control self-organized biological dynamics for therapeutic purposes.

18.8 Conclusions and outlook

The observations and ideas outlined in this book suggest a new research agenda for biology and medicine. While important steps have been taken toward answering some of the questions posed in the Introduction to this book, many questions remain open and even more are raised by the new findings. Several of the reported observations still need to be confirmed and the general applicability of some of the theoretical proposals needs to be validated by future work. This state of the science should not serve as a criticism, but it is a testament to the promise and vitality of the work at this new research frontier. Most importantly, the tools and ideas from nonlinear biodynamics have motivated researchers to design experiments and to research and develop diagnostic and therapeutic modalities that never would have been considered otherwise. The time and opportunity has come for biomedical scientists to adopt an information-based, whole-systems approach to biological understanding and in the development of advanced biomedical technology. The rewards for medicine could be immeasurable.

References

Adachi, Y., Kindzelskii, A. L., Ohno, N. and Petty, H. R. (1999) Amplitude and frequency modulation of metabolic signals in leukocytes: synergistic role of IFN-gamma in IL-6 and IL-2 mediated cell activation. *J. Immunol.* **163**: 4367–4374.

Arkin, A., Ross, J. and McAdams, H. H. (1998) Stochastic analysis of developmental pathway bifurcation in phage λ-infected *Escherichia coli* cells. *Genetics* **149**: 1633–1648.

Astumian, R. D. (1997) Thermodynamics and kinetics of a Brownian motor. *Science* **276**: 917–922.

Astumian, R. D. and Moss F. (eds.) (1998) *Focus Issue: The Constructive Role of Noise in Fluctuation Driven Transport and Stochastic Resonance. Chaos* **8**: 533–664.

Bassingthwaighte, J. B., Liebovitch, L. S. and West, B. J. (1994) *Fractal Physiology.* New York: Oxford University Press.

Berridge, M. J. and Gallione, A. (1988) Cytosolic calcium oscillators. *FASEB J.* **2**: 3074–3082.

Bezrukov, S. M. and Vodyanoy, I. (1995) Noise-induced enhancement of signal transduction across voltage-dependent ion channels. *Nature* **378**: 362–364.

Bezrukov, S. M. and Vodyanoy, I. (1997) Stochastic resonance in non-dynamical systems without response thresholds. *Nature* **385**: 319–321.

Cannon, W. B. (1929) Organization for physiological homeostasis. *Physiol. Rev.* **9**: 399–431.

Chiou-Tan, F. Y., Magee, K., Robinson, L., Nelson, M., Tuel, S., Krouskop, T. and Moss, F. (1996) Enhancement of subthreshold sensory nerve action potentials during muscle tension mediated noise. *Int. J. Bifurc. Chaos* **6**: 1389–1396.

Collins, J. J., Imhoff, T. T. and Grigg, P. (1996) Noise-enhanced tactile sensation. *Nature* **383**: 770.

Dolmetsch, R. E., Xu, K. and Lewis, R. S. (1998) Calcium oscillations increase the efficiency and specificity of gene expression. *Nature* **392**: 933–936.

Eichwald, C. F. and Walleczek, J. (1998) Magnetic field perturbations as a tool for controlling enzyme-regulated and oscillatory biochemical reactions. *Biophys. Chem.* **74**: 209–224.

Förster, A., Hauck, T. and Schneider, F. W. (1994) Chaotic resonance of a focus and entrainment of a limit cycle in the periodically driven peroxidase–oxidase reaction in a CSTR. *J. Phys. Chem.* **98**: 184–189.

Förster, A., Zeyer, K.-P. and Schneider, F. W. (1995) Chemical resonance and chaotic response induced by alternating electrical current. *J. Phys. Chem.* **99**: 11889–11895.

Garfinkel, A., Spano, M. L., Ditto, W. L. and Weiss, J. N. (1992) Controlling cardiac chaos. *Science* **257**: 1230–1235.

Glanz, J. (1997) Mastering the nonlinear brain. *Science* **277**: 1758–1760.

Glass, L. and Mackey, M. C. (1979) Pathological conditions resulting from instabilities in physiological control systems. *Ann. NY Acad. Sci.* **316**: 214–235.

Gluckman, B. J., Netoff, T. I., Neel, E. J., Ditto, W. L., Spano, M. L. and Schiff, S. J. (1996) Stochastic resonance in a neuronal network from mammalian brain. *Phys. Rev. Lett.* **77**: 4098–4101.

Goldberger, A. L. (1997) Fractal variability versus pathological periodicity: complexity loss and stereotypy in disease. *Persp. Biol. Med.* **40**: 543–561.

Goldbeter, A., Dupont, G. and Berridge, M. J. (1990) Minimal model for signal-induced Ca^{2+}-oscillations and for their frequency encoding through protein phosphorylation. *Proc. Natl. Acad. Sci. USA* **87**: 1461–1465.

Haag, J. and Borst, A. (1996) Amplification of high-frequency synaptic inputs by active dendritic membrane processes. *Nature* **379**: 639–641.

Hively, L. M., Gailey, P. C. and Protopopescu, V. A. (1999) Detecting dynamical change in nonlinear time series. *Phys. Lett. A* **258**: 103–114.

Lehnertz, K. and Elger, C. E. (1998) Can epileptic seizures be predicted? Evidence from nonlinear time series analysis of brain electrical activity. *Phys. Rev. Lett.* **80**: 5019–5022.

Li, W. H., Llopis, J., Whitney, M., Zlokarnik, G. and Tsien, R. Y. (1998) Cell-permeant caged $InsP_3$ ester shows that Ca^{2+} spike frequency can optimize gene expression. *Nature* **392**: 936–941.

Mackey, M. C. and Glass, L. (1979) Oscillations and chaos in physiological control systems. *Science* **197**: 287–289.

Mayr, E. (1982) *The Growth of Biological Thought*. Cambridge, MA: Harvard University Press.

Mayr, E. (1988) *Toward a New Philosophy of Biology*. Cambridge, MA: Harvard University Press.

Meyer, T. and Stryer, L. (1991) Calcium spiking. *Annu. Rev. Biophys. Biophys. Chem.* **20**: 153–174.

Milton, J. and Black, D. (1995) Dynamic diseases in neurology and psychiatry. *Chaos* **5**: 8–13.

Poon, C.-S. and Merrill, C. K. (1997) Decrease in cardiac chaos in congestive heart failure. *Nature* **389**: 492–495.

Prigogine, I. and Nicolis, G. (1971) Biological order, structure and instabilities. *Quart. Rev. Biophys.* **4**: 107–148.

Richet, C. (1900) *Dictionnaire de Physiologie*. Paris.

Schiff, S. J., Jerger, K., Duong, D. H., Taeun, C., Spano, M. L. and Ditto, W. L. (1994) Controlling chaos in the brain. *Nature* **370**: 615–620.

Schrödinger, E. (1944) *What is Life?* Cambridge: Cambridge University Press.

Steinbock, O., Zykov, V. and Müller, S. C. (1993) Control of spiral-wave dynamics in active media by periodic modulation of excitability. *Nature* **366**: 322–324.

Xie, T. D., Chen, Y.-D., Marszalek, P. and Tsong, T. Y. (1997) Fluctuation-driven directional flow in biochemical cycle: further study of electric activation by Na,K-pumps. *Biophys. J.* **72**: 2496–2502.

Xie, T. D., Marszalek, P., Chen, Y.-D. and Tsong, T. Y. (1994) Recognition and processing of randomly fluctuating electric signals by Na,K-ATPase. *Biophys. J.* **67**: 1247–1251.

Index

acetylcholine, 148, 220
acoustic stimuli, 102, 301, 311, 321
action potential, 55, 153, 240–244, 257, 350,
  365–368, 388
  timing of, 154
activation barrier, 259, 303, 318
aging, 78–80, 87–93, 248
alamethicin, 258, 267–271, 274, 275
allosteric effector, 116, 121
allosteric regulation, 46, 115
amplification factor, 200, 209
amplification properties, 6, 135, 200, 206–212, 415;
  see also biochemical amplification; signal
  amplification
amplitude control, 205
ampullae of Lorenzini, 157
analog circuits, 116
aperiodic oscillations, 19, 260, 276, 351; see also
  chaos
Aplysia, 154
Arnold tongue, 22, 23; see also resonance horn
arrhythmias, see heart arrhythmias
astrocyte, 252
ATP, 5, 123, 175, 177, 180, 181, 185, 186, 205, 295,
  297, 311, 312, 318
atrial fibrillation (AF), 330, 355–362, 369, 410
autocatalysis, 47, 251, 388, 410
autocatalytic process, 24, 47, 388, 391, 394
autocorrelation function, 75, 100, 101, 262

B12-dependent ammonia lyase, 195, 199–201
band-pass filter, 6, 135, 136, 138, 272
Belousov–Zhabotinsky (BZ) reaction, 328–339,
  387–407
  chaos control, 337–339
  electric field control, 394–399
  frequency control, 402–406
  light control, 399–406
  photosensitive, 251, 391, 392
  stratification, 393
  subexcitable, 251
  wave formation, 387–392, 407, 415; see also
    spiral wave

Bénard instability, 2
bifurcation
  analysis, 125, 126
  diagram, 50, 59, 60
  global, 25
  Hopf, 21, 24, 27, 40, 375, 380
  local, 21
  parameter, 21, 50
  period-doubling, 21, 332, 334, 337
  point, 31, 40, 330, 332, 411
  saddle-node, 21, 104
  tangent, 21
biochemical amplification, 163, 193, 200, 201, 204,
  206–212; see also amplification factor;
  peroxidase–oxidase reaction; zero-order
  ultrasensitivity
biochemical oscillations, 33, 44, 46, 128–129, 188,
  204–212; see also Ca^{2+} oscillations; glycolytic
  oscillations; NADH oscillations; neutrophil
  oscillations; peroxidase–oxidase reaction
biochemical reaction networks (BRN), 4–7, 9,
  112–115, 126, 127, 139–141, 208–212; see also
  genetic regulatory networks; regulatory
  motifs
biochemical switching, 6, 117, 123–126, 131
biodynamics, 1–10, 409, 418
bioelectromagnetics, 8, 211, 212; see also electric
  field; magnetic field
biological signaling, 4, 5, 9, 134, 194, 204, 302; see
  also cellular signal transduction; biochemical
  reaction networks
biological state dependence, 194
birhythmicity, 46
bistability, 2, 16, 46, 47, 124, 125, 380
Boltzmann function, 159, 266, 294
Boltzmann constant, 286
brain defibrillator, 376, 382, 383, 413
brain dynamics, 7, 47, 54, 139, 374, 412; see also
  epilepsy; hippocampus; psychophysics
Brownian motion, 73, 281, 286–289, 295, 297

Ca^{2+} ATPase, 317
Ca^{2+} channel, 58

421

Ca^{2+} oscillations, 5, 7, 16, 34–39, 41, 45, 137, 188,
    416
  lymphocytes, 137, 205, 208
  modeling, 34–39; *see also*
    Goldbeter–Dupont–Berridge model
  neutrophils, 173
Ca^{2+} wave, 45, 252–254
cancer, 5
cardiac arrhythmias, 44, 93, 350, 356; *see also* heart
    arrhythmias
cardiomyopathy, 342
categorical perception, 102, 110; *see also* speech
    perception
cell cycle, 5, 112, 118, 128, 133
cell-cycle oscillator, 118, 128; *see also* mitotic
    oscillator
cell-to-cell communication, 34, 235, 257, 323
cellular
  control, 5, 112, 410
  division, 5, 133, 173; *see also* cell cycle
  oscillations *see* Ca^{2+} oscillations; cell-cycle
    oscillator; glycolytic oscillations; neutrophil
    oscillations
  regulation, 5
  reliability, 138
  signal transduction, 5, 33–39, 128, 173, 180, 194,
    301, 409, 416; *see also* biochemical reaction
    networks; biological signaling; signal
    transduction
central nervous system (CNS), 51, 68, 87, 90, 154,
    157, 162, 163, 166, 248, 253; *see also* brain
    dynamics; hippocampus
cGMP, 225, 229
chaos, 2, 7, 9, 17–21, 31, 33, 139, 328–334, 341, 342
  atrial fibrillation, 355–362
  Belousov–Zhabotinsky reaction, 337–339
  cardiac arrhythmias, 350–355
  epilepsy, 53, 54, 60, 61
  hippocampal brain tissue, 362–364
  peroxidase–oxidase reaction, 46–51
chaos control, 9, 40, 41, 328–339, 341, 344, 369
  atrial fibrillation, 330, 355–362
  Belousov–Zhabotinsky reaction, 337–339
  cardiac arrhythmias, 350–355
  hippocampal brain tissue, 218, 330, 343, 362–364
  spatiotemporal systems, 41, 338
chaos maintenance, 363, 369
chemical
  circuits, 112–126, 136; *see also* circuits;
    regulatory motifs
  filter, 129, 134–138; *see also* frequency filter
  kinetics, 32, 56, 133, 140, 210; *see also* enzyme
    kinetics
chemotaxis, 139, 176, 187, 188
Cheyne–Stokes frequency, 78, 92, 93
circadian rhythm, 33, 128, 406
circuits
  analog, 116
  asynchronous, 117–119
  digital, 116–123
  electrical, 6, 114–123
  synchronous, 117–119

see also biochemical reaction networks; chemical
    circuits
clinical therapy, *see* therapy
coexistence, 19, 25, 103, 375, 379, 381
coherence, 7, 31, 162, 251
  abnormal, 51; *see also* epilepsy
  long-range, 17, 145, 147, 149, 162–167
  noise-mediated, 251; *see also* stochastic
    resonance
  quantum, 196
  resonance, 31
  spin, 197; *see also* radical pair mechanism
concentric circle, 388; *see also* spiral wave
context effect, 107, 110; *see also* speech perception
continuous-flow stirred tank reactor, 209, 330, 337,
    338
control parameter, 7, 21, 102, 104, 330, 332, 337,
    360, 375, 412; *see also* bifurcation parameter
control, *see* chaos control; electric field control;
    feedback control; frequency control; light
    control; magnetic field control; nonlinear
    control
convection cells, *see* Bénard instability
cooperative enzyme activity, 116
cooperativity, 15, 116, 117, 138
cornu ammonis (CA3), 56, 58, 221, 226, 363; *see
    also* hippocampus
coupled lattice model, 60, 61
current noise, 157

Dale's Law, 55
defibrillator, 153, 376, 382, 383, 413
Degn–Olsen–Perram (DOP) model, 49, 50, 60
delayed recurrent loop, 377–379
deterministic chaos, *see* chaos
detrended fluctuation analysis (DFA), 73–76, 80–89
Δg-mechanism, 198, 199
diagnosis, 8, 10, 66, 78, 114, 212, 236, 343, 418; *see
    also* dynamical diagnosis
dichlorophenol, 47
dielectrophoretic potential, 296
digital circuits, 116–123
dimensionality
  correlation, 19
  fractal, 19, 21, 68, 343
  Hausdorf, 19
  information, 19
diseases
  and chaos, 46, 411; *see also* chaos; chaos control
  and stochastic resonance, see stochastic
    resonance
  atrial fibrillation, 355–362, 369
  cancer, 5
  Ebola virus infection, 188
  epilepsy, 51–55, 374–383; *see also* epileptic
    seizures
  fetal distress syndrome, 92
  heart arrhythmias, 9, 44, 93, 327, 343, 350–362,
    413
  Huntington's disease, 88, 90, 93
  in aging, 78, 80, 87
  inflammatory, 186, 189

neurodegenerative, 87, 90
Parkinson's disease, 44, 90
pyoderma gangrenosum, 186
susceptibility, 188
ventricular fibrillation, 342–350, 369
ventricular tachycardia, 344
Wiskott–Aldrich syndrome, 184, 187
see also diagnosis; dynamical diseases; prognosis;
  therapy
dissipation, 23, 32, 120, 281, 282, 283
DNA
  damage, 184
  inversion reaction, 112
  methylation, 112
  promoter, 129
  replication, 5
dynamical assay, 93
dynamical biology, 10; *see also* biodynamics
dynamical diagnosis, 412
dynamical diseases, 9, 44, 51, 54, 218, 374–376, 412,
  418
dynamical equilibrium, 16

Ecto-ATPase, 318
elasmobranch fish, 148
electric dipole, 150, 248, 267, 291, 294
electric field
  amplification, 151, 152
  biological interactions, 147–168, 173–187,
    246–251, 301–323, 341–369
  control, 167, 182–184, 350–369, 394–399, 415
  detection, 147–168, 184, 185, 318, 319
  endogenous, 148
  external, 148–168, 305, 368, 394, 395
  extremely-low-frequency (ELF), 152, 318
  low-frequency, 152, 153, 158
  of the Earth, 149
  phase-matched, 184, 186
  pulsed, 182–184
  randomly fluctuating, 297, 305–311, 313–320
  sensitivity, 148, 247
  static (DC), 154
  stimulation, 185, 220, 313
  therapy, 147, 151, 162, 168
  vector, 395, 398, 399
electrical filter, 135
elecrocardiogram (ECG), 78 80, 346
electroconformational coupling (ECC), 301,
  308–320
electroencephalogram (EEG), 44, 52–54, 66, 216,
  374, 380, 412
electromigration, 399
electromyography, 249, 414
electroporation, 152
electroreceptor, 246, 247
emergent behavior, 56
emergent complexity, 5
emergent phenomena, 2, 7, 17, 51, 417
emergent properties, 2, 4, 6
enforced conformation oscillation, 305; *see also*
  electroconformational coupling
enhanced contrast effect, 102, 106, 107; see also
  speech perception
entropy, 3, 302
  Kolmogorov, 19
enzymatic futile cycle, 123, 125
enzyme–substrate complex, 49, 202, 203, 318
enzyme dynamics, 7, 415
  magnetic field effects, 204–208, 210–212
  see also peroxidase–oxidase reaction
enzyme kinetics, 44, 46, 47
  electric field effects, 301–323
  magnetic field effects, 193, 199–203, 229
  see also biochemical reaction networks; chemical
    circuits
ephaptic signaling, 148, 231
epigenetic control, 112
epilepsy, 44, 45, 51–55, 61, 66, 92, 167, 216, 218,
  223, 231, 232, 369, 413
  modeling, 55–60, 374–378
  multistability, 374–378
  see also hippocampus
epileptic seizures, 44, 45, 52, 61, 218, 330, 377, 382,
  412, 413; *see also* partial seizures
epileptiform activity, 154, 218, 223–225, 365
  magnetic field effects, 216, 225, 231
  nitric oxide, 222–225
  see also rhythmic slow activity
epistemology, 416
equilibrium, 130, 131, 258, 281, 294, 409, 410
  constant, 180, 181, 268, 292, 293, 303, 306
  cooperativity, 15
  phase transition, 15
  state, 16, 416
Escherichia coli, 114, 133
excitability, 2, 7, 35, 251, 344, 349, 393
  cellular, 55, 153
  chemical, *see* Belousov–Zhabotinsky reaction
  membrane, 54, 153, 258, 267, 269
  threshold, 251
  tissue, 153, 154, 338
excitation, 7, 29, 45, 55, 59, 231
  period, 403
  wave, 380, 388–406; *see also* Ca^{2+} wave; rotor;
    spiral wave
excitatory neuron, 55, 59, 163, 377–379

Faraday induction, 151, 248
Farey construction rule, 23, 25
feedback, 5, 61, 86, 185, 205, 338, 381
  control, 8, 129, 180, 329, 343, 377; *see also*
    biochemical reaction networks; chaos control
  excitatory, 47
  loop, 47, 92, 209, 355, 378
  low-gain, 377
  negative, 47, 61, 377, 410
  neuronal, 377
  positive, 47, 61, 124, 125, 410
  stabilization, 139, 381
  strength, 124
  time-delayed, 327, 376–379; *see also* time delay
    effects
fetal distress syndrome, 92
FitzHugh–Nagumo model, 35, 36

fixed point (FP), 19, 21, 27, 31, 104, 344, 358, 389
   driven, 24, 28, 29, 31, 35, 37
   unstable (UFP), 22, 27, 344, 353, 355, 358, 359, 360, 362
fluctuation-driven transport, 9, 295–297, 414; *see also* molecular motor; ratchet; stochastic resonance
fractal
   analysis methods, 10, 68–76; *see also* rescaled range analysis
   complexity, 341, 412
   dimension, 19, 21, 343
   dynamics, 67, 76, 80, 86, 87, 91, 92
   gait rhythm analysis, 80–91
   heart rate analysis, 76–80
   scaling analysis, 73, 78–90
   torus, 50, 51, 60
fractals, 7, 18, 68
fractional dimensionality, 68
fractional Gaussian noise, 98, 100
frequency control
   Belousov–Zhabotinsky reaction, 402–406
   biological activity, 204, 208, 212
   gene expression, 41, 205, 416
   Na^+,K^+-ATPase, 311–313
   peroxidase–oxidase reaction, 204–208
frequency encoding, 34, 416
frequency filter
   band-pass, 6, 135, 136, 138, 272
   chemical, 129, 134–138
   electrical, 135
   high-pass, 135
   low-pass, 135, 136, 202, 203
   notch, 137
frequency sensitivity, 8, 31, 203, 207, 318, 415, 416; *see also* frequency control

γ-aminobutyric acid (GABA), 55, 225, 229
$GABA_A$ receptor, 225
gait dynamics, 66, 87–92; *see also* walking
galvanotaxis, 183
gap junctions, 161, 162
gene expression, 9, 41, 112, 129–134, 144, 188, 416
   stochastic processes, 133, 134
gene networks, 120, 126, 132
genetic control, 129
genetic regulatory circuits, 5, 7, 129–134
genome projects, 6, 114, 141
glutamate, 55, 224
glycolysis, 46
glycolytic oscillations, 16, 128, 173, 180
glycosylphosphatidylinositol-linked membrane protein, 174, 186
Goldbeter–Dupont–Berridge (GDB) model, 35, 36

heart
   arrhythmias, 9, 44, 93, 327, 343, 350–362, 413
   beat regulation, 67, 76–78, 91, 330, 412
   canine, 343–350, 355
   chaos control, 350–362
   failure, 78–80, 92, 93
   human, 9, 66, 344

rabbit, 344, 346, 350, 354
   rate, 66, 68, 71, 78, 80, 91, 412; *see also* fractal heart rate analysis
   sinus rhythm, 76
high-pass filter, 135
hippocampus, 54–61, 154, 212, 330, 343, 356, 377, 383, 413, 415
   chaos control, 362–364
   electric field sensitivity, 217, 218, 364–369
   magnetic field sensitivity, 216–232
   stochastic resonance, 364–368
   see also cornu ammonis; epileptiform activity; rhythmic slow activity
histone acetylation, 132
holism, 417, 418
Holter monitor, 83
homeodynamics, 410, 411
homeostasis, 91, 140, 410, 411, 416
hormone, 66, 138, 173, 188
horseradish peroxidase, *see* peroxidase–oxidase reaction
Huntington's disease, 88, 90, 93
Hurst exponent, 73, 101
hydrodynamic hair receptor, 241–246
hyperfine coupling (HFC), 198, 199
hyperpolarization, 155, 218
hysteresis, 6, 16, 21, 22, 25, 102, 106, 124

immune defense, 5
inflammatory disease, 186, 189; *see also* neutrophil oscillations
information
   dimension, 19
   encoding, 25, 34, 40, 161, 166, 188, 204, 260
   processing, 5, 9, 16, 34, 40, 166, 168, 196, 364, 365, 409, 410, 416
   ratchet, 287, 290–294
   storage, 16, 27
   theory, 283
   transfer, 27, 39, 40, 115
inhibitory interneuron, 56–59, 377, 378
initial state sensitivity, 8, 21, 138, 328, 332, 341, 342, 343
instability, 21, 32, 410, 411
   spatial, 16
   spatiotemporal, 17, 41
   temporal, 16
   see also Bénard instability; bifurcation; transitions
integrin, 125, 174, 179, 180, 181, 186, 187
interictal bursting, 223, 224
interictal spikes, 363
intermediary metabolism, 114
intermittency, 17, 25, 78
ion channel, 7, 61, 128, 240, 301, 313, 368, 382, 414
   noise, 157, 158–162
   stochastic resonance, 165, 265–278
   voltage-gated, 148, 153, 158–166, 257–278, 414
ion pump, *see* Ca^{2+}-ATPase; Ecto-ATPase; mitochondrial ATPase; Na^+,K^+-ATPase

K^+ channel, 58, 153, 267, 269

keratinocyte, 148
kinases, 123, 124, 174, 175, 180, 186, 187
  pp125 focal adhesion kinase (FAK), 125
  protein kinase C (PKC), 180
Kirchoff's Law, 120
Kolmogorov entropy, 19

light control, 244, 391, 400–406, 415
light stimuli, 7, 129, 244–246, 251, 252, 286, 297
limit-cycle oscillator, 16, 17, 21–28, 53, 54, 123, 154, 381, 382
  biophysical modeling, 33–39
  driven, 21–32
  see also active oscillator; Van der Pol oscillator
lipid bilayer, 152, 258, 268, 273, 301
local activity, 17
low-pass filter, 135, 136, 202, 203
Lyapunov exponent, 19, 21, 343
lymphocyte, 137, 184, 194, 205, 208

magnetic field
  biological interactions, 8, 147, 193–212, 216–232
  control 204–212
  detection, 195, 217, 320, 321, 322
  of the Earth, 193, 197
  oscillating, 151, 193, 201–212, 216–232, 329
  pulsed, 202
  radical pair mechanism, 196–199, 229
  sensitivity, 193, 205, 216, 320, 415
  static, 193, 199–201, 204–212, 219, 225, 226, 229
  therapy, 194, 204, 212, 217
magnetic resonance imagining (MRI), 217
magnetite, 193, 194, 229
magneto-orientation, 193
magnetochemistry, 194, 196
magnetoencaphlography (MEG), 167, 216, 230
magnification factor, 68, 69, 70; *see also* fractal analysis methods
Maxwell's demon, 282, 283, 286, 287, 295
McCulloch–Pitts artificial neuron, 127
mechanical stimuli, 7, 8, 16, 34, 216, 251, 254, 311, 313
mechanoreceptor, 239, 240–243
memory, 51, 112, 216, 218
  long-persistent, 101
  long-range, 92, 98, 100, 101, 110
  short-range, 81, 101
mitochondria, 128, 311
mitochondria ATPase, 311, 312, 320
mitotic oscillator, 5; *see also* cell-cycle oscillator
molecular biology, 1, 4, 417
molecular motor, 128, 290–298; *see also* fluctuation-driven transport; ratchet
morphogenesis, 16
Morris–Lecar model, 56, 57
multistability, 25, 103, 327, 374–383, 412
myocyte, 153–155, 161, 162, 356

Na$^+$ channel, 153, 159, 267, 269
Na$^+$,K$^+$-ATPase, 297, 307, 311–320, 414
NADH oscillations, 46, 173
NADPH oscillations, 5, 173, 175–188, 204

NADPH oxidase, 180, 181
network analysis, 114
neuronal amplification, 415
neuronal assemblies, 5, 110
neuronal bursting, 218, 223, 226, 227, 230, 356, 379
neuronal circuitry, 86, 376
neuronal control, 66, 76, 80, 92, 412
neuronal dynamics, 47, 55–62, 380; *see also* epilepsy; hippocampus
neuronal modeling, 55–60, 374–378
neuronal network, 127, 166, 364–368
neuronal noise, 244, 365
neuronal oscillations, 4, 51
neuronal processing, 149, 167
neuronal stochastic resonance, 364–368
neuronal subnetwork model, 60
neuronal synchronization, 377
neutrophil oscillations
  actin assembly, 173, 185
  adhesive events, 180
  cell shape change, 173, 177
  clinical abnormalities, 186–188
  electric field control, 182–184
  intracellular Ca^{2+}, 173
  intracellular NADPH, 173, 175–188, 204
  membrane potential, 173
  migration velocity, 173
  receptor interactions, 173–175, 180
  respiratory burst, 173
nitric oxide (NO), 216, 219, 222–225, 229–231
nitric oxide synthase (NOS), 222, 223, 225
noise-induced switching, 380, 381
noise
  $1/f$, 75, 76, 78, 88, 157
  current, 157
  electrical, 157, 248, 249, 313–317, 319, 320, 414
  external, 35, 40, 156, 166, 167, 244, 251, 257–259, 270–278
  filter, 119, 135, 138, 276
  fractional Gaussian, 98, 100
  internal, 31, 35, 40, 138, 156–167, 240, 241, 244, 249, 264, 365
  ion channel, 157, 158–162
  muscle-tension mediated, 249
  network 252, 253
  random telegraph, 313
  shot, 157
  spatiotemporal, 251
  thermal, 113, 133, 148, 196, 240, 281–298
  thermal voltage, 157
  white, 71, 75, 76, 79, 81, 83, 93, 272, 311, 319, 320
  see also fluctuation-driven transport; stochastic resonance
nonequilibrium chemical reactions, 251, 290; *see also* Belousov–Zhabotinsky reaction
nonequilibrium fluctuations, 298, 414; *see also* fluctuation–driven transport; ratchet
nonlinear biochemical amplification, *see* biochemical amplification
nonlinear control
  goals, 6

nonlinear control (*cont.*)
  methods, *see* chaos control; feedback control;
    frequency control
notch filter, 137
Nyquist frequency, 239

oscillator
  active, 22, 28, 29, 35, 36, 38, 39
  chaotic, *see* chaos
  damped, 28, 46, 377; *see also* fixed point
  Duffing, 24
  limit-cycle, 16, 17, 21–28, 53, 54, 123, 154, 381,
    382
  mixed-mode, 50, 51, 60
  overdamped, 103
  passive, 24, 28, 29, 35
  self-sustained, 22, 33, 204; *see also* limit-cycle
    oscillator
  Van der Pol, 23, 25; *see also* limit-cycle oscillator
  see also Belousov–Zhabotinsky reaction;
    biochemical oscillations; cellular oscillations;
    heart arrhythmias; hippocampus
oscillatory activation barrier (OAB) model, 318,
  319
Ott–Grebogi–Yorke (OGY) method, 329, 343; *see*
  also chaos control

paramagnetism, 195, 197, 199; *see also* radical pair
  mechanism
Parkinson's disease, 44, 90
partial seizures, 44, 52, 55, 60; *see also* epileptic
  seizures
pattern formation, 16, 103, 327, 399, 411, 417
  spatial, 387
  spatiotemporal, 2, 393, 415; *see also* Ca^{2+} wave;
    excitation wave; rotor; spiral wave
Pauli exclusion principle, 198
perceptual category, 104; *see also* speech
  perception
peroxidase–oxidase (PO) reaction, 44, 46–51, 128,
  139, 162, 208–211
  amplitude control, 210
  chaos, 46–51
  electric field control, 415
  frequency control, 415
  magnetic field control, 210
  modeling, 46–51
  nonlinear amplification, 208–210
perturbations
  acoustic, *see* acoustic stimuli
  electromagnetic, *see* electric field; light stimuli;
    magnetic field
  mechanical, *see* mechanical stimuli
  stochastic, *see* noise; nonequilibrium
    fluctuations; stochastic resonance
phagocytosis, 188
phase
  locking, 21, 50, 60, 154, 155, 158, 167, 174, 368,
    404, 406
  plane diagram, 19, 22, 25
  singularity, 348
  space, 21, 53, 54, 341, 358

synchronization, 377
tracking, 226–228, 230
transition, 4, 15, 29, 38, 149, 163, 165, 166, 167;
  see also bifurcation; transitions
phosphatase, 123, 124, 175, 180, 187
photoreceptor, 244
photosensitivity, 251, 388, 391, 393, 399–406
Poincaré map, 16, 343, 344, 353, 354, 360, 362; *see*
  also phase plane diagram
Polydon spathula (paddlefish), 246
population cycles, 33
power law, 71, 74, 78, 81, 100, 101, 252, 253
  anti-correlations, 75
  long-range correlations, 75, 78, 92
power spectral analysis, 66, 76, 86
power spectral density, 75, 100, 260, 261, 270, 272,
  276
power spectrum, 19, 25, 75, 83, 102, 239, 241, 244,
  262, 271
preturbulence, 17
Procambarus clarkii, 241
prognosis, 8, 78, 80, 92
proprioceptive neuron, 248–251
psychophysics, 7, 246, 414; *see also* sensorimotor
  coordination; speech perception
pyoderma gangrenosum, 186; *see also* neutrophil
  oscillations

quantum, 1
  coherence, 196; *see also* spin coherence
  mechanics, 120, 259
quasiperiodicity, 19, 46, 138

radical pair mechanism (RPM), 194–199, 202, 229;
  see also enzyme dynamics; enzyme kinetics
random fluctuations, *see* noise
random telegraph noise (RTN), 313
random walk analysis, 73, 159; *see also* detrended
  fluctuation analysis
randomly fluctuating force field, *see* electric field;
  noise
ratchet, 281–298
  electric, 310, 311; *see also* electroconformational
    coupling
  energy, 287–290, 294, 295
  Feynman's, 283–285, 291, 295
  flashing, 295, 296, 298
  information, 287, 290–294
  macroscopic, 283
  microscopic, 283
  molecular, 286, 297, 414
reactive oxygen metabolites (ROM), 177–181, 184,
  188
reductionism, 1, 10, 51, 416–418
regulatory motifs, 116, 126, 127–138, 140; *see also*
  biochemical reaction networks; chemical
  circuits; genetic regulatory circuits
REM sleep, 167
rescaled range analysis, 101; *see also* Hurst
  exponent
resonance
  coherence, 31

diagram, 23
drift, 405, 406
energy transfer (RET), 174
horn, 23, 24, 27; *see also* Arnold tongue
metabolic, 182–185, 187
stochastic, *see* stochastic resonance
subharmonic, 22, 25, 29, 40
superharmonic, 22, 25, 29, 40
whole-body, 151
rhythmic slow activity (RSA), 216, 218–226, 231;
   see also
epileptiform activity; hippocampus
   magnetic field effects, 218–222
   nitric oxide, 222, 223
   phase tracking, 226–228
ribosome, 128, 129, 133, 134
RNA polymerase (RNAP), 129–134
RNA transcription, 5, 129–134, 141, 414; *see also*
   gene expression; genetic regulatory circuits
rotating spiral wave, 344, 388, 394, 400; *see also*
   rotor; spiral wave
rotor, 344–350

self-organization, 1–10, 17, 34, 92, 149, 393, 407
   biological, 10, 406
   chemical, 387, 393, 406
   hydrodynamic, 2
   in health and disease, 418
   paradigm of, 417, 418
self-organized criticality, 110, 252
self-organized dynamics, 8, 15, 44, 97, 102, 196, 204,
   212, 410, 418
self-similar time series, 69, 73, 74
self-similarity, 68–75
   parameter, 69, 71, 73, 74, 75
sensitivity to frequency, *see* frequency sensitivity
sensitivity to initial conditions, *see* initial state
   sensitivity
sensorimotor coordination, 97–102
shot noise, 157
signal amplification, 31, 38–40, 121, 135, 196, 209,
   229, 230, 276, 277
signal detection, 147, 149, 157, 411, 414
   array averaging, 156
   frequency, 196, 229; *see also* frequency control;
      frequency filter
   long-range coherence, 162–166
signal processing, 112–141, 157, 165, 239, 244, 245
signal transduction, 34, 35, 120, 194, 257–278, 281,
   302, 323
   noise-improved, 257–278; *see also* stochastic
      resonance
   see also biological signalling; cellular signal
      transduction
signal-to-noise ratio (SNR), 38, 156–160, 239, 241,
   249–251, 258, 259, 260, 263, 264, 272–278, 319,
   367, 414
signals
   aperiodic, 31, 40, 276
   coherent, 35, 38, 240, 246, 251
   external, 15, 16, 21, 27, 31, 34, 35, 38, 40, 286
   noise-enhanced, 249; *see also* stochastic

resonance
   noisy, 31, 40, 134, 138, 244, 367
   oscillatory, 129, 160, 173–189
   periodic, 31, 40, 134, 138, 270, 276, 301, 365,
      367
   static, 31, 32
   subthreshold, 237, 238, 239, 249, 364, 365, 366,
      367; *see also* stochastic resonance
   see also stimuli
sonoluminescence, 16
spectral maximum likelihood estimator, 98, 102
speech perception, 102–110
spin coherence, 197
spin-correlated radical pair, 197, 229; *see also*
   radical pair mechanism
spin-orbit coupling, *see* Δg-mechanism
spiral
   annihilation, 398
   core, 389, 400, 406
   meandering, 389, 403
   tip, 388, 389, 395, 398, 400–406
spiral wave
   Belousov–Zhabotinsky reaction, 251, 252, 337,
      387–407
   Ca^{2+} concentration, 45, 252–254
   cardiac muscle, 344–350
   collision, 395–398
   formation, 347
spiraling wave, 15; *see also* spiral wave
stability analysis, 19
stimuli
   auditory, 98, 218
   electromagnetic, 8, 211
   external, 6, 7, 16, 18, 22, 24, 29, 32, 34, 196, 204,
      410, 416, 418
   hydrodynamic, 244, 245
   visual, 5, 218
   see also acoustic stimuli; electric field; light
      stimuli; magnetic field; mechanical stimuli;
      signals
stochastic fluctuations, *see* noise; nonequilibrium
   fluctuations; stochastic resonance
stochastic resonance (SR), 7, 9, 31, 35, 38, 40, 138,
   149, 165, 236–254, 257–278, 311, 369, 414
   basic principles, 237–239
   hippocampus, 364–368
   medical science, 248–251
   Na^+,K^+-ATPase, 319, 320
   sensory biology, 239–248
   spatiotemporal systems, 251–254
   voltage-gated ion channels, 257–278
stoichiometric network analysis, 126
strange attractor, 19, 53, 334, 335; *see also* chaos
stratification, 393
substrate inhibition kinetics, 205
synchronization, 2, 3, 7, 27, 117–119, 128, 149, 154,
   162–166
   error time series, 98–102, 110; *see also*
      sensorimotor coordination
   neuronal, 5, 31, 52, 56, 218, 230, 231, 232, 363,
      367, 377, 402, 414
synergetics, 17, 34

therapy, 8, 9, 10, 162, 418
  drug, 187
  dynamical, 218, 375
  electric field, 147, 151, 168
  epilepsy, 374, 382, 383
  gene, 1
  magnetic field, 194, 204, 212, 217
  stochastic resonance, 236, 249
  ventricular fibrillation, 343
  see also defibrillator
thermal activation, 238, 259, 281
thermal energy, 162, 196, 302, 320
thermal gradient, 2, 282, 286
thermal noise, 113, 133, 148, 157, 196, 282–298
thermal voltage noise, 157
thermodynamics, 281, 307, 411, 417
  second law of, 3, 282
threshold
  crossing, 124, 238, 239, 240, 292
  effect, 16, 27, 185, 222, 229, 237, 251, 296, 362, 381
  firing, 154, 159, 240, 248, 378
  function, 117, 239
  model, 239
  potential, 59, 153, 154
  relative spread of, 160
  stimulation, 160, 358
time delay effects, 18, 21, 34, 35, 36, 39, 118, 158, 159, 376–379, 410
tissue repair, 194, 212
transcranial magnetic stimulation (TMS), 217
transition-state theory, 259
transitions
  chaos-to-order, 7, 31, 210, 412; *see also* chaos control
  noise-induced, 31; *see also* noise-induced

switching; stochastic resonance
  order-to-order, 31
  synchronized, 162, 165
  thermal noise-activated, 281; *see also* fluctuation-driven transport
  see also bifurcation; phase transition
transmembrane potential, 152, 258, 265, 270, 311, 312, 344, 347, 365
transporter, *see* ion channel; ion pump
turbulence, 16, 17, 18, 41, 92, 328

unstable fixed point (UFP), 22, 27, 344, 353, 355, 358, 359, 360, 362
unstable periodic orbit (UPO), 40, 51, 344, 358
urokinase receptor, 174

vagal nerve stimulation, 61, 383
Van der Pol oscillator, 23, 25; *see also* limit-cycle oscillator
Van't Hoff–Arrhenius Law, 259
ventricular fibrillation (VF), 342–350, 369
ventricular tachycardia, 344

walking, 80–90
  metronomic, 86
  see also gait dynamics
wavebreak, 346, 348
wavefront, 346, 348, 399
waves, *see* Ca^{2+} wave; excitation wave; rotor; spiral wave
white noise, 71, 75, 76, 79, 81, 83, 93, 272, 311, 319, 320
Wigner spin conservation rule, 197
Wiskott–Aldrich syndrome (WAS), 184, 187
wound healing, 147

zero-order ultrasensitivity, 124